The Golgi Apparatus

The Golgi is What?

Hilton H. Mollenhauer

In days long gone
the Golgi was what?

Why, a stringy thing,
or so Camillo said.

Oh no, said some,
just dots – not strings.

You are wrong said others,
not real at all.

But, now we have worked and now we all know
that the Golgi is, well, you know.

A bunch of flat plates
stacked one on another.

With vesicles close by
all positioned in space.

And a pile of small tubules
that spread out and about.

And with buds of small tubules
important no doubt.

But how do we know that all is now clear?
What have we done to bring this to fore?

Why, we have seen it you dummy
with EM's galore.

We have chopped it to bits
to see what's inside.

We fed it some markers
to make it abide.

So now we know all, there's no reason to doubt.
That what we have now is a Golgi for sure.

But, somehow, I am still not so sure,
that all is clear as some might suppose.

Many still fret about this and about that,
about strings, about dots, and about tubules galore.

Perhaps it is time to look once again
to consider the past and look on ahead.

Perhaps this time we might actually know.
The Golgi is what? – and be reasonably sure.

The Golgi Apparatus

The First 100 Years

D. JAMES MORRÉ
Medicinal Chemistry and Molecular Pharmacology
Purdue University, West Lafayette, IN, USA

HILTON H. MOLLENHAUER
College Station, TX, USA

 Springer

D. James Morré
Medicinal Chemistry and Molecular
   Pharmacology
Purdue University, West Lafayette,
   IN, USA
201 South University St
West Lafayette 47907
e-mail: morre@pharmacy.purdue.edu

Hilton H. Mollenhauer
College Station, TX, USA
1208 Ridgefield Circle
College Station 77840

ISBN: 978-0-387-74346-2   e-ISBN: 978-0-387-74347-9

Library of Congress Control Number: 2008933895

Printed on acid-free paper

9 8 7 6 5 4 3 2 1

springer.com

Preface

The Golgi Apparatus: The First 100 Years traces the first 100 years of Golgi apparatus discovery from the first published accounts from Pavia, Italy in 1898 to the Centenary Celebration in Pavia, Italy in 1998 and into the decade beyond. It is not intended, however, to be a comprehensive survey but rather to present the perspectives of the authors to summarize their contributions over the past 50 years in parallel with the modern era of Golgi apparatus discovery initiated in 1954 and made possible by the advent of the electron microscope. Included are methods of cell fractionation and biochemical analysis leading up to the present where efforts focus heavily on molecular biology.

Topics where the authors and their colleagues have made substantial and/or pioneering contributions are emphasized including Golgi apparatus morphology and structural organization and function (especially in plants), the existence and importance of cisternal tubules, development of methods of plant and animal Golgi apparatus isolation and subfractionation, biochemical analyses of highly purified plant and animal Golgi apparatus fractions in comparison to equally highly purified reference fractions. The use of such fractions in cell-free system analyses of membrane trafficking, the concept of Golgi apparatus function as part of an integrated system of internal endomembranes (the Endomembrane System), evidence for differentiation of membranes across the stacks of Golgi apparatus cisternae, and flux of membrane constituents along the polarity gradient defined by membrane differentiation all culminating in the membrane maturation or flow-differentiation model of Golgi apparatus function. More recent contributions to Golgi apparatus in cell growth (enlargement) and to cancer are summarized in the final chapters.

The authors' view of the dynamic working of the Golgi apparatus were based initially on static electron micrographs of the maize root tip generated by Hilton Mollenhauer in the Cell Research laboratory of the University of Texas in Austin, then under the direction of the late W. Gordon Whaley. Subsequent quantitative studies in the laboratory of D. James Morré at Purdue University suggested that massive amounts of membrane were moved to the plasma membrane at the cell surface in the discharge of secretory products in the outer cap cells of the maize root. The logical source of this membrane was the endoplasmic reticulum. The concept was further fueled by observations of Stanley Grove, then a graduate student in the laboratory of Charles Bracker at Purdue University. Grove's investigations with the Golgi apparatus of a fungus clearly demonstrated a gradient of membrane morphology across the stacked cisternae from endoplasmic reticulum-like on one face to plasma membrane-like on the opposite face. A membrane composition of Golgi apparatus intermediate between that of the endoplasmic reticulum and the plasma membrane was determined using isolated cell fractions from both rat liver and mammary gland in collaboration with Thomas W. Keenan also from Purdue University. The actual concept of the dynamic passage of membrane material from the endoplasmic reticulum to the plasma membrane (i.e., membrane flow) was first tested experimentally in the laboratory of Werner Franke in the Department of Peter Sitte at the University of Freiburg. With the assistance of Barbara Deumling, Ernst Jarash, Jürgen Kartenbeck, Ronald Cheetham, and Hans-Walter Zentgraf, rats were pulse-labeled with ^{14}C-leucine, the livers were excised, purified fractions of endoplasmic reticulum, Golgi apparatus, and plasma membrane were isolated and stripped to remove extrinsic and sectretory proteins, and the residual intrinsic membrane proteins were analyzed for specific radioactivity. The pulse-chase kinetics were consistent with passage of membrane proteins from the endoplasmic reticulum to the plasma membrane via the Golgi apparatus. The membrane flow concept seemed secure. Unfortunately, alternative views based on recycling models soon prevailed and dominated the literature for more than two decades nearly up until the centenary year of Golgi apparatus discovery in 1998. That year marked a renewed appreciation for the dynamic, cisternal maturation model of Golgi apparatus function. The new resurgence of interest in the potential for membrane flux or flow across the stacked cisternae of the Golgi apparatus has provided the impetus to complete this monograph on the Golgi apparatus with emphasis on the dynamic aspects of Golgi apparatus underrepresented in all but the most recent Golgi apparatus literature.

We express our appreciation to the many colleagues, postdoctorals, graduate students, undergraduate assistants, and technicians whose invaluable assistance made possible the experimental studies and the even greater numbers who challenged and criticized the work to force us to work even more diligently to distinguish among possible interpretations of the findings. We thank Janet Sweet, Sarah Craw, Keri Safranski, and Peggy Runck for assistance with manuscript preparation covering at least four different versions over several years and Matthew Miner for assistance with preparation of the figures. We are especially indebted to the unwavering support of Dorothy Morré, Barbara Mollenhauer, the Morré children, Connie, Jeffrey, and Suzanne, and the Mollenhauer children, Paul, John, and David, some of whom still sport electron micrographs of negatively stained Golgi apparatus on their refrigerator doors.

D. James Morré
Hilton H. Mollenhauer

The authors, Hilton H. Mollenhauer (left) and D. James "Jim" Morré (right) celebrating Camillo Golgi's birthday on July 8, 1972. Electron Microscope Laboratory of the Cell Research Institute, The University of Texas, Austin.

Contents

Preface   v

1. Discovery and Rediscovery **1**

   1.1. The Discovery of the Golgi Apparatus   1
   1.2. The Controversy   4
   1.3. Modern Rediscovery   6
   1.4. Summary   7

2. Structure **9**

   2.1. Cisternae   10
   2.2. The Cisternal Stack or Dictyosome   17
   2.3. Golgi Apparatus (Denotes either Singular or Plural)   28
   2.4. Golgi Apparatus Functioning as Part of an Integrated
        Endomembrane System   29
   2.5. Associations with Other Organelles and Cell Components   31
   2.6. Vesicles of the Golgi Apparatus   32
          2.6.1. COP (COPII)-Coated Transition Vesicles   32
          2.6.2. Clathrin-Coated Vesicles   32
          2.6.3. Secretory Vesicles   32
          2.6.4. Secretion Granules and Condensing Vacuoles   33
          2.6.5. Fusiform Vesicles and Cisternal Remnants   35
          2.6.6. Trans Golgi Apparatus Network   36
          2.6.7. Cis Golgi Apparatus Network (Intermediate Compartment)   37
   2.7. Summary   37

3. Isolation and Subfractionation **39**

   3.1. Golgi Apparatus Isolation   39
          3.1.1. Procedure for Rodent Liver   40
          3.1.2. Preparation of Reference Fractions from Rodent Liver   44
          3.1.3. Isolation of Golgi Apparatus Fractions from Plant Cells   49
          3.1.4. Isolation of Golgi Apparatus from Mammalian Cells
                Grown in Culture   52
   3.2. Subfractionations of Golgi Apparatus based on Density   53
   3.3. Golgi Apparatus Subfractionation by Free-Flow Electrophoresis   57
   3.4. Summary   61

4. Tubules 63

4.1. Function of Golgi Apparatus Peripheral Tubules in Delivery
of Cargo from Endoplasmic Reticulum to the Golgi Apparatus 66
   4.1.1. Liver Parenchyma and Intestinal Absorptive Cells 66
   4.1.2. Peripheral Tubule Function in Acinar Cells of Pancreas
      and Parotid Gland and Chromaffin Cells of the Adrenal Medulla 71
4.2. Golgi Apparatus Buds – Vesicles or Coated Ends of Tubules? 71
   4.2.1. Isolation of Tubule-Enriched Fractions 74
4.3. Summary 76

5. Endomembrane Biogenesis 77

5.1. Role of Endoplasmic Reticulum in Membrane Biogenesis 78
5.2. Biosynthetic Capabilities of Golgi Apparatus Relevant
to Membrane Biogenesis 79
5.3. Biosynthesis of Membrane Lipids 80
5.4. Biosynthesis of Membrane Sterols 83
5.5. Biosynthesis of Membrane Proteins 84
   5.5.1. The Origins of Golgi Apparatus Proteins 85
5.6. Glycosylation of Membrane Glycoproteins 88
5.7. Glycosylation of Membrane Glycolipids 88
5.8. Formation of Sugar Nucleotides and Other Active
Intermediates of Glycosylation Reactions 88
5.9. Sulfation Reactions 90
5.10. Distribution of Glycosyltransferases Across the Polarity
Axis of the Golgi Apparatus 90
5.11. Summary 91

6. Function in the Flow-Differentiation of Membranes 93

6.1. Morphological Evidence for Membrane Differentiation
within the Golgi Apparatus 94
   6.1.1. Measurements of Membrane Thickness 94
   6.1.2. Organization of Membrane Constituents 96
   6.1.3. Evidence from Cytochemistry 99
6.2. Biochemical Evidence for Membrane Differentiation
within the Golgi Apparatus 101
   6.2.1. Survey of Biochemical Constituents Common
      to All Endomembranes 102
   6.2.2. Changes in Constituents Concentrated in a Particular
      Membrane Compartment 104
6.3. Immunological Manifestations of Endomembrane Differentiation 110
   6.3.1. Evidence from Induced Systems 112
6.4. Mechanisms of Membrane Differentiation 118
   6.4.1. Biosynthetic Contributions to Membrane Differentiation 118
   6.4.2. Selectivity of Membrane Differentiation Mechanisms 119
   6.4.3. Golgi Apparatus Polyribosomes – A Means to Achieve
      Selective Addition of Proteins? 119

6.4.4. Examples of Selective Enzyme Deletion 120
6.4.5. Summary of Flow-Differentiation Mechanisms 122
6.5. Functional Significance of Flow-Differentiation of Membranes 124
6.5.1. Cell, Tissue and Organ Differentiation 125
6.6. Dynamic Aspects of the Flow-Differentiation of Membranes –
Membrane Flow 125
6.6.1. General Morphological Basis for Membrane Flow 126
6.6.2. Kinetics of Membrane Flow 130
6.6.3. Bulk Flow of Membrane Lipids 133
6.6.4. Evidence from Induced Systems 134
6.6.5. Energetics of Membrane Flow-Differentiation
and Problems of Regulation 135
6.7. Summary 135

7. Biochemistry **137**

7.1. Introduction 137
7.2. Enzymology of the Golgi Apparatus 138
7.3. Glycosphingolipid Synthesis 143
7.4. Nucleotide Sugar Transporters 146
7.5. Golgi Apparatus Markers for Medial Cisternae 147
7.6. Lipid Composition 147
7.7. Phospholipid Biosynthesis 148
7.8. Protein Composition of the Golgi Apparatus 149
7.9. Summary 152

8. Function in Secretion **155**

8.1. Role of the Golgi Apparatus in Secretion 155
8.2. A General Model for Golgi Apparatus Functioning in Secretion 156
8.3. Specific Examples of Golgi Apparatus Secretion 158
8.3.1. Enzyme and Proenzyme Secretion by Acinar Cells
of Pancreas and Parotid Gland 158
8.3.2. Secretion of Lipoprotein Particles by Liver Parenchymal
Cells and Adsorptive Cells of the Small Intestine 161
8.3.3. Mucin Secretion 162
8.3.4. Cell Walls and Cell Wall Units 166
8.3.5. Hormones 173
8.3.6. Fat-Soluble Vitamins and Essential Oils 173
8.3.7. Protein Secretion 175
8.3.8. Simple Sugars, Ions and Other Small Molecules 175
8.4. Processing of Large Molecules: An Integral Aspect of Golgi
Apparatus Function in Secretion 175
8.4.1. Glycosylation of Glycoproteins 176
8.5. Signal Hypothesis 178
8.6. Control of Secretion 179
8.7. Segregation of Lysosomal Enzymes 181
8.8. NSF, SNAPS, and SNARES in Membrane Fusion
and the Regulation of Membrane Traffic 183
8.9. Summary 185

9. Replication 187

9.1. A Mechanism of Golgi Apparatus Multiplication 187
    9.1.1. Extension of Forming Face Regions 187
    9.1.2. Appearance of Cisternae with Twice Normal Diameters
          at the Forming Face 189
    9.1.3. Replicating Forms having Two Stacks of Cisternae with Normal
          Dimensions on Top of a Single Stack of Cisternae with Twice
          Normal Dimensions 189
    9.1.4. Separation into Two Stacks having Normal Dimensions 191
    9.1.5. Control of Multiplication of Golgi Apparatus Stacks 191
9.2. Precisternal Stages of Golgi Apparatus Ontogeny 191
9.3. Golgi Apparatus Fragmentation and Reformation during Mitosis 194
9.4. Experimental Golgi Apparatus Fragmentation and Reformation 194
9.5. Summary 196

10. Cell-Free Analysis 197

10.1. Cell-Free Systems Development 197
10.2. Cell-Free Transfer Assay Development 198
10.3. Reconstitution of Transitional Endoplasmic Reticulum
     to Golgi Apparatus Transfer 198
    10.3.1. Fidelity and Efficiency of Cell-Free Transfer in Rat Liver:
           Comparison to Studies with Liver Slices and Tissues 203
    10.3.2. Donor and Acceptor Specificity 203
    10.3.3. Temperature Dependence and 16° C Temperature Block 203
    10.3.4. Lipid and Protein Cotransfer 204
    10.3.5. Processing of Transferred Constituents: Evidence for Functional
           Fusion of Donor and Acceptor Compartments 205
    10.3.6. Lipid Processing 205
    10.3.7. Glycoconjugate Processing 207
10.4. Nucleoside Triphosphate Dependence of Endoplasmic
     Reticulum to Golgi Apparatus Membrane Transfer 208
10.5. Reconsititution of Golgi Apparatus to Plasma Membrane Transfer 211
    10.5.1. Cell-Free Transfer in Cultured Cells 211
    10.5.2. Cell-Free Transfer in Yeast 213
    10.5.3. ATP-Independent Vesicle Budding 214
10.6. Cell-Free Membrane Transfer in Plants 214
10.7. Model for ATP-Dependent Vesicle Budding based on the Rat
     Liver System 215
    10.7.1. Retinol Stimulation of Vesicle Budding in Rat Liver 218
10.8. Summary 219

11. Growth and Cell Enlargement 221

11.1. Golgi Apparatus and Growth 222
    11.1.1. Inhibitor Studies 222
11.2. Evidence from Tip-Growing Cells for a Role of Golgi
     Apparatus Activity in Cell Enlargement 228

11.3. Physical Membrane Displacement................................ 228
    11.3.1. Membrane Budding 229
11.4. Energy Requirements for Physical Membrane Displacement........... 233
11.5. Summary 237

12. Cancer **239**

12.1. The Ultrastructural Cancer Phenotype of the Endomembrane System.... 239
    12.1.1. Rough (with attached ribosomes) Endoplasmic Reticulum...... 239
    12.1.2. Smooth Endoplasmic Reticulum 241
    12.1.3. Golgi Apparatus 241
    12.1.4. Plasma Membrane 245
12.2. Role of Endomembranes in Signal Transduction and Oncogene
    Expression in Cancer.. 248
12.3. Summary 251

Epilogue **253**

Appendix Tables **257**

References **271**

Index **301**

Discovery and Rediscovery

The era of Golgi apparatus discovery may be divided conveniently into three phases – the initial discovery (1865–1925), the controversy (1925–1955), and the modern rediscovery (1955–1963) (see chronology of events, Appendix Table 1 p. 257). Although doubtless seen by others earlier, the discovery of the Golgi apparatus is ascribed to the Italian cytologist Camillo Golgi (1898) who described an *apparato reticolare interno* (internal reticular apparatus) in his now famous "Sur la structure des cellules nerveuses." The description was based on light microscopy of nerve cells of the barn owl and cat made possible through experiments in specimen preparation, where tissues were placed in a silver nitrate bath after preliminary fixation in a solution of bichromate. This method, in modified form, persists today as the "Golgi–Cox" method of preparing nervous tissue for light microscopic examination. Various elements of the nervous system (i.e., Purkinje cells) are rendered dark brown or black against an almost clear background. The modern equivalent of what Golgi initially described is still not clear. It was a darkly staining internal reticular apparatus (Fig. 1). There is little doubt that the region of the cell presently equated with the Golgi apparatus was included in Golgi's observation but the possibility remains that other cell components, such as portions of the endoplasmic reticulum which may be darkly stained by the method, also were included.

Many workers, in the period 1915–1945 and beyond, applied Golgi's methods or variations thought to stain an equivalent region of the cell. They equated a variety of structures with the apparatus of Golgi but not in all cell types and tissues. Various misidentifications were major contributions to an element of doubt as to the reality, generality, and function of the Golgi apparatus. This era between 1925 and 1955 is often referred to as the era of *Golgi apparatus controversy*. It was not until the advent of the electron microscope and the publications of Dalton and Felix in 1953 and of Sjöstrand and Hanzon in 1954 that the modern era of Golgi apparatus discovery would begin.

1.1. The Discovery of the Golgi Apparatus

Camillo Golgi, for whom the Golgi apparatus is named, was born on July 9 (July 8 by some accounts), 1843 in the Italian town of Corteno in the province of Lombardy. The son of a physician, he studied medicine, and accepted a position at the University of Pavia where he worked, with two interruptions, from 1865

D. James Morré and Hilton H. Mollenhauer, *The Golgi Apparatus.*
© Springer 2009

Camillo Golgi   Santiago Ramon y Cajal

Fig. 1.1. Camillo Golgi and Santiago Ramon y Cajal.

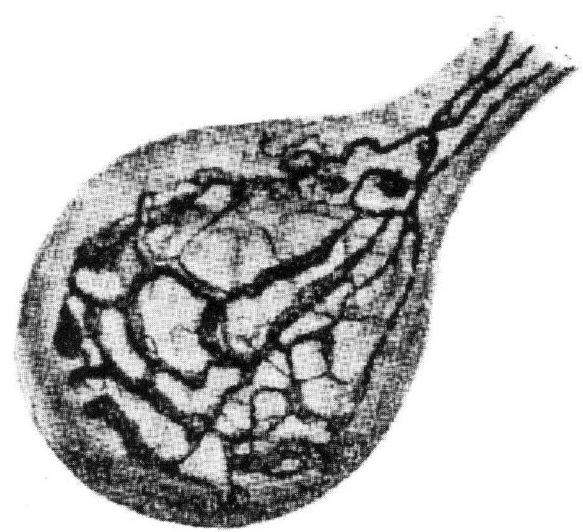

Fig. 1.2. An original drawing from the work of Golgi (1898) of the internal reticular apparatus as seen in a Purkinje cell of a barn owl.

until shortly before his death in 1926. It was on one of his absences from Pavia (1872–1875) while resident physician at the Home for Incurables at Abbiategrosso near Pavia that he discovered his chromate of silver method (*la reazione nera*) that was to revolutionize the study of the nervous system and eventually made possible the observation of the *apparato reticolare interno*.

Much is known about events in Golgi's life preceding the discovery of the apparatus and a modern account is given in the book by Whaley (1975). Most Golgi followers find their first reading of the original account a disappointment. What is presented is a brief description of what Golgi observed along with a pen and ink drawing. This is entirely in character with Golgi's style of writing which was to limit himself to a concise description of the morphology of the nervous system. It is generally agreed that Golgi and his followers considered that they had discovered a new cell component and were aware of its potential importance to secretion but most was left for contemporaries to sort out.

One such contemporary of Golgi was Santiago Ramón y Cajal (1852–1934), a Spanish physician, who, with Golgi, shared a Nobel prize in 1906 for work on the nervous system (Cajal, 1923). The views of Cajal differed from those of Golgi. Cajal viewed each nerve cell as a separate entity whereas Golgi's view of the nervous system was that of a continuous network. As to the internal reticular apparatus, Cajal accepted Golgi's view, refined Golgi's impregnation methods, established the generality of the structure, and contributed substantially to the beginnings of a functional understanding. A major account, published in 1914, marks the real beginning of early Golgi apparatus discovery so much so that Cajal might be justly credited as a co-discoverer. It is in Cajal's writings that subsequent generations found the impetus to probe deeper. Cajal, perhaps more than Golgi, established that the internal reticular apparatus was a new cell component, that it existed in diverse cell types, and that its form and appearance changed during differentiation of the cell and with changes in metabolism. He noted the relationship between the apparatus and the region of the cells that contain the centrioles, and its changes with activity of secretory cells. Basically, he proposed that the apparatus enclosed materials that were consumed during periods of activity and that accumulated during the quiescent phases. He was aware of the appearance of the apparatus and its changes under many conditions and came very close to deducing its correct role as a component of the cell's secretory apparatus. Cajal's papers might be profitably read even today in terms of certain aspects of Golgi apparatus dynamics and susceptibility to postmortem change often overlooked in the design of modern biochemical and molecular investigations.

A role of the Golgi apparatus in secretion was implicit in most of the early accounts. The work of Nassonov (1923, 1924) following that of Negri on Golgi apparatus function in the parotid and pancreas (see Whaley, 1975) was important in that a consistent association between secretory products and Golgi apparatus was noted in addition to staining reactions common to the two (Bowen, 1929). Also, important was the paper by Fuchs (1902) concerning the epididymal epithelium of the mouse. He cautioned correctly, for example, that in certain instances the Golgi apparatus might serve only to function as an intermediate between synthesis and final discharge of secretory materials.

The value of Nassonov's contributions were heightened by the fact that he was able to reach certain conclusions bearing correctly on Golgi apparatus function. He noted that secretory granules made their initial appearance within the Golgi apparatus meshwork. As a second step, he concluded that the granules upon reaching a certain size were released from the apparatus and collected near the luminal surface of the cell. He also noted that in different cells different materials were processed through the apparatus. Finally, he concluded that formation of the granules containing secretory materials and the eventual discharge via the granules of the secretory materials from the cell were separable events. Granule formation was ascribed to an immediate activity of the Golgi apparatus per se, while the exteriorization process was considered to proceed without further Golgi apparatus involvement.

Much of the early work leading to concepts of considerable importance would be virtually inaccessible to the hurried contemporary cell biologist were it not for Bowen's epic review "The Cytology of Glandular Secretion" published in 1929. Bowen concluded both from his own work and his extensive familiarity with the literature that secretion was a cellular process whereby products destined for export were collected or built up in the Golgi apparatus and then separated from it. As these materials were being discharged to the cell's exterior, new secretory granules would take their place through continued activity of the Golgi apparatus. The Golgi apparatus always appeared to remain intact during successive cycles of secretion and details of the process varied from one cell type to another. Bowen was also to set the stage for the concept of Golgi apparatus function as an integral part of a complex system of internal cytomembranes (the Endomembrane System of Chapter 2). This is found in his comment, "apparently the Golgi apparatus plays some immediate role in the process of accumulation and final synthesis of the secretion products, but the concomitant changes in other cellular structures suggest that all parts of the cell contribute in some way."

1.2. The Controversy

Much of what was to be learned in the light microscopy era [see book by Wilson (1925), for example] was discovered by 1925 and can be found in Bowen's (1929) review. Subsequent investigations only served to complicate the problem to the extent that were it not for the coming of modern electron microscopy techniques, the Golgi apparatus might have been relegated to the category of artefact. To appreciate this, one must realize that study of the Golgi apparatus in all of its variations from cell to cell and even within a single cell type using techniques then available to light microscopists was a formidable task. Only with information derived using much more adequate instrumentation was it possible to move ahead. Some of the controversial literature arising in the 1930s and 1940s became so involved as to defy resolution even with modern techniques.

Exemplary in this regard was the analogy between the Golgi apparatus and the Canals of Holmgren (1902) which led to assumptions concerning the equivalence of Golgi apparatus and elements of the vacuome (vacuolar apparatus/vacuoles). An extreme view elaborated by Parat and Painlevé (1924a, b) was that all animal and plant cells have two fundamental and independent morphological elements, the vacuome and the chondriome (mitochondria). The vacuome was regarded as an aqueous phase stained specifically with neutral red, while the chondriome was a lipid phase stained by lipid-soluble dyes. The reticular apparatus of Golgi and the Canals of Holmgren were initially regarded as artefacts produced by precipitation of silver or osmium at the surface of, inside, or between the vacuoles. The whole subject of neutral red cytology was attacked by Gatenby in 1931 but some aspects of the arguments of Parat may have been correct. Elements of the chondriome are now equated with mitochondria and the possibility remains that the Holmgren canals were endoplasmic reticulum.

Yet another aspect of the controversy was contributed over the period 1944 to1963 by Baker and colleagues at Oxford who denied the existence of the apparatus and explained their observations and those of others to modification of spherical bodies which were stained with Sudan black, a lipid stain. Mostly they were unable to find any signs of a network using standard impregnation techniques and phase contrast microscopy. The situation was further clouded by the inability of various workers to visualize the typical network of fixed preparations in unfixed living cells. No matter that Hirsch (1939) pointed out that the inability to see the apparatus in living cells was no indication that it was not present since it had a refractive index identical with that of the cytoplasm. In fact, Ludford (1925) published photographs of living tissue culture cells stained in vitro with methylene blue in which a distinct Golgi apparatus area was visible near the nucleus. The area appeared somewhat reticular. Ludford also presented images of cells photographed in ultraviolet light where the Golgi apparatus region occupied a half-moon area around the nucleus and appeared reticular. In the same year, Strangeways and Canti (1927) denied categorically that the Golgi apparatus existed.

Doubtless much artefact contributed to the controversies of the light microscope era. Palade and Claude (1949a, b) unequivocally showed that Golgi apparatus-like structures could be created in cells by the fixatives used in the Golgi impregnation techniques. These were actually myelin figures probably produced from phospholipids.

The situation in plant cells became hopelessly complex. Bowen first developed an idea that the plastids might equate to the plant Golgi apparatus. He later returned to the question and correctly described structures, termed osmiophilic platelets, which he equated to the animal Golgi apparatus. Strong evidence in support of this concept was provided by Beams and King (1935), who used centrifugations of whole tissues to demonstrate that the Golgi apparatus of animal cells and the osmiophilic platelets of plant cells were displaced to the same level. However, the majority of plant papers were concerned with other approaches. In the absence of definitive correlative information Nahm (1940) concluded, based on a thorough review of the literature, that a Golgi apparatus equivalent did not exist in plant cells.

It is impractical to mention all of the various aspects of the period of Golgi apparatus controversy much less analyze the vast literature dealing with the subject. This brief summary may provide some indication of the level of confusion that existed prior to the electron microscope era. An excellent review is that of Bourne and Tewari (1964).

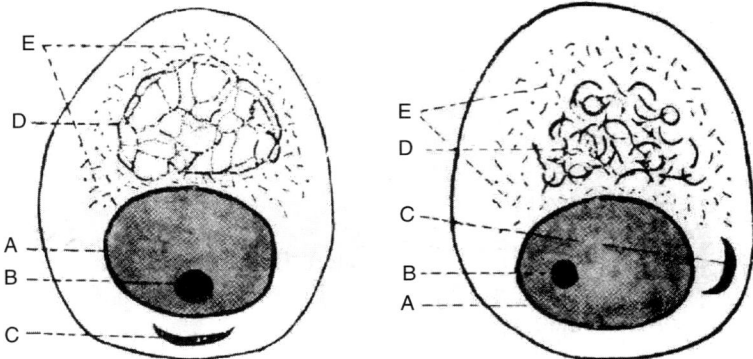

Fig. 1.3. Original diagram of Perroncito (1910) illustrating dictyokinesis and the elements to which he applied the term dictyosome. *Left*: A cell before dictyokinesis begins. A = nucleus, B = nucleolus, C = centrosome, D = Golgi apparatus, E = chondrosomes of Meves (mitochondria). *Right*: Dictyokinesis in progress. A = nucleus, B – nucleolus, C = centrosome, D = dictyosomes, E = chondrosomes of Meves (mitochondria).

1.3. Modern Rediscovery

The first micrographs correctly identifying the Golgi apparatus in the electron microscope were published by Dalton and Felix (1953). These micrographs were available to Gatenby for his 1955 review and were accompanied in 1954 by papers by Sjöstrand and Hanzon describing the Golgi apparatus of mouse pancreas by electron microscopy. The structures took the form of a mixture of large vacuoles, flattened sacs, and groups of vesicles. In 1956, Dalton and Felix showed that all these parts of the Golgi apparatus reduced osmium tetroxide and helped to form the classic reticulum of Golgi. The Golgi apparatus equivalents in invertebrates, the osmiophilic platelets or dictyosomes, were also shown by electron microscopy to be homologous with the Golgi apparatus of vertebrate cells (Dalton and Felix, 1956; Beams and colleagues, 1956). Similar structures were subsequently found in a variety of plant cells (Porter, 1957; Buvat, 1957a, b; Heitz, 1957a, b, c; Perner, 1957, 1958; Sitte, 1958; Dalton and Felix, 1957; Charder and Rouiller, 1957; Sager and Palade, 1957). A portion of a Golgi apparatus stack (dictyosome) is illustrated but not identified in a 1956 electron microscope study by Hodge et al. (1956) with *Nitella*.

Thus, within a span of about 2 to 4 years, the advanced technology afforded by the electron microscope erased nearly 40 years of controversy generated from light microscopy. There were some who resisted. Kanwar (1961–1962) argued that the so-called "Golgi apparatus" of electron microscopy was not homologous to the light microscope "apparato reticolare interno" of Golgi and not until 1963 did Baker reluctantly become a follower.

1.4. Summary

The "apparato reticolare interno" discovered by Camillo Golgi and published in the late 1800's resulted from work with silver impregnation techniques developed to visualize neuronal networks. The 50 years following the first report were years of controversy. Even the reality of the structure was openly questioned by some while others proceeded with structural and functional studies. It remained for electron microscopy to end the "Golgi controversy" at the beginning of the 1950s with the description of a morphological entity with the requisite characteristics to be called the Golgi apparatus. In the 50 years that followed the modern rediscovery, rapid progress was made in the cytochemical differentiation of the cisternal stack, and elucidation of its role in secretion and in post-translational modifications including demonstrations by autoradiography and by classical biochemical analyses applied to isolated Golgi apparatus. The roles of the Golgi apparatus in secretion were firmly established. Development of cell-free systems and of immunochemical and molecular probes of Golgi apparatus function provided the beginnings for the modern era.

2

Structure

With no cell compartment or organelle has morphology served such a pivotal role in its discovery and investigation as with the apparatus of Golgi. The original description of the "apparato reticulo interno" (internal reticular apparatus) now known as the apparatus of Golgi or Golgi apparatus was based on light microscopy (Chapter 1). Both classical Golgi apparatus study prior to 1953 and the modern rediscovery due to the advent of the electron microscope all were based on morphology. An understanding of Golgi apparatus architecture was one of the more important early developments resulting from electron microscopy.

Morphology was the sole basis to guide early attempts at Golgi apparatus isolation and, until its isolation in the mid to late 1960s and early 1970s, and the introduction of autoradiography (Peterson and LeBlond, 1964), morphology was the only basis for investigation of this complex cellular component. Even today, morphology remains as a major criterion by which Golgi apparatus are defined.

Biochemical definitions of Golgi apparatus are complicated by the fact that Golgi apparatus (either singular or plural) are transitional cell components sharing many biochemical characteristics with either the endoplasmic reticulum or the plasma membrane or both. Unlike the situation with true organelles such as chloroplasts or mitochondria, there have been virtually no biochemical markers common to all cell types and not shared with either endoplasmic reticulum or plasma membranes. Additionally, the functioning of Golgi apparatus in secretion and other activities is highly dependent upon the co-participation of endoplasmic reticulum, secretory vesicles, and other structures. Thus, the Golgi apparatus exists in the cell as a component within a highly integrated endomembrane system (Morré and Mollenhauer, 1974) often with biochemical characteristics intermediate between those of the endoplasmic reticulum and those of the plasma membrane, but with a characteristic and easily recognized pattern or morphology that unambiguously distinguishes the Golgi apparatus from all other cell components.

The most common form of the Golgi apparatus, exemplified by most mammalian cells, consists of side by side piles or stacks of smooth membrane cisternae lacking ribosomes (Fig. 2.1). The stacks are seen to be distributed and located within a special region of cytoplasm called the Golgi apparatus zone or Golgi apparatus matrix (Fig. 2.2). In plants, and in many animal cells as well, individual stacks of cisternae may be spread more or less evenly, through the entire cytoplasm. These stacks, even though dispersed, function synchronously indicating that they are functionally and perhaps structurally interconnected.

D. James Morré and Hilton H. Mollenhauer, *The Golgi Apparatus.*
© Springer 2009

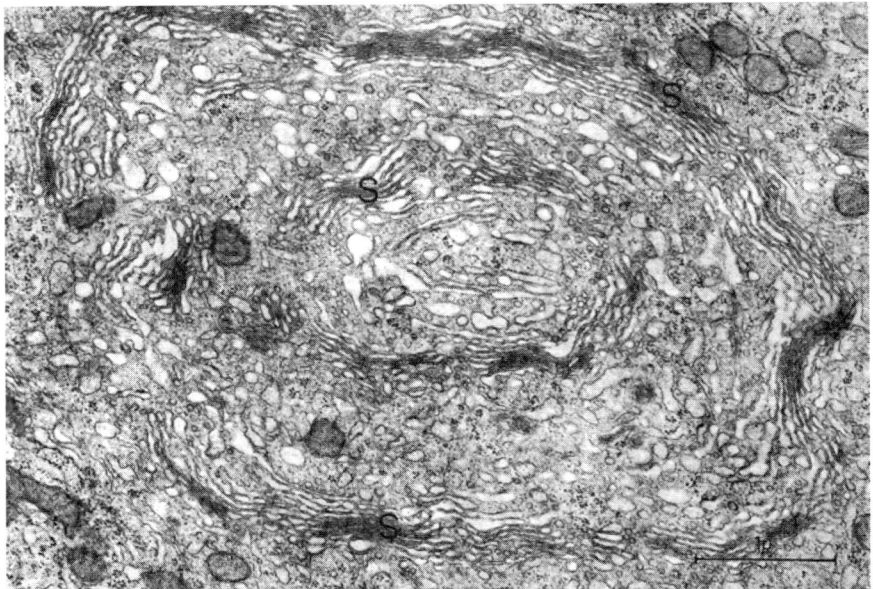

Fig. 2.1.   Portion of the Golgi apparatus of rat epididymis showing numerous closely-spaced stacks (S). Karnovsky's fixative. From Flickinger (1969b). Electron micrograph courtesy of Dr. C. J. Flickinger, University of Virginia Medical School, Charlottesville. Reprinted by permission of Academic Press, Inc., New York.

Vesicles and tubules either may be attached to or associated with various parts of the Golgi apparatus. Structural features at each level of organization will vary depending on cell type, method of fixation, and of specimen preparation and the physiological state of the cell. However, those emphasized may be observed as consistent features of most electron microscope preparations.

2.1. Cisternae

By definition, a cisterna (plural = cisternae) is a sac or cavity within a cell or organism, usually filled with fluid (i.e., a cistern). The term was used originally in electron microscope morphology to describe one of the interconnected vesicles, lamellae or tubules comprising the endoplasmic reticulum but has served the same purpose for the Golgi apparatus. Each cisterna of the Golgi apparatus consists of a lumen or central cavity surrounded by a membrane lacking ribosomes (one of the several so-called "smooth" membranes that frequently contribute to the smooth microsome fraction of cell homogenates). Cisternae will differ in architectural detail depending on cell type and their position in the stack (Figs. 2.3 to 2.7). Perhaps no two are exactly alike. Many cisternae especially from the mid-region of the stack (i.e., the intercalary cisternae) often have

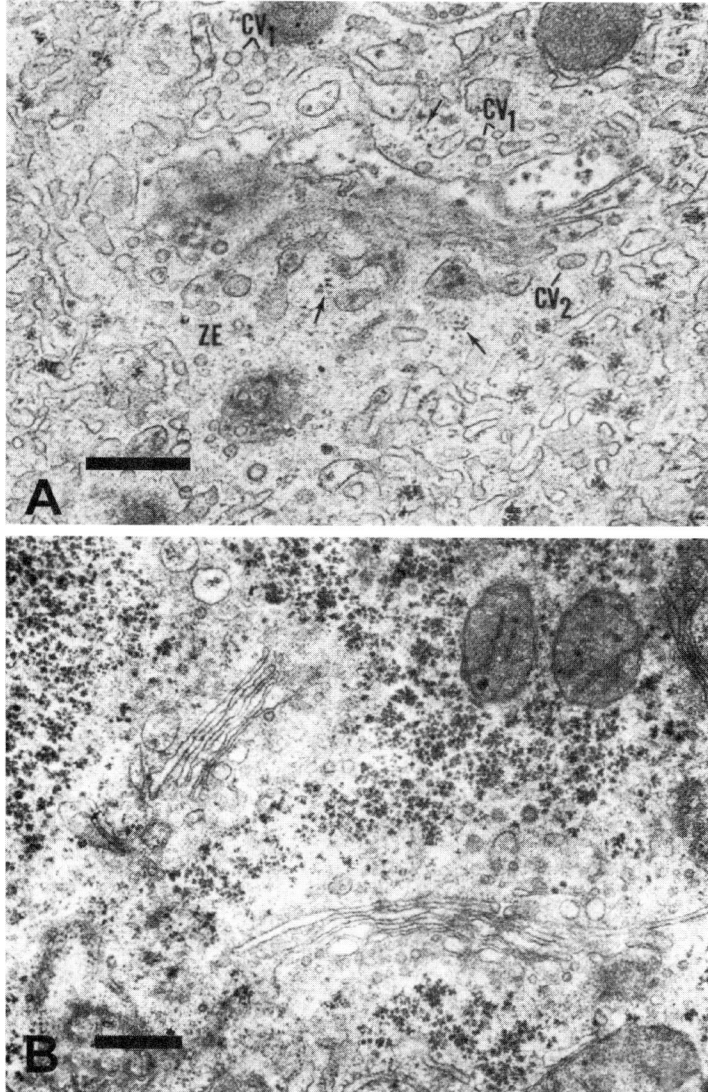

Fig. 2.2. Golgi apparatus regions showing the zone of exclusion or Golgi matrix. Coated vesicles (CV) of the Golgi apparatus regions are restricted to this zone. Elements of endoplasmic reticulum entering the zone are smooth (lacking attached ribosomes). Free polyribosomes (Golgi apparatus polyribosomes) are frequently observed within the zone. (A). Normal rat liver. (B). Rat hepatomas induced by N-2-fluroenylacetamide (FAA). The hepatoma cells contain a Golgi apparatus with dispersed stacks but still each stack is surrounded by a zone of exclusion. Reproduced from Mollenhauer and Morré, 1978b with permission from Springer Science+Business Media. Scale bar = 5 μm.

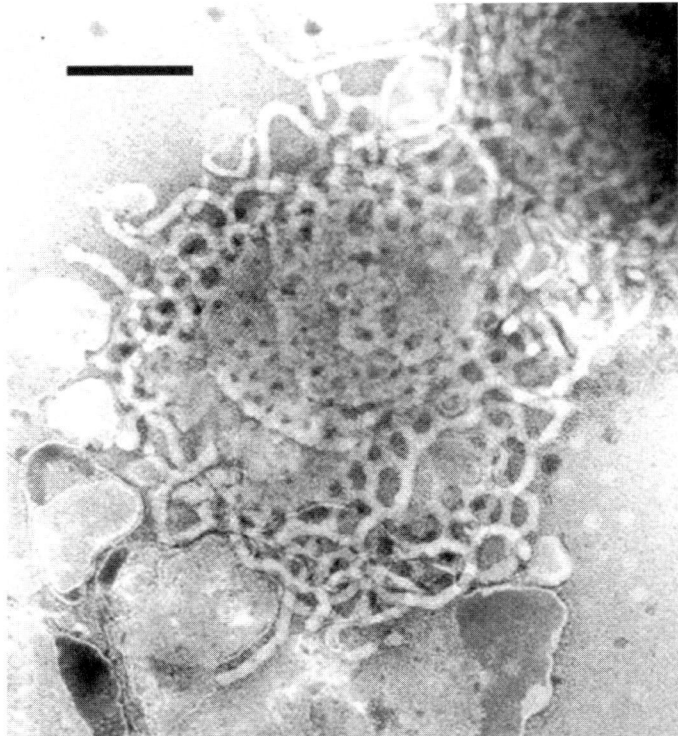

Fig. 2.3. A Golgi apparatus stack from radish root isolated and viewed in negative contrast directly on an electron microscope grid. The stack has been partially unstacked to reveal a progression change diagrammed in Figure 2-4 from the cis (top) to trans (bottom) face of the stack. Reproduced from Mollenhauer and Morré (1966) from the Journal of Cell Biology, 1966, 29:373-376. Copyright 1966b The Rockefeller University Press. Scale bar = 0.2 μm.

small perforations or fenestrae at the margins. This central portion of the cisternal stack is referred to also as the central *saccule* or simply as the saccule. Normally, 4 to 6 such saccules are arranged one on top of the other, in parallel array, to yield the characteristic "stack" or dictyosome as one of the more obvious features used to identify Golgi apparatus both in situ and isolated in cell fractions (Chapter 3). The plate-like regions, up to the fenestrated margins, are typically 0.5 to 1 μm in diameter (Fig. 2.5).

Continuous with the central plate like region is a complex system of tubules and secretory or coated vesicles (Fig. 2.9). Secretory vesicles lack coats or are partially coated over their surfaces with spiny clathrin coats. Vesicles of the Golgi apparatus zone coated either with clathrin or coatomer proteins (COP) may be either free or attached to tubule ends (Mollenhauer and Morré, 1966b). The connecting tubules that attach secretory vesicles to saccules are short (Fig. 2.7) and begin as narrow partitions extending beyond the fenestrated peripheries of the central plates. Longer tubules may continue for several microns and follow

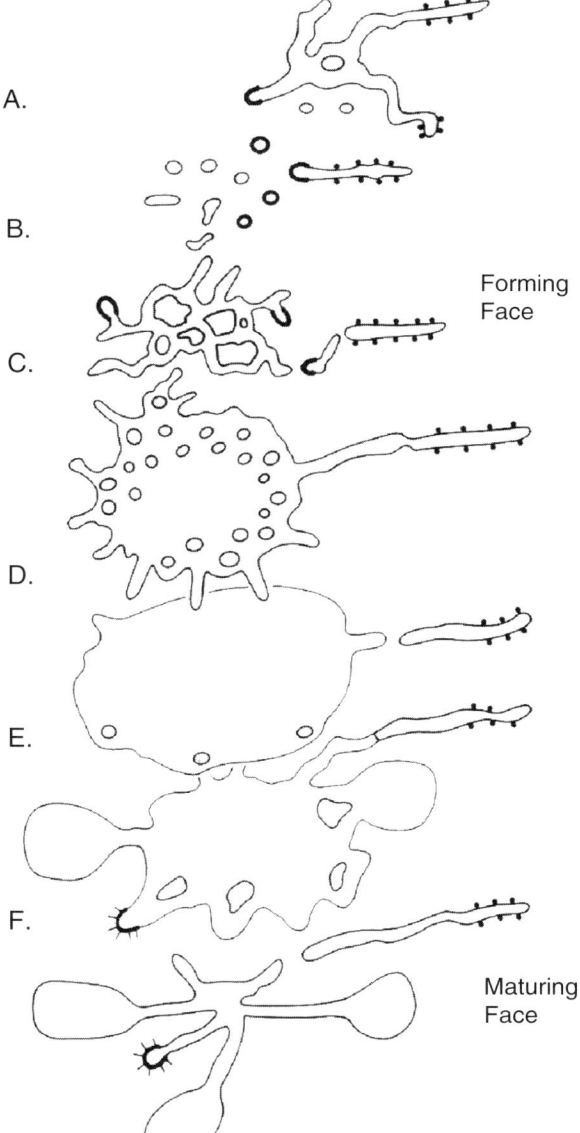

A.

B.

C.   Forming
                                                                             Face

D.

E.

F.

                                                                             Maturing
                                                                             Face

Fig. 2.4.   Diagrammatic representation of successive cisternae within a single stack as partially revealed in Figure 2-3. (A). Part rough (with ribosomes) -part smooth (lacking ribosomes) transitional endoplasmic reticulum with (B). COPII—coated buds and vesicles which (C). coalese to form the largely tubular first Golgi apparatus cisternae of the forming face. (D). Continued delivery of ER-derived material results in fenestrated cisternae with a central plate-like region and tubular peripheries. (E). As cisternal maturation proceeds, plate-like regions dominate. (F). Secretory vesicles connected to plate-like regions by tubules characterize the maturing face. (G). The post Golgi structures or trans Golgi network = cisternal remnants with clathrin-coated membrane and vesicles represent the final stage of cisternal maturation and utilization in secretory vesicle formation.

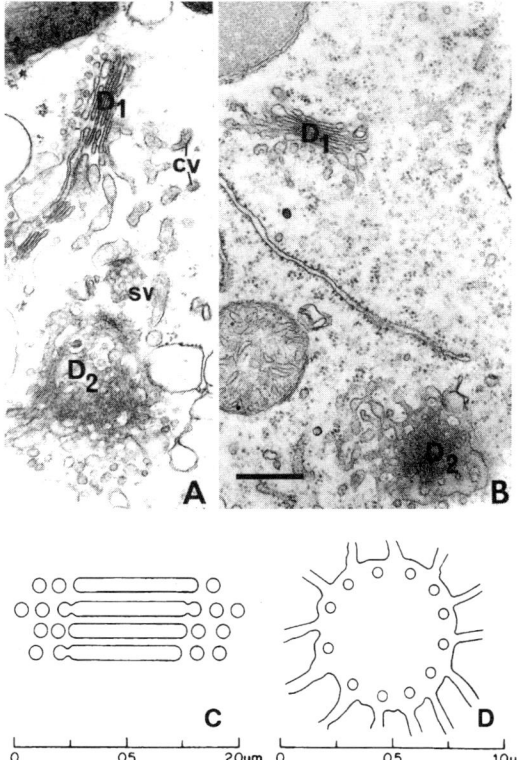

Fig. 2.5. Golgi apparatus stacks isolated (A) and in situ (B) from soybean illustrating the complementary cross sectional (D_1) and face (D_2) views. SV = Secretory vesicle. CV = clathrin-coated vesicle. Reproduced from Morré, 1977a with permission from the author. Scale bar = 0.5 μm. C, D. Diagram illustrating the correspondence of the cross-sectional and face view images of stacked Golgi apparatus cisternae. The fenestrated (with openings or holes) and/or tubular peripheries seen in face view (D) are represented by small vesicles or interruptions in the flattened saccules when viewed in cross section (C). Reprinted from Morré and Ovtracht, 1981, Copyright 1981 with permission form Elsevier.

an irregular course through the cytoplasm. The full extent of such tubules, or whether they interconnect to distant stacks, has not been determined.

In intercalary cisternae, the solid plate like regions seem to dominate (Figs. 2.4 and 2.5). Exterior cisternae tend to be more fenestrated with a dominance of tubular elements. These differences in morphology observed within a single stack contribute to the polarity of the dictyosomal stacks discussed in the next section. The interconnected system of plates, tubules, and vesicles that constitute Golgi apparatus cisternae allows for considerable subcompartmentation and restriction of functional activities to specific regions even within a single cisterna.

In the diagram of Fig. 2.4 based on analysis of negatively stained partially unstacked plant (Fig. 2.3) and animal (Fig. 2.10) preparations as well as serial section analysis (e.g., Brown and Arnott, 1971; Fig. 2.8), predominantly tubular cisternae are present at the pole proximal to the nuclear envelope (cis face) while

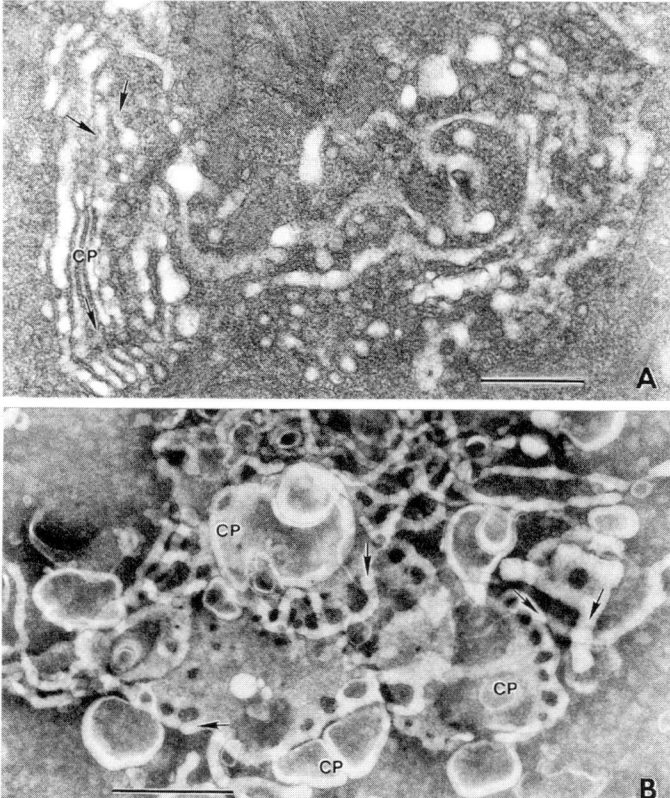

Fig. 2.6. Tubular peripheries of Golgi apparatus cisternae (arrows) of cat trachael epithelia fixed in situ with a tannic acid-containing fixative (A) and seen by negative staining in an isolated Golgi apparatus preparation (B). CP = central plate-like portion of cisternae. Reproduced from: Tandler and Morré, 1983 with permission from Springer-Wien. Scale bar = 0.5 μm.

plate-like regions become dominant toward the center of the stack. Cisternae at the pole most distal from the nuclear envelope (trans face) may again present a more tubular aspect although the form of the tubules is quite distinct from those at the cis face.

The extent of the system of peripheral tubules associated with the Golgi apparatus first became evident from preparations of plant Golgi apparatus stabilized with aldehyde fixatives during isolation and visualized by negative staining with phosphotungstic acid (Mollenhauer and Morré, 1966a; Cunningham et al., 1966; Fig. 2.6B) and later extended to animal cells (Mollenhauer et al., 1967; Morré and Ovtracht, 1981; Fig. 2.10). The tubules, which are 300–500 Å in diameter, may serve to connect cisternae of adjacent dictyosomes (Fig. 2.11), function as attachment sites for secretory vesicles (Figs. 2.7 and 2.12), and seem to facilitate direct connections between Golgi apparatus cisternae and smooth portions of the endoplasmic reticulum (Fig. 2.13). Some authors have argued that the tubules

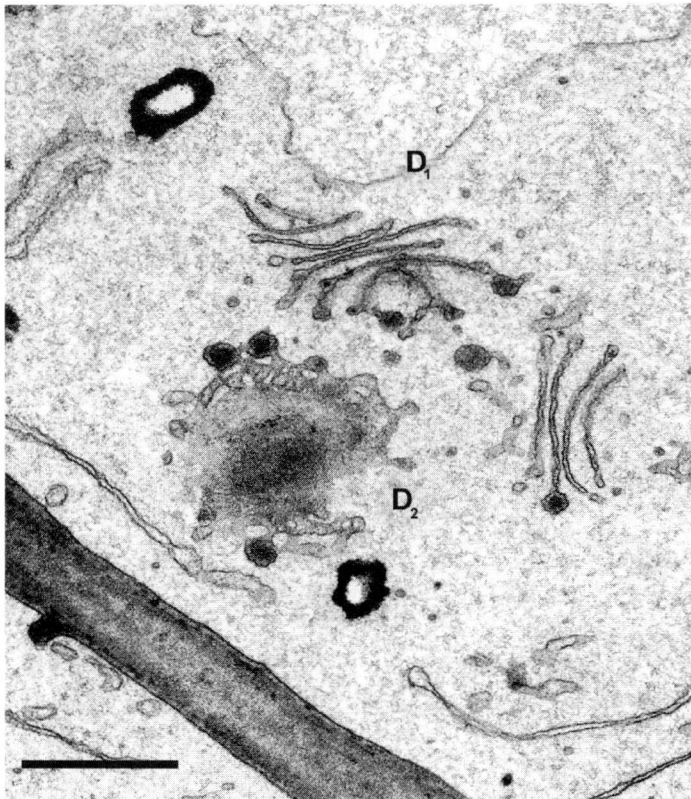

Fig. 2.7.   Dictyosomes (cisternal stacks) from epidermal cells of maize root fixed with potassium permanganate and contrasting cross-sectional (D_1) and face (D_2) views. In face view, the cisternal peripheries are seen to exhibit the tubular structures characteristic of other fixatives. Reproduced from Mollenhauer and Morré, 1994 with permission from Springer-Wien. Scale bar = 0.5 μm.

and fenestrae may be artefacts due to extraction of proteins from the Golgi apparatus membranes during preparation of the specimens for staining (Cunningham et al., 1974). Clearly, prolonged contact between Golgi apparatus and any concentrated salt solution will result in extraction of lipids and proteins which may result in morphological modifications. However, cisternae tubules are observed *in situ* following fixation with glutaraldehyde–osmium tetroxide (Fig. 2.5), glutaraldehyde–tannic acid–osmium tetroxide (Fig. 2.6), osmium tetroxide alone (not shown), or potassium permanganate (Fig. 2.7). They are present in freeze–fracture–etch preparations in which no fixatives are involved (Fig. 2.14). While care must be exercised in the interpretation of negatively stained images, especially of small or fragmented structures, Golgi apparatus tubules in situ and in isolated preparation, negatively stained, however, are similar in extent and appearance (Tandler and Morré, 1983; Fig. 2.6) which argues against them being solely artifactual.

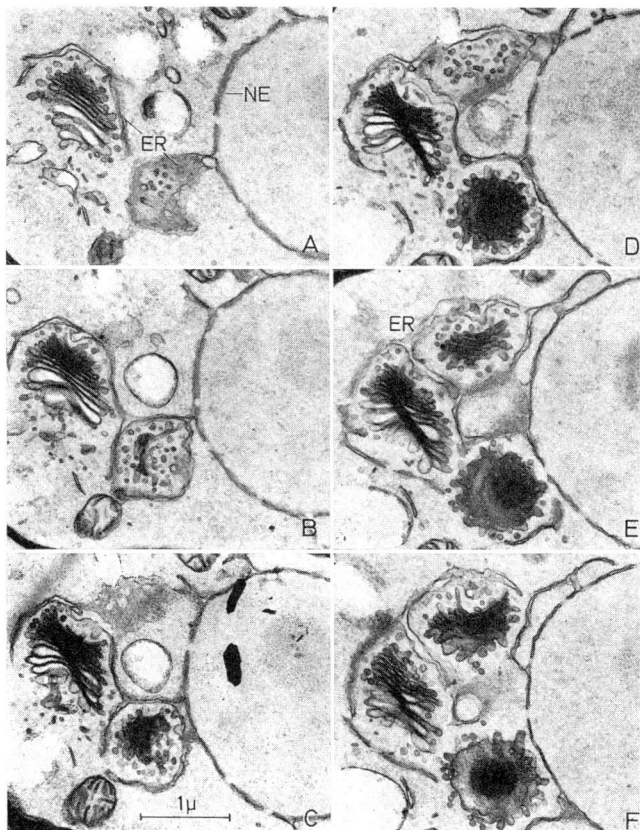

Fig. 2.8. Serial sections through a portion of the endomembrane system of the alga *Tetracystis excentrica* including three adjacent stacks. Sheets of endoplasmic reticulum (ER) continuous with the nuclear envelope (NE) and known as the amplexus, surround each stack, except for a region at the maturing face through which secretory vesicles are discharged. The stack at the lower center is sectioned tangentially and shows successive cisternae in face view beginning with the endoplasmic reticulum at the forming face in A. To the left, a second stack is sectioned transversely and shows the entire stack in cross section. A third stack becomes evident in D–F at upper center and C provides a tangential view of the sheet of endoplasmic reticulum associated with the proximal face of the stack. $KMnO_4$ fixation. Unpublished electron micrographs courtesy of Drs. R. Malcolm Brown, Jr. and Howard J. Arnott, Cell Research Institute, The University of Texas, Austin. From: Morré et al., 1971c. Reproduced by permission of Elsevier. Scale bar = 1 μm.

2.2. The Cisternal Stack or Dictyosome

Golgi apparatus cisternae are usually organized into stacks of five to eight cisternae (Table 2.1). The stacks also are known as *dictyosom*es (Mollenhauer and Morré, 1966a). Twenty or more cisternae per dictyosomal stack are not unusual among lower organisms such as *Euglena gracilis* (Mollenhauer, 1974).

The term dictyosome (from the Greek word *dictyos* meaning net or network = net body) was used originally by Perroncito (1910, see Whaley, 1975, Chapter 1)

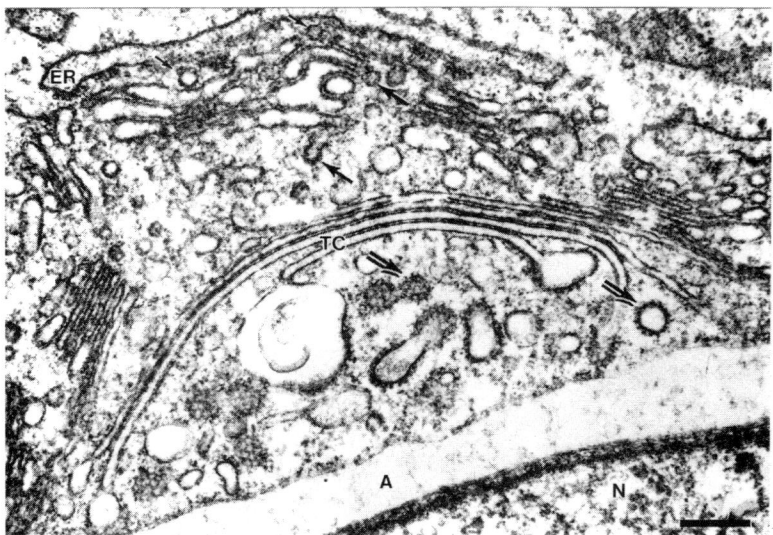

Fig. 2.9.   Golgi apparatus region of a rat spermatid adjacent to the forming acrosome (A) fixed with 37% glutaraldehyde – 2% paraformaldahyde – 10% saturated picric acid containing 1% tannic acid to enhance vesicle coats. Illustrated are COP-coated endoplasmic reticulum-derived transition vesicles at the forming face (single arrows) and clathrin-coated membranes and vesicles of the maturing face (double arrows). Also present is an elaborate post Golgi structure consisting of elongated cisternae with thick membranes (TC). N = nucleus. Reproduced from Mollenhauer, Hass, and Morré, 1976 with permission from the Société Française de Microscopie Electronique. Scale bar = 0.5 μm.

to designate a component of the Golgi apparatus that was visible following cell division and that had a definite pattern of distribution in the daughter cells. The term was also used to indicate a form of Golgi apparatus characteristic of invertebrates which appeared more as discrete units than the complex reticular apparatus first described by Camillo Golgi. The modern usage of the term dictyosome differs little from the historical, and "net body" accurately describes the modern concept of Golgi apparatus structure.

A major difference among species and cell types in regard to Golgi apparatus organization is the distance by which the individual stacks or dictyosomes are separated. There are approximately 500 such stacks in a typical plant or animal cell. These may be arranged side by side as an almost continuous ribbon as seen in many mammalian cells (Figs. 2.1 and 2.15) or they may be so widely separated as to appear as discrete units as in most plants and invertebrates. Generally, the dispersed arrangement is more characteristic of undifferentiated cells and tissues or differentiated cells not involved in protein secretion whereas the compact arrangement becomes most evident in cells specialized for protein secretion.

Dictyosomes are polarized structures in that cisternae at one pole or face of the cisternal stack differ from those at the opposite pole or face (Fig. 6.2). In many cells of animals, algae, and fungi, one pole of each dictyosome is associated with the nuclear envelope or endoplasmic reticulum in a characteristic manner. This pole or "face" of the Golgi apparatus stack and of the Golgi apparatus per se

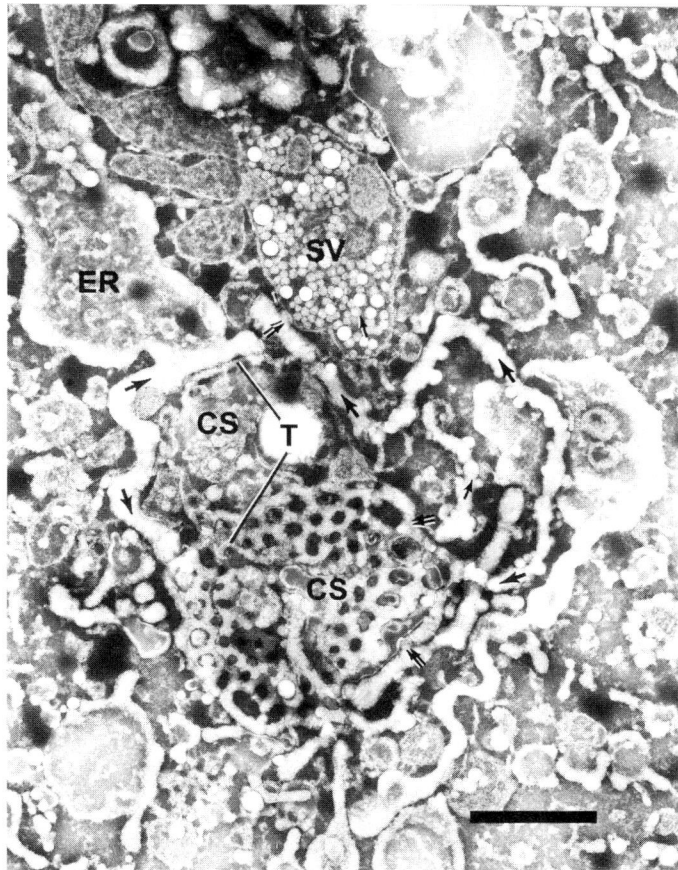

Fig. 2.10. Tubules (T) of the Golgi apparatus (boulevard périphérique) connecting (large arrows) endoplasmic reticulum (ER) and lipoprotein particles (small arrows) containing secretory vesicles (SV) of a Golgi apparatus preparation isolated from rat liver. CS = central saccule. Reprinted from Morré and Ovtracht, 1981, Ultrastruct. Res., 74, 284-295, Copyright 1981 with permission from Elsevier. Scale bar = 0.5 μm.

is referred to as the pole proximal to endoplasmic reticulum or proximal pole, or as the "cis" face. For biogenetic considerations, it has also been referred to as well as the "forming" face. The opposite pole or face of the dictyosome is the distal pole, also known as the mature(ing), secreting, or trans face. Membranes of the proximal pole or cis face cisternae are morphologically and cytochemically similar to the membranes of endoplasmic reticulum (Chapter 6). Toward the opposite pole or trans face, the morphology and staining characteristics of the cisternae become progressively more like those of plasma membrane.

Associations between endoplasmic reticulum (or nuclear envelope) and Golgi apparatus take many forms. One of the first to be noted was that of an endoplasmic (nuclear envelope) cisterna lying parallel to the cis face of the Golgi apparatus (Fig. 2.16). Attached ribosomes were present on the cytoplasmic

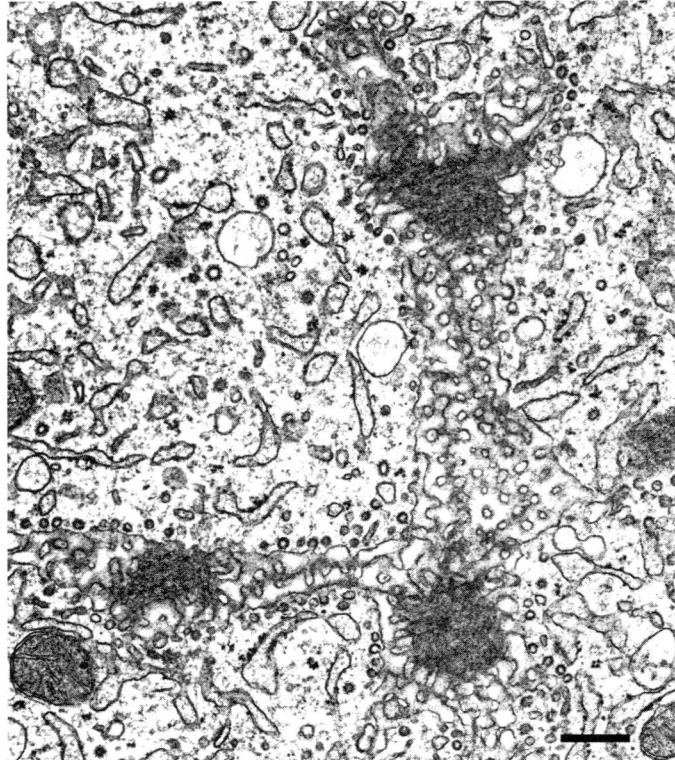

Fig. 2.11. Golgi apparatus stacks from a rat epidermal principal cell illustrating tubular connection between stacks. These and images such as that of Fig. 2-10 suggest that most, if not all, of the stacks of the Golgi apparatus are interconnected and perhaps function synchronously. Reproduced from Mollenhauer and Morré, 1998 with permission from Springer Science + Business Media. Scale bar = 1 μm.

surface of the endoplasmic reticulum (nuclear envelope) on the side opposite the compact Golgi apparatus ribbons or of dispersed Golgi apparatus stacks and in regions of the endoplasmic reticulum not associated with the Golgi apparatus. However, ribosomes were absent from the endoplasmic reticulum (or nuclear envelope) membranes immediately adjacent to the Golgi apparatus. This smooth portion of the endoplasmic reticulum (nuclear envelope) was characterized also by an appearance of forming small evaginations or "blebs" of the cisternal membrane that extended into the intercisternal space between the endoplasmic reticulum and the Golgi apparatus. Approximately 50 nm diameter transition vesicles were observed in the space between the subtending endoplasmic reticulum or nuclear envelope cisternae and the most proximal Golgi apparatus cisternae. The transition vesicles resembled the blebs and were either uncoated or coated with a nap-like material or material seen in the electron microscope (Fig. 2.9) and later noted as having simple spines subsequently identified as coatomer protein. These

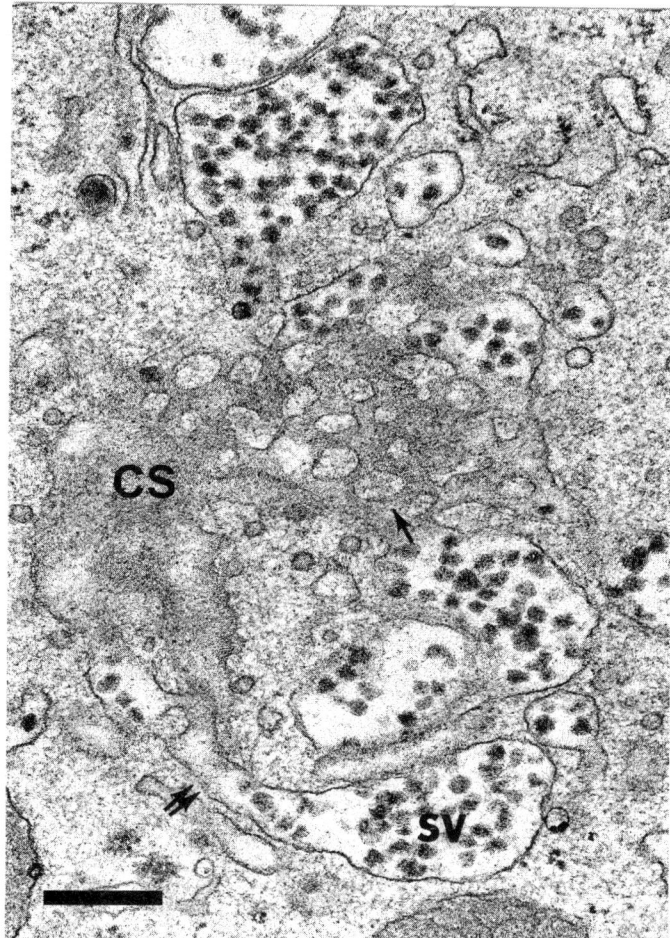

Fig. 2.12. Portion of a rat hepatocyte Golgi apparatus illustrating the attachment of lipoprotein-filled secreting vesicles (SV) to the central saccules (CS) via peripheral tubules. Reprinted from the Academic Press, J., Morré and Ovtracht, 1981, Ultrastruct. Res., 74, 284-295, Copyright 1981 with permission from Elsevier. Scale bar = 0.2 μm.

observations provided one of the morphological bases for a dynamic concept of Golgi apparatus function in which small transition vesicles were originally suggested to bleb from the endoplasmic reticulum and migrate to the Golgi region to coalesce to form new Golgi apparatus cisternae (Dalton and Felix, 1956).

Extensive direct connections between Golgi apparatus and endoplasmic reticulum are observed with plant cells grown at low temperatures (Fig. 2.17) but are much less common at temperatures optimal for Golgi apparatus functioning. The distal pole or trans face of Golgi apparatus in cells actively engaged in secretion is characterized by the presence of secretory vesicles (Figs. 2.12 and 2.13; Figs. 8.5 and 8.16) and/or condensing vacuoles (Figs. 2.18 and 8.3). These vesicles dissociate

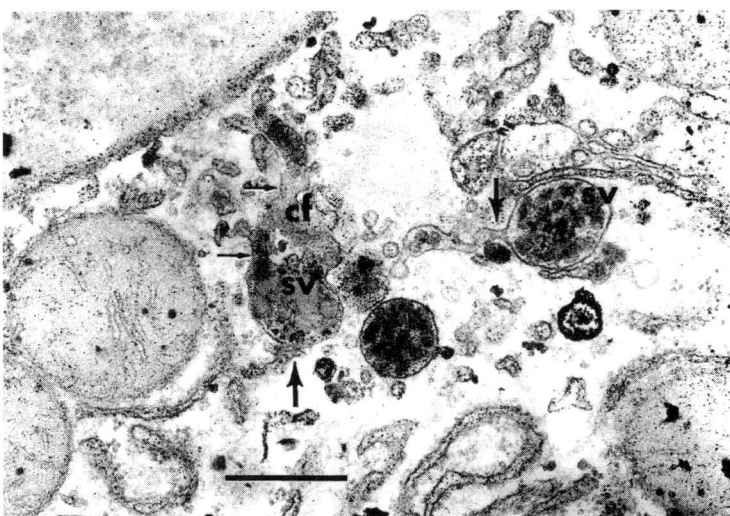

Fig. 2.13. Portion of a hepatocyte Golgi apparatus illustrating connections, in thin section, between lipoprotein-containing endoplasmic reticulum tubules and the secretory vesicles (sv) of the Golgi apparatus (large arrows) and between these tubules and portions of a Golgi apparatus cisternae (cf) seen in flat section (small arrows). The tubules connect directly with forming secretory vesicles. Lipoprotein particles are almost never seen within the flattened portion of the cisternae suggesting a peripheral route of transfer of secretory lipoproteins from smooth endoplasmic reticulum to the forming vesicles of the Golgi apparatus. Reprinted from Morré and Ovtracht, 1981, Copyright 1981 with permission from Elsevier. Scale bar = 0.5 μm.

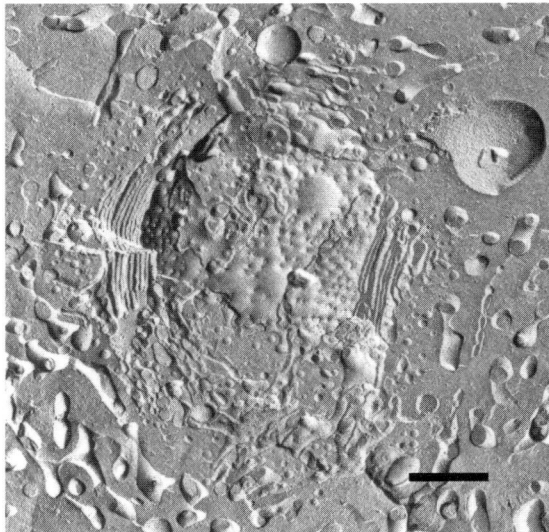

Fig. 2.14. Stacked cisternae of the Golgi apparatus from guinea pig testes seen after freeze fracture. Fenestrated cisternae with tubular peripheries were similar to those observed in chemically-fixed preparations. Unpublished electron micrograph provided by Hilton H. Mollenhauer with permission from the author. Scale bar = 0.25 μm.

Table 2.1.   Summary of structural similarities and differences comparing
plant and animal Golgi apparatus.

Characteristic	Animal	Plant
Cisternal stacks (Dictyosome subunits)	+	+
Stacked cisternae	+	+
Av. 5 cisternae/stack	+	+
Polarity	+	+
Membrane differentiation	+	+
Product transformations	+	+
Change in luminal thickness	±	+
Flattened cisternae with fenestrated margins		
continuous with tubules	+	+
Central plate (0.5 to 1 µm diameter)	+	+
Peripheral tubules (ca. 50 nm diameter)	+	+
Secretory vesicles	+	+
Transition vesicles	+	+
Trans Golgi network	+	+
Cis Golgi network	+	+
Coated vesicles and membranes	+	+
Intercisternal region; "zone of exclusion"	+	+
Golgi apparatus-associated polysomes	+	+
Intercisternal elements	−	+

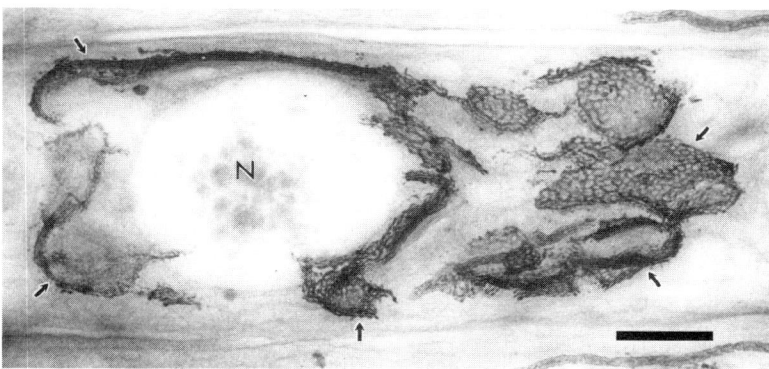

Fig. 2.15.   Golgi apparatus from multifid gland of the snail *Helix pomatia*, treated so as to precipitate osmium tetroxide, thick sectioned (1 µm), and examined in an electron microscope operating at 2.5 meV accelerating potential. The Golgi apparatus, which courses through much of the cell, is outlined by arrows. The tubular constituents are substantial and appear to interconnect the stacks to form a single large organelle. N: Nucleus. Courtesy of P. Favard, N. Carasso and L. Ovtracht, CNRS, Saclay, France. Reproduced from Mollenhauer and Morré, 1998 with permission from Springer Science + Business Media. Scale bar = 0.5 µm.

from their connections to cisternae and migrate to the cell surface where the vesicle membranes coalesce (fuse with) the plasma membrane to effect the discharge of the vesicle contents to the cell's exterior as well as the addition of the vesicle membrane to the plasma membrane. The formation and release of secretory vesicles is thought to result in a loss of cisternae from the trans face that compensates for the formation

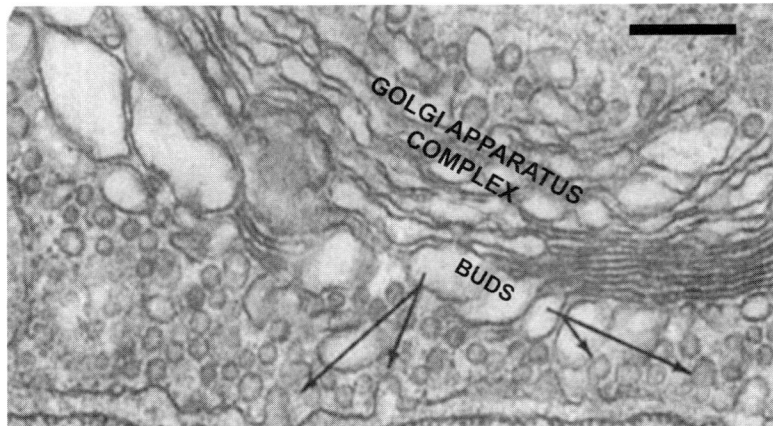

Fig. 2.16.   Blebbing and fusion profiles in the region between part rough and part smooth regions of endoplasmic reticulum and the forming or cis face of the Golgi apparatus stack of the Brunner's gland of the mouse. Reproduced from Friend, 1965, Journal of Cell Biology, 1965, 25:563-576. Copyright 1965 The Rockefeller University Press. Scale bar = 0.5 μm.

of new cisternae at the cis face. The release of mature cisternae as secretory vesicles has been observed directly with the light microscope in certain favorable unicellular organisms (Brown, 1969; Schnepf, 1969).

Within each Golgi apparatus stack (= dictyosome), the cisternae are separated from one another by a minimal space of 100 to 150 Å. In plant cells, a single layer of parallel rod-like elements or fibers, named intercisternal elements, are present within the intercisternal region midway between the surfaces of adjacent cisternae (Mollenhauer, 1965; Turner and Whaley, 1965; Fig. 2.19). Intercisternal elements appear to be unique to plants and have not been observed in Golgi apparatus of other organisms (Table 2.1). Nothing is known about their function and/or composition. A role in shaping secretory vesicles has been indicated especially in the elaboration of nonspherical (cylindrical or elongated) vesicles (Mollenhauer and Morré, 1975). As far as can be determined, animals and fungi lack them but still maintain a constant minimal spacing of 100 to 150 Å between adjacent cisternae. Plant cells also show a consistent and marked narrowing of the cisternal lumina. Cisternal lumens are widest at the cis face and narrowest at the trans face. This cisternal narrowing follows approximately the gradient of intercisternal elements across the stack. Animals and fungi that lack the intercisternal elements fail to reveal the marked gradient in cisternal narrowing suggesting the presence of intercisternal elements may somehow by related to the constriction of cisternal lumina characteristic of the Golgi apparatus of plant cells.

Cisternal stacks are surrounded by a differentiated region of cytoplasm where rough endoplasmic reticulum, glycogen, and mitochondria are scarce or absent. This region has been called the Golgi apparatus zone of exclusion (= Golgi ground substance of Sjöstrand and Hanzon, 1954; Fig. 2.2; see Whaley,

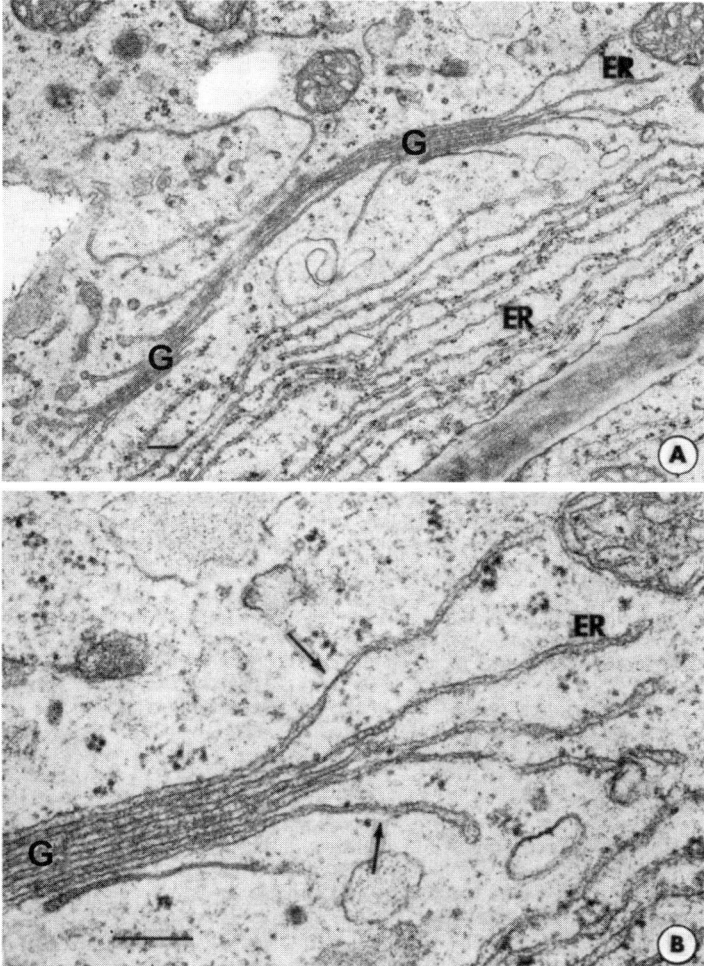

Fig. 2.17. Endoplasmic reticulum (ER)-Golgi apparatus (G) continuities (arrows) in outer cap cells of the maize root tip of plants grown at 4 °C for 12 h. The endoplasmic reticulum may be associated with cisternae at any level within the stacks. In these micrographs, which are exceptional, segments of endoplasmic reticulum are continuous with the peripheral portions of at least five cisternae. The junctions between Golgi apparatus cisternae and endoplasmic reticulum occur via a transition region (arrows). Reprinted from Mollenhauer and Morré, 1975, Copyright 1975 with permission from Elsevier. Scale bar = 0.2 μm.

1975; Mollenhauer and Morré, 1978b) or Golgi matrix (Shorter and Warren, 2002; Barr and Short, 2003). Endoplasmic reticulum entering the zone of exclusion becomes smooth surfaced (lacking ribosomes) and the small-coated vesicles of the Golgi apparatus zone are restricted to this zone usually at the trans Golgi apparatus face (Chapter 5).

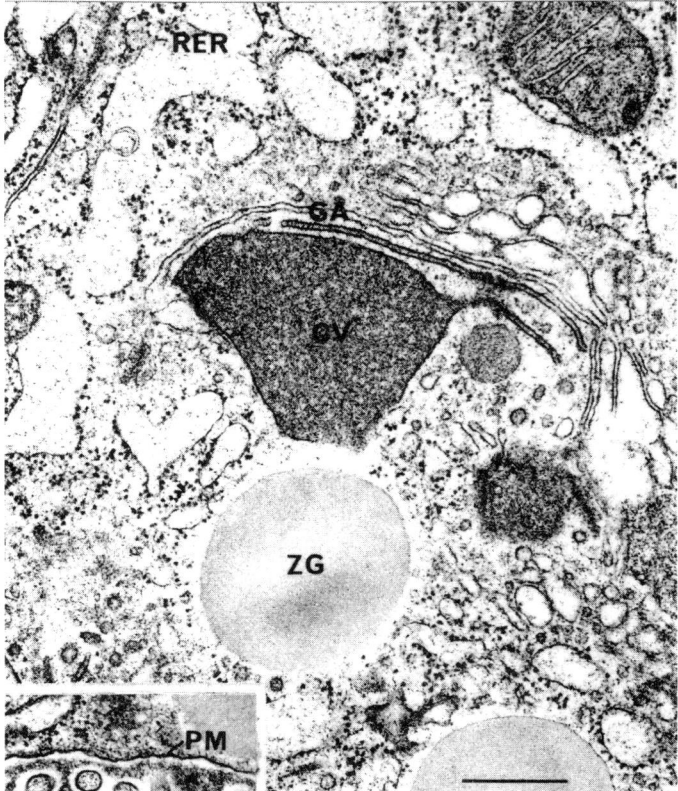

Fig. 2.18.   A forming zymogen granule (ZG) or condensing vacuole (CV) of the Golgi apparatus of the rat pancreas attached to a trans-most cisternae of a Golgi apparatus (GA). Also illustrated is a frequently observed pattern of endomembrane differentiation observed across the stacked Golgi apparatus cisternae from endoplasmic reticulum-like at the cis face to plasma membrane (PM)-like (insert), that is thicker and more darkly stained, at the trans face. RER = rough endoplasmic reticulum. Reprinted from Morré and Ovtracht, 1977, Acad. Press, Int. Rev. Cytol. Suppl., 5, 61-188, Copyright 1997 with permission from Elsevier. Scale bar = 0.5 μm.

Polarity of Golgi apparatus stacks (dictyosomes) is expressed also in terms of membrane differentiation. This differentiation is observed as a progressive change in staining intensity, membrane thickness, or other properties. Invariably, Golgi apparatus membranes at the cis, proximal, or forming face resemble endoplasmic reticulum, and at the trans, distal, or maturing face, they resemble plasma membrane (Chapter 6).

In Golgi apparatus of many cells, manifestations of secretory activity also contribute to polarity, as do materials within the cisternal lumens including enzyme proteins. Some enzyme proteins, demonstrable by enzyme cytochemistry such as thiamine pyrophosphatase, are located mainly in distal cisternae whereas in rat liver, at least, NADPase is mostly concentrated in the intercalary (middle) cisternae. Secretory materials accumulated within attached Golgi apparatus vesi-

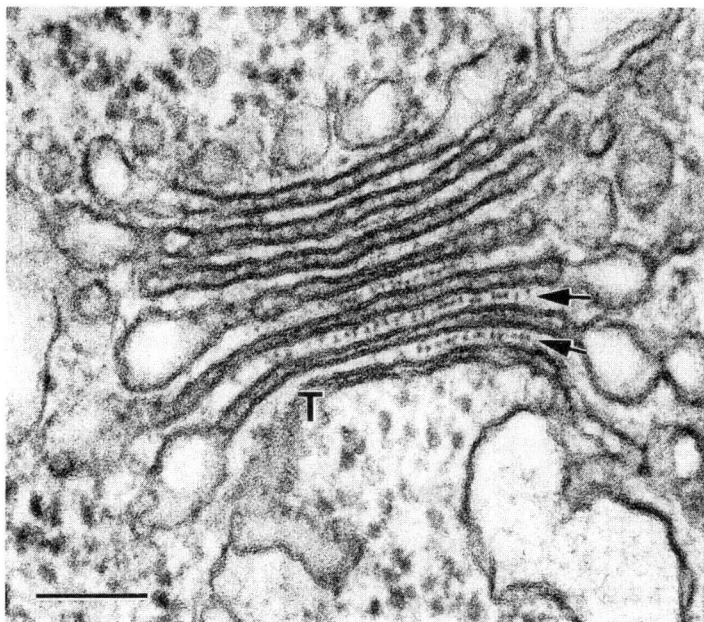

Fig. 2.19. Golgi apparatus stack from a cortical cell of the maize (corn) root with intercisternal element viewed end on. The filaments are always aligned in the same direction. The number of filaments increases toward the trans pole (T) of the dictyosomes but then decrease before the trans cisternae are released. The cis to trans narrowing of saccules characteristic of plant cells follows approximately the gradient of intercisternal elements across the stack. Reproduced from Mollenhauer and Morré, 1991 with permission from John Wiley and Sons, Inc. Bar = 0.1 μm.

cles usually show a gradient across the stack with greater concentrations toward the trans face. These observations will be emphasized in more detail in later chapters dealing with secretory activities of the Golgi apparatus.

Various impregnation methods similar to the original heavy metal stains utilized by Golgi have been used widely to demonstrate Golgi apparatus polarity. Osmium and silver salts have been traditionally employed but osmium–zinc iodide has also been used. Generally with osmium impregnation, the endoplasmic reticulum and forming or immature Golgi apparatus cisternae become filled with electron dense reduced osmium (Fig. 2.20) whereas Golgi apparatus cisternae at the trans face are unreactive. Using this method, Rambourg and collaborators (1969) have demonstrated that the cis faces of the collective dictyosomes of very complex animal Golgi apparatus consist of a ribbon of fenestrated or tubular plates interconnected by contiguous peripheral cisternal tubules.

The number of stacks per cell ranges from none (certain fungi where single cisternae carry out the Golgi apparatus function as Golgi apparatus equivalents i.e. Franke et al. 1971a and in prokaryotes where Golgi apparatus equivalents are completely lacking) to over 25,000 (algal rhizoids). A plant or animal cell usually contains 500 or more discrete cisternal stacks.

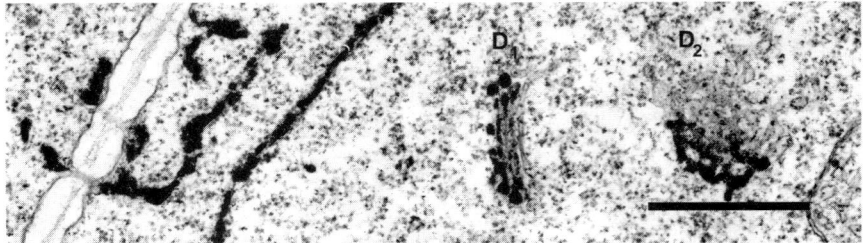

Fig. 2.20. Golgi apparatus stack of the maize root fixed in osmium tetroxide at elevated temperature. The reduced osmium marks primarily the trans most cisternae seen in cross section (D_1) and in face view (D_2). Reproduced from Mollenhauer and Morré, 1991 with permission from John Wiley and Sons, Inc. Scale bar = 0.5 µm.

2.3. Golgi Apparatus (Denotes either Singular or Plural)

The collective system of cisternal stacks of most cells appear to function synchronously suggestive of strong functional inter-associations. In most differentiated mammalian cells what is commonly recognized as the Golgi apparatus is a system of inter-associated cisternal stacks. In a few algal species, when the Golgi apparatus is represented by a single stack or dictyosome, that stack or dictyosome is the Golgi apparatus. In fungal cells, single cisternae or tubules function as Golgi apparatus equivalents and a system of stacked cisternal elements is entirely lacking. The yeast, *Saccharomyces*, is intermediate in that in certain developmental stages stacked cisternae are absent whereas at other times they may be present.

While usage among different authors will vary, the term "Golgi apparatus" may be most appropriately used to denote the totality of stacked cisternae within a cell that function together (Morré et al., 1971c). Based on analyses of inter-connections among stacks, some cells seem to have but a single Golgi apparatus composed of many stacks or dictyosomes. In Urodele sperm development (Werner, 1970), an example was reported where more than one Golgi apparatus was observed to occur in a single cell. However, what Werner observed as the second system of Golgi apparatus stacks was subsequently identified as structures unrelated to Golgi apparatus and termed dictyosome-like structures (Mollenhauer and Morré, 1977; Fig. 2.21). Frequently, what may appear as two or more Golgi apparatus in thin sections are two or more parts of the same apparatus curving in and out of the plane of the section.

Since the individual stacks that comprise the Golgi apparatus are polar structures, they impart a polarity to the entire Golgi apparatus. This is especially evident in the condensed forms of the Golgi apparatus where the stacks are closely connected with their cis and trans faces in register (Fig. 6.4). Because of this arrangement of Golgi apparatus stacks, one face of a Golgi apparatus is comprised of the aggregate of all the forming cisternae of its component dictyosomes and is designated as the cis, proximal, immature, or forming face. The opposite face is designated as the trans, distal, mature, secreting, or exit face. Other des-

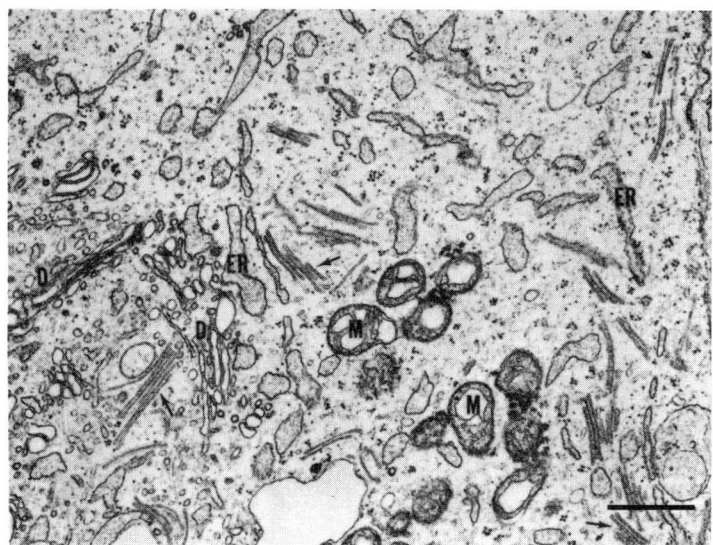

Fig. 2.21. Thin section of a portion of guinea pig primary spermatocyte showing stacks or dictyosome (D) of conventional Golgi apparatus compared to dictyosome-like structures (arrows). The dictyosome-like structures or DLS consist of stacks of several saccules are distributed throughout the cytoplasm. Single DLS saccules also are common. Mitochondria (M); endoplasmic reticulum (ER). Reproduced from Mollenhauer and Morré, 1977 with permission from John Wiley and Sons, Inc. Scale bar = 1 μm.

ignations such as convex and concave or inner and outer may be useful in some types of mammalian cells but these terms are not generally applicable to Golgi apparatus of all types of cells. In general, the concave, inner surface corresponds to the trans face while the convex, outer surface corresponds to the cis face. The relationship between surface curvature and the functional face is not absolute since curvature is not always in the same direction and other, more objective, criteria must also be applied. Golgi apparatus, while observed to be in a constant state of motion in living cells, are sometimes fixed in a configuration where the convex outer face corresponds to the trans face.

2.4. Golgi Apparatus Functioning as Part of an Integrated Endomembrane System

In recognition of the interdependence of intracellular membranes in membrane biogenesis and other functional activities, the concept of an endomembrane system was proposed (Morré and Mollenhauer, 1974). The system is viewed as a functional continuum of membrane types in which processes of membrane biogenesis, membrane differentiation, and membrane flow are combined. It accounts for Golgi apparatus origins and function in secretion as well as the biogenesis and differentiation of internal membranes. Hence, only the endoplasmic reticulum and/or nuclear envelope could be considered as even a semiautonomous organelle with respect to biogenesis of endomembrane constituents. Within this concept, functional activities of the Golgi apparatus in glycosylation of glycoconjugates, secretion, etc., and perhaps even the very existence of the cell component, per se, would depend upon an obligatory association and integration with other parts of the endomembrane system.

Included within the endomembrane system were the nuclear envelope, rough and smooth endoplasmic reticulum, Golgi apparatus, and vesicles derived from the above. Plasma membranes, vacuolar membranes, and lysosomes were considered as end products of the system. Organelles such as mitochondria and chloroplasts were not included as part of the endomembrane system even though their outer membranes may be closely associated with, or on occasion, even directly connected to, the endoplasmic reticulum (Franke and Kartenbeck, 1971; Bracker and Grove, 1971; Morré et al., 1971b).

The major parts of the system are illustrated diagrammatically in Fig. 2.22. While each part differs from the other in structure, position, and composition,

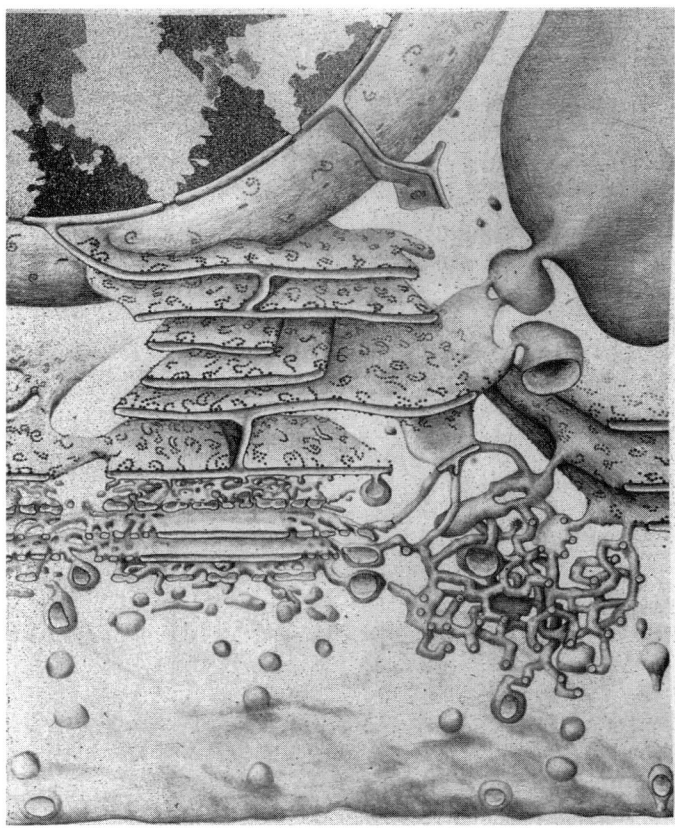

Fig. 2.22. Generalized interpretation of a portion of the endomembrane system characteristic of eukaryotic cells. Reproduced from Morré, 1994a with permission from the American Chemical Society.

they have many properties in common (Chapter 7). Most important, however, is that Golgi apparatus function, perhaps even the very existence of the Golgi apparatus, per se, may depend on the integrated activities of the entire endomembrane system.

Evidence for a functional continuum among endomembrane components has come from autoradiographic and, more recently, kinetic analyses with isolated cell fractions (Chapter 6). Such a continuum is more or less implicit in the consistent positional relationships among various endomembrane constituents. Obvious direct connections exist between nuclear envelope and rough endoplasmic reticulum and between rough and smooth endoplasmic reticulum. Vesicular traffic provides continuity between the Golgi apparatus and the plasma membrane and between the endoplasmic reticulum and the Golgi apparatus.

At the cell surface, the membranes of the secretory vesicles fuse with the plasma membrane and the contents of the secretory vesicles are discharged from the cell. This route provides for the direct transfer of plasma membrane, or plasma membrane constituents, from the Golgi apparatus to the cell surface. In fast growing cells such as neurons or tips of fungal hyphae or pollen tubes, much or all of the plasma membrane needed to keep pace with rapid cell extension appears to be derived from the Golgi apparatus via this route (Pfenninger and Bunge, 1974; Morré and VanDerWoude, 1974; Grove et al., 1970; Fig. 8.12).

2.5. Associations with Other Organelles and Cell Components

Additional opportunities for Golgi apparatus–endoplasmic reticulum communication are at the Golgi apparatus periphery. In liver, direct connections have been suggested to function as a "boulevard périphérique" to deliver very low density lipoprotein particles to forming secretory vesicles (Chapter 8; Figs. 8.5 and 8.6; Morré and Ovtracht, 1981; Figs. 2.12 and 2.13).

Golgi apparatus have been investigated in the formation of primary lysosomes and, possibly through participation in membrane recycling, in the numerous activities attributed to secondary lysosomes. A specialized structure known as GERL (Golgi apparatus–endoplasmic reticulum–lysosome) now known as the trans Golgi network (TGN), located at the mature or trans face of some complex Golgi apparatus (Novikoff, 1964), has long been associated with endocytic function (Novikoff et al., 1971). Lysosomal enzymes have been associated with the breakdown of secretory products packaged into Golgi apparatus vesicles but where degradation, rather than secretion, occurs. This type of abortic secretory event has been demonstrated in a variety of secretory activities including lipoproteins, serum albumin, viruses, and mucopolysaccharides.

The lysosomal equivalent in plant cells is the tonoplast or vacuole. An origin of vacuole membranes from vesicles of the Golgi apparatus has been suggested by numerous investigators. However, most authors consider the endoplasmic reticulum as the primary source of vacuolar membranes (see Morré, 1975; Morré and Mollenhauer, 1976, for literature). Participation of Golgi apparatus of elongating plant cells in the delivery of membranes to vacuoles which increase in surface to nearly the same extent as the plasma membrane remains largely uninvestigated.

Microtubules frequently occur within the Golgi apparatus zone. Additionally, close associations between dictyosomes and the microtubules of centrioles, flagellar bases, and rhizoplasts are common in certain cell types.

Drugs targeted to microtubules such as colchicine, colcemid, vinblastine, vincristine and griseofulvin alter Golgi apparatus mediated secretion, and microtubules have long been regarded as potential guide elements to direct the vectorial migration of secretory vesicles. In some cells, where secretion is not inhibited by microtubule poisons, the microfilament-directed drugs, such as the cytochalasins are effective.

There is little or no information implicating any unique relationship of the Golgi apparatus to mitochondria, plastids, or peroxisomes. Positional relationships among Golgi apparatus and mitochondria and/or plastids are observed in certain algae but always cisternae of endoplasmic reticulum and endoplasmic reticulum-associated membrane blebs are observed in the space between the mitochondrial or plastidal outer membrane and the proximal cisterna of perimitochondrial or periplastidal dictyosomes (Franke et al., 1972; Peyriére, 1975). Thus, these relationships emerge as manifestation of the well-known associations between mitochondria or plastids and endoplasmic reticulum and the association of the latter with the Golgi apparatus.

A consistent relationship is found between peroxisomes and Golgi apparatus in some, but certainly not all, cells. A good example is liver where nearly every Golgi apparatus seen in thin section in the electron microscope will have one or more associated peroxisomes.

2.6. Vesicles of the Golgi Apparatus

2.6.1. COP (COPII)-Coated Transition Vesicles

The 30 to 60 nm transition vesicles apparently bleb off the nuclear envelope or rough endoplasmic reticulum and, presumably, coalesce to form new Golgi apparatus cisternae. These vesicles frequently, but not necessarily, are covered with a nap-like coating composed of coatomer proteins (Fig. 2.9). Useful fractions of transition vesicles have been isolated from rat liver (Chapter 10).

2.6.2. Clathrin-Coated Vesicles

Clathrin coated vesicles tubules and membranes at the trans Golgi apparatus face (Croze et al., 1982) have the typical complex spiny coats associated with the major coat protein, clathrin (Pearse, 1976) a property associated with the involvement of such vesicles in functional specialization, transport, and/or fusion.

2.6.3. Secretory Vesicles

In continuously secreting cells, secretory vesicles are formed at the ends of tubules which, in turn, are attached to flattened cisternae (Figs. 2.7 and 2.12).

Secretory vesicles are sometimes referred to as dilated rims of cisternae. Even when giving that appearance in cross-section, the mode of attachment most frequently observed in the connection is via short intervening tubules at the fenestrated cisternal peripheries. In some cells, the entire cisternae round up and become the secretory vesicles (Brown, 1969).

Once fully formed, mature secretory vesicles detach from the cisternae and migrate. In continuously secreting cells, the detached vesicles move to the cell surface to fuse with the plasma membrane. This type of vesicle is characteristic of liver, mammary gland, most plant cells, and tip growing cells. The composition and morphological appearance of the vesicle membrane resembles closely those of the plasma membrane. The membranes of mature secretory vesicles, i.e., those that fuse with the plasma membrane, are morphologically and biochemically similar to the plasma membrane. As the membranes of the vesicles fuse with the plasma membrane, any content is discharged into the extracellular space, probably within milliseconds, and the vesicle membranes are incorporated, at least for a time, into the plasma membrane. In this manner, secretory vesicles provide functional continuity between the plasma membrane and the rest of the endomembrane system. The secretory vesicle is the most extensively documented example of a cell component derived from one membrane structure (the Golgi apparatus), which migrates to and coalesces with another membrane structure (the plasma membrane) to effect the physical transfer of membrane (membrane flow).

Mature secretory vesicles frequently exhibit clathrin coats in the vicinity of the Golgi apparatus or at the cell surface but not during migration across the cytoplasm. These observations suggest a special role for clathrin coats of secretory vesicles in the detachment and/or fusion process or in facilitating short distance migration.

2.6.4. Secretion Granules and Condensing Vacuoles

In contrast to continuously secreting cells, in discontinuously secreting cells, secretions accumulate in the cytoplasm as granules (Figs. 2.18 and 2.23). Exocytosis is slow or infrequent in the absence of stimulation, and the rate-limiting step to secretion becomes the coalescence of granules with the plasma membrane. In these cells, discharge requires extracellular calcium and an appropriate hormonal or secretagogue drug stimulus. Examples include the classical endocrine and exocrine gland cells such as the pancreatic beta cells, pancreatic acinar cells, zymogen cells of the parotid gland, adrenal medulla and neurohypophysis, and goblet cells of the intestine. Secretion in these cells, i.e., fusion with the plasma membrane and the discharge of the secretory product, is not continuous but in response to the specific secretagogue stimulus. Once stimulated, discharge is synchronous and the cells degranulate rapidly. The membranes of the mature granules are morphologically and biochemically dissimilar from the plasma membrane and do not appear to make permanent contributions to the plasma membrane. Much of the excess membrane discharged in response to a secretory stimulus appears eventually to become internalized and degraded and/or the components are recycled to be utilized in successive rounds of secretion.

Secretory granules described above originate from condensing vacuoles (Figs. 2.23 and 8.3), i.e., vesicles discharged from the Golgi apparatus perhaps

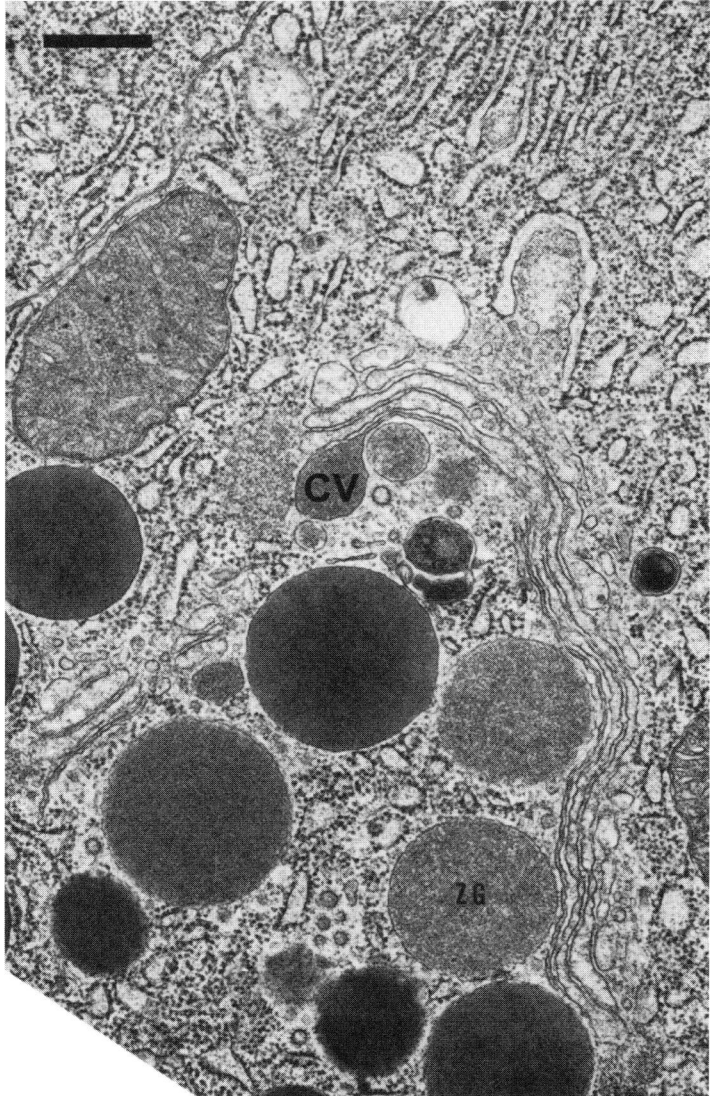

Fig. 2.23. Electron micrograph of the apical portion of an acinar cell of rat pancreas illustrating mature condensing vacuoles and zymogen (Z) granules in the vicinity of the Golgi apparatus. Reproduced from Morré and Mollenhauer 1978b with permissions from Springer Science + Business Media. Scale bar = 0.5 μm.

in much the same manner as simple secretory vesicles. Once released from the immediate vicinity of the Golgi apparatus, condensing vacuoles function for a time in the cytoplasm in a process whereby the various products destined for secretion are concentrated to form secretion granules.

2.6.5. Fusiform Vesicles and Cisternal Remnants

Some vesicles released from Golgi apparatus appear to lack content and most probably serve as specialized sources of plasma membrane. A classical example is the single intact cisterna or saccule ("fusiform vesicle") released from the mature face of Golgi apparatus of urothelial cells (Fig. 2.24; Porter et al., 1967; Hicks, 1966; Alroy et al., 1982). These vesicles give rise to the specialized luminal membrane of the epithelium of the urinary tract. Located on the inner membrane lumens are thickened plaques composed of hexagonal lattice of dodecameric subunits separated by thinner, unstructured, and narrow bands. This specialized surface membrane, easily recognized in the electron microscope, seems to be elaborated by the Golgi apparatus and provides a clear example of direct Golgi apparatus participation in the synthesis and export of specific plasma membrane domains (Fig. 2.24).

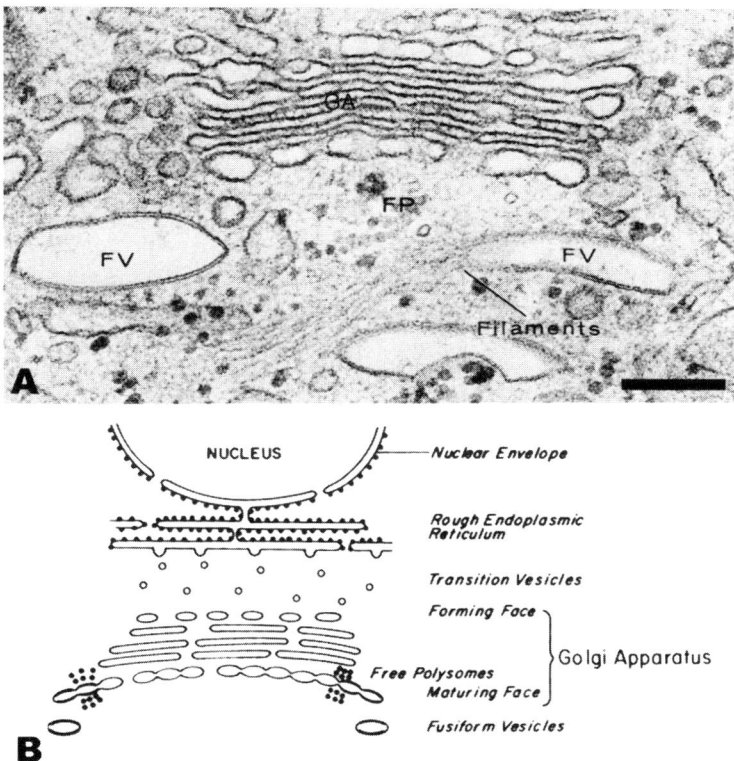

Fig. 2.24. A. Furiform vesicle (FV) formation in canine urinary bladder illustrating Golgi apparatus involvement. Microfilaments are associated with fusiform vesicles adjacent to Golgi apparatus (GA). Scale bar = 0.2 μm. B. Schematic diagram of the formation site of fusiform vesicles in urothelial cells. Free polyribosomes are adjacent to the lateral region of the maturing face. The membrane plaques which characterize the fusiform vesicles and represented by thickened lines are first associated with immature fusiform vesicles that have not yet detached from the Golgi apparatus. Reproduced from Alroy, Merk, Morré and Mollenhauer, 1982 with permission from John Wiley and Sons, Inc.

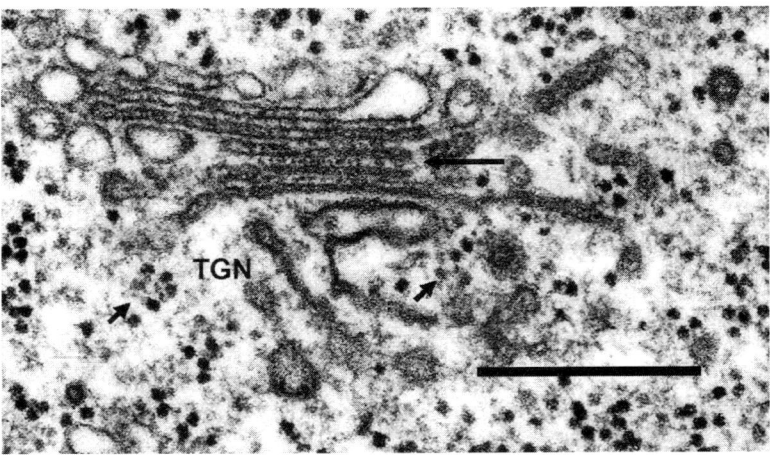

Fig. 2.25. Golgi apparatus stack from a cortical cell of the maize (corn) root with 2 intercisternal filamentous elements viewed end-on (large arrow). The filaments are always aligned in the same direction. Intercisternal substance of the Golgi apparatus-associated polyribosomes (small arrows). TGN = trans Golgi network. Reproduced from Mollenhauer, Morré and Totten, 1973 with permission from Springer-Wien. Scale bar = 0.1 μm.

Fusiform vesicles have been described, as well, in certain developmental stages of algae (Falk, 1969). As with the fusiform vesicle of the urothelium, these vesicles are of interest since they appear to involve no special secretory products and only the delivery of preformed plasma membrane units.

Cisternal remnants or discoid vesicles (cisternae) have been suggested to be released into the cytoplasm concomitant with the discharge of secretory vesicles by both plant and animal Golgi apparatus during the normal course of Golgi apparatus functioning (Fig. 2.25). Some of these structures may be synonymous with the trans Golgi network (TGN). The fate of these elements has not been conclusively determined although a contribution to the cell surface should not be excluded. Release and migration to the cell surface of entire Golgi apparatus cisternae is a common occurrence in many scale-forming algae (Fig. 8.13; Brown, 1969; see Morré and Mollenhauer, 1976, for references and for additional examples).

2.6.6. Trans Golgi Apparatus Network

The trans Golgi apparatus network is a morphologically distinct, often ridged-appearing tubular-vesicular, component located at the trans most aspect of the Golgi apparatus (Griffiths and Simons, 1986; Geuze and Morré, 1991). Its membranes are frequently clathrin-coated and secretory vesicles and products are not in evidence. This structure may correspond to the cisternal remnant of Mollenhauer et al. (1991) and others.

2.6.7. Cis Golgi Apparatus Network (Intermediate Compartment)

Some authors have distinguished a subset of membranes at the cis Golgi apparatus pole apparently formed from transitional endoplasmic reticulum by fusion of transition vesicle that has been suggested to function as an intermediate compartment between endoplasmic reticulum and the cis Golgi apparatus (e.g., Saraste and Svensson, 1991; Schweizer et al., 1991). Not all cells exhibit such an intermediate compartment. One possibility is that the formation occurs under conditions, where transition vesicle formation exceeds the need for membrane replacement at the Golgi apparatus and that the intermediate compartments form as a repository for the excess membranes.

2.7. Summary

Early electron micrographs established the Golgi apparatus as a system of saccules or cisternae flattened and stacked together in a characteristic structure, the Golgi apparatus stack or dictyosome. The stack of saccules and associated vesicles appeared as an oriented structure with one pole or "face" adjacent to endoplasmic reticulum. The opposite pole was associated with the secretory vesicles that carried sequestered or elaborated products to various intra- or extracellular compartments. This oriented appearance of the Golgi apparatus, with endoplasmic reticulum at one pole and mature secretory vesicles at the opposite pole, led to the concept of an input face (cis face, forming face, or proximal pole) and an exit face (trans face, maturing face, or distal pole). A morphological basis for endoplasmic reticulum–Golgi apparatus continuity was provided by small (ca. 50 nm) transition vesicles with a nap-like covering of coatomer COPII proteins usually aligned between a part rough, part smooth cisterna of endoplasmic reticulum and the Golgi apparatus at the cis face.

An anastomotic tubular network of the Golgi apparatus periphery, shown first by negative staining of isolated plant and animal dictyosomes in the 1960s, has since been found to be a feature of at least some cisternae of all Golgi apparatus. These tubules may serve to connect adjacent cisternal stacks of the Golgi apparatus, to connect Golgi apparatus cisternae and secretory vesicles, and, as transition elements, to connect the Golgi apparatus and smooth elements of the endoplasmic reticulum. The extent of the tubular interconnections that typify the Golgi apparatus has been elegantly demonstrated by high voltage electron microscopy after impregnation with osmium tetroxide.

Numerous other structures participate in the vesicular traffic of the Golgi apparatus, especially between the Golgi apparatus and the cell surface. These include secretory vesicles, coated membrane elements such as the clathrin containing small coated vesicles of the trans Golgi apparatus face or COPI-coated membrane structures, condensing vacuoles and various forms of secretion granules as well as microtubules and various filament systems. An association between Golgi apparatus, endoplasmic reticulum, and lysosomes, designated as GERL, as well as other "thick" or discoid cisternae or cisternal remnants may be found associated with the mature or trans Golgi apparatus face. Distal elements are often organized as a distinctive trans Golgi apparatus network.

Golgi apparatus occupy a special region of cytoplasm, the Golgi apparatus zone, frequently in association with a unique class of free polyribosomes (the so-called Golgi apparatus-associated polyribosomes) and smooth elements of endoplasmic reticulum (the boulevarde périphérique). Vesicles derived from plasma membrane through recycling also appear to enter the Golgi apparatus zone, as do various other elements of the endosome–lysosome–vacuolar system.

Thus the Golgi apparatus emerges as one of the morphologically more complex portions of the cell. As a direct result of this complexity, terminology tends to proliferate. In 1920, Bowen listed more than 600 different terms applied to Golgi apparatus (Whaley, 1975). This number has probably doubled since. Appendix Table 2 summarizes some of the terms in current usage.

3

Isolation and Subfractionation

Among the more compelling evidences in support of the reality of the Golgi apparatus was the demonstration in 1953 by Schneider and coworkers (Schneider et al., 1953; Schneider and Kuff, 1954; Dalton and Felix, 1954; Kuff and Dalton, 1959) of recognizable Golgi apparatus material from epididymal homogenates by a density gradient technique with light microscopy to evaluate the procedure. Cells were disrupted in a sucrose medium containing 0.34 M (2%) sodium chloride using a loose-fitting homogenizer. Golgi apparatus ranging in size from 0.5 μ upward (average 1.2 μ) had isopycnic densities between 1.09 and 1.13. The possibility of contamination was recognized. In a subsequent study (Kuff and Dalton, 1959), the procedure was modified to include flotation of the membrane material to its equilibrium position and the product was examined by electron microscopy. Unfortunately, these preparations were insufficiently characterized to provide new information on Golgi apparatus structure and function but they paved the way for future efforts in 1964 with the isolation of Golgi apparatus from plant cells (Morré and Mollenhauer, 1964) and culminating in 1968 with the isolation of intact Golgi apparatus from rodent livers in sufficient yield and fraction purity to permit definitive biochemical studies (Morré et al., 1968a, b).

Categorically, the Golgi apparatus was the last major cell component to be isolated from mammalian cells. A factor was the lack of a unique biochemical function or property that could be ascribed to the cell component to provide an independent biochemical marker to aid in its isolation. As detailed in Chapter 2, the definition of Golgi apparatus was, and continues to be, based on morphology. Therefore, it is logical that the basis for their initial isolation, however tedious, was by necessity based on morphology (Hamilton et al., 1967; Ovtracht et al., 1969; Morré et al., 1970a). A summary of the development of Golgi apparatus isolation methods and the many contributions from all different laboratories is given in Chapter 7 as they relate to biochemical studies of the Golgi apparatus.

3.1. Golgi Apparatus Isolation

The isolation of Golgi apparatus allows a direct approach to the study of its physical, chemical and enzymatic properties. However, it is essential that the preparations are free of contamination by other membrane components and that they be representative of the state of the Golgi apparatus in situ.

D. James Morré and Hilton H. Mollenhauer, *The Golgi Apparatus.*
© Springer 2009

The procedures described in this chapter fractionate decisively, yet are sufficiently gentle to minimize gross modifications or loss of individual Golgi apparatus components. The electron microscope remains indispensable in the quantitative and qualitative assay of isolated Golgi apparatus fractions but morphological evaluations are best carried out in conjunction with assays for enzymatic activities now known to concentrate in Golgi apparatus and through comparisons with analyses of marker enzymes to help monitor contamination by other organelles and cell components (Morré et al., 1979a).

3.1.1. Procedure for Rodent Liver

The procedure which follows is simplified somewhat from that described originally (Morré et al., 1968a, b, 1969, 1970a) and yields intact Golgi apparatus stacks (dictyosomes) ready for analysis within about 1 h from the time the animal is sacrificed (Morré et al., 1972) (Fig. 3.1). It works equally well for rat or mouse liver. It is applicable to livers of other species as well as to kidney, mammary gland and other nonhepatic tissues (Morré, 1976).

Cells are first disrupted by low-shear homogenization and the Golgi apparatus are concentrated by differential centrifugation. Degradation by lysosomal enzymes is minimized by the presence of dextran in the medium and the rapidity with which homogenization and centrifugation are carried out. Structural preservation is favored by the use of concentrated homogenates and a sucrose gradient where the Golgi apparatus collect at the gradient/homogenate interface without actually entering the sucrose layer. The development of the method and references are given by Morré (1971) and Morré et al. (1972).

(a) Homogenization medium: Detailed instructions for preparation of stock solutions are given by Morré (1971).

37.5 mM Tris maleate, pH 6.5
0.5 M Sucrose
1% Dextran
5 mM $MgCl_2$ protease inhibitors or sulfhydryl protectants optional.

(b) Homogenization procedure: One-half to 1 rat liver (5 to 10 g liver) in the ratio of 1 g tissue per 2 ml of medium for 40 sec at 10,000 rpm with the Polytron 20ST (Kinematica, Lucerne, Switzerland, distributed by Brinkman Instruments, Westbury, NY, USA).

(c) Low-speed differential centrifugation: The homogenate is transferred to 50-mL lusterloid tubes for the Sorvall HB-4 rotor (or any appropriate tube for a refrigerated centrifuge fitted with a swinging bucket rotor and centrifuged for 15 min at 5,000 X g (5,500 rpm, HB-4) to concentrate the Golgi apparatus. Most of the supernatant fluid is removed into another tube and saved for isolation of rough endoplasmic reticulum and mitochondria and also to resuspend the crude Golgi apparatus.

The yellow-brown phase of the pellet (upper 1/2 to 2/3) which lies above the red to pink (containing whole cells and nuclei) and dark brown (containing large mitochondria and endoplasmic reticulum fragments) layers is resuspended

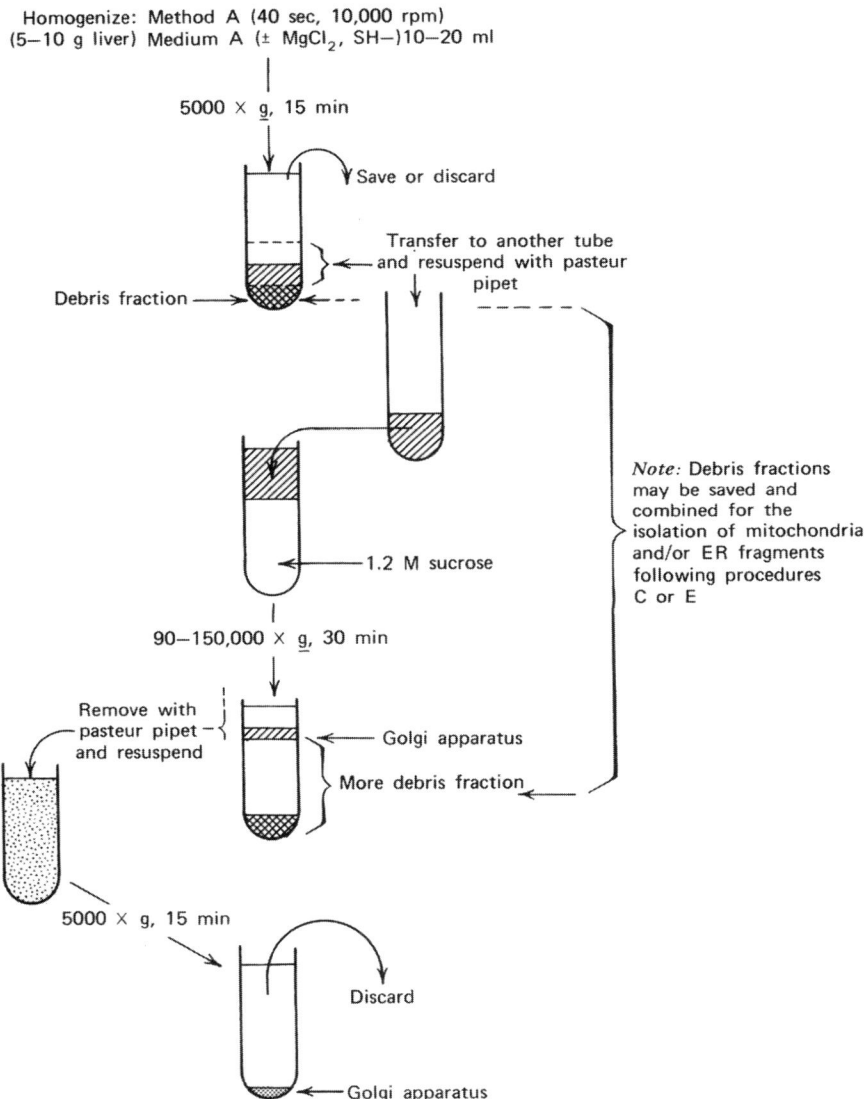

Homogenize: Method A (40 sec, 10,000 rpm)
(5–10 g liver) Medium A (± MgCl$_2$, SH–)10–20 ml

5000 × g, 15 min

Save or discard

Transfer to another tube
and resuspend with pasteur
pipet

Debris fraction

1.2 M sucrose

90–150,000 × g, 30 min

Remove with
pasteur pipet
and resuspend

Golgi apparatus

More debris fraction

5000 × g, 15 min

Discard

Golgi apparatus

Note: Debris fractions
may be saved and
combined for the
isolation of mitochondria
and/or ER fragments
following procedures
C or E

Fig. 3.1. Diagrammatic representation of a rapid protocol for isolation of Golgi apparatus from rodent liver. Reproduced from Morré, 1973 with permission from John Wiley & Sons, Inc.

in a portion of the supernatant (final volume about 6 ml per 10 g liver) by several excursions through a large-bore Pasteur pipette. Use of all-glass or glass-teflon homogenizers at this step will tend to cause unstacking.

    (d) Sucrose gradient centrifugation: The suspension of crude Golgi apparatus is then layered on 1.5 to 2 volumes of 1.2 M sucrose and centrifuged for 30 min at 90,000 to 150,000 X g. After centrifugation, Golgi apparatus are collected as a "rug" from the 1.2 M interface using a Pasteur pipette. The 1.2 M sucrose layer should be clear at the end of the centrifugation. Occasionally, it is cloudy with microsomal fragments. Under these conditions, special care must be taken to avoid removing any of the 1.2 M sucrose layer with the Golgi apparatus rug. By squeezing the bulb to dispel air beforehand and by placing the tip of the pipette one or two millimeters above the rug, the rug can be lifted off the 1.2 M sucrose layer without removal of the underlying solution.

The purified Golgi apparatus are resuspended in about 5 ml of one of the following:

1. Clear supernatant from the sucrose gradient for optimum preservation of morphology.
2. Distilled water for the highest fraction purity (to vesiculate ER fragments).
3. Homogenization medium or enzyme assay "cocktails" for the preservation of enzymatic activities.

The Golgi apparatus are then collected by centrifugation at 5,000g for 15 to 20 min.

(e) Yield: 4 to 8 mg Golgi apparatus protein from 10 g liver.

(f) Recovery: Based on estimates of galactosyltransferase activity (Morré et al., 1969), the recovery of Golgi apparatus ranges from 25 to 70% (average 40%) depending on degree of homogenization. Inadequate homogenization results in losses through unbroken cells to the low-speed pellet while over homogenization results in losses through unstacking and fragmentation to the low speed supernatant (microsomal fraction).

(g) Purity: When properly carried out, the product is intact portions of the Golgi apparatus which sediment at low centrifugal forces. With intact preparations (Fig. 3.2), the identification of the isolated Golgi apparatus can be based on their morphology which is characteristic and serves as a reliable marker. Once unstacking and vesiculation of individual Golgi apparatus elements have occurred, they become more difficult to identify on morphological characteristics alone. Intact Golgi apparatus elements can also be identified by the rapid technique of negative staining (see Morré, 1971, for details). This method does not require the usual time-consuming steps of fixation, embedment in plastic and thin sectioning and can be employed to monitor fractionations without interruption of the experiment. Quantitative analyses of electron micrographs are facilitated by various sterological procedures (Loud, 1962; Weibel et al., 1969).

Information from morphological analyses is confirmed by analyses of marker enzymes (Table 3.1). Contamination by endoplasmic reticulum, mitochondrial inner membrane, mitochondrial outer membrane, plasma membranes, and microbodies is estimated by assaying glucose-6-phosphatase, succinate dehydrogenase, monoamine oxidase, 5'-nucleotidase and uric acid oxidase, respectively. The estimates of contamination of the Golgi apparatus by

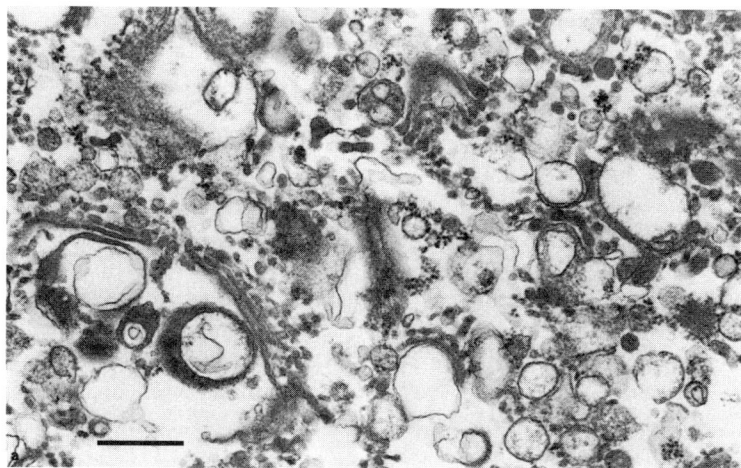

Fig. 3.2. Electron micrograph of isolated Golgi apparatus fraction from rat liver consisting primarily of intact stacks of cisternae that resemble closely those found in intact hepatocytes. Reproduced from Morré, Morré and Heidrich, 1983b, with permission from Wissenschaftliche Verlagsgesellschaft mbH. Bar = 0.5 μm.

Table 3.1. Enzymatic activities of Golgi apparatus-rich fractions from rat liver.

Enzyme[a]	Specific activity		
	Homogenate	Golgi apparatus	Relative specific activity
5'-Nucleotidase	7.1	11.4	1.6
Glucose-6-phosphatase	6.7	4.0	0.6
Succinate-INT-reductase	2.6	0.2	0.08
Monoamine oxidase	0.8	0.5	0.6
Acid phosphatase	1.0	3.4	3.4
Galactosyltransferase	2.4	220.0	91.5

Source: From Huang and Morré (unpublished); Merritt and Morré (1973); Cheetham et al. (1970)
[a] Units of specific activity are micromoles of inorganic phosphorus formed per hour per milligram of protein for glucose-6-phosphatase, 5'-nucleotidase, and acid phosphatase; micromoles of INT reduced per hour per milligram of protein for succinate-INT-reductase; micromoles of benzaldehyde formed per hour per milligram of protein for monoamine oxidase; and millimicromoles of glucosamine-dependent hydrolysis of UDP-galactose for galactosyl-transferase

endoplasmic reticulum and plasma membrane observed from biochemical measurements must be considered as upper limits, since both 5'-nucleotidase and glucose-6-phosphatase are enzymes endogenous to the Golgi apparatus albeit at lower specific activities (see Chapter 6; Morré et al., 1979a).

By morphology, the fractions are judged to be more than 90% Golgi apparatus derived and greater than 80% by enzymatic analysis. The discrepancy between these two methods of estimation results principally from overestimates of levels of endoplasmic reticulum and plasma membrane from biochemical measurements.

(h) Golgi apparatus marker enzyme: Uridine-5′-diphosphate galactose: N-acetylglucosamine galactosyltransferase (galactosyltransferase) is a convenient marker enzyme for Golgi apparatus (Fleischer et al., 1969; Morré et al., 1969; Fleischer and Fleischer, 1970). This activity is enriched 90- to 100-fold in purified Golgi apparatus relative to the total homogenate (Fig. 3.2).

3.1.2. Preparation of Reference Fractions from Rodent Liver

In order to properly evaluate the composition and purity of Golgi apparatus fractions and to provide functional comparisons with other parts of the endomembrane system (Cheetham et al., 1970, 1971), it is necessary to have available a wide variety of reference fractions, especially endoplasmic reticulum and plasma membrane, prepared from the same tissue and at levels of fraction purity exceeding 90%. This is readily accomplished for rodent liver (Fig. 3.3) and is reasonably expected for other tissues as well.

(a) Endoplasmic reticulum: Large sheets of rough endoplasmic reticulum are obtained most readily if the homogenate is undiluted. However, mitochondria surrounded by large ER sheets also frequent these preparations. As an expedient, the large sheet preparations are resuspended in dilute sucrose solution to vesiculate the ER, mitochondria are removed by sedimentation at $10,000g$ for 15 min and the vesicles derived from large rough ER sheets are finally purified by sucrose gradient centrifugation as summarized in Fig. 3.4.

The homogenization medium may be the same as isolation of the Golgi apparatus except that 5 mM $MgCl_2$ is included to retard loss of ribosomes from the rough membranes. A convenient source of rough ER as a reference fraction in the study of Golgi apparatus is the initial Golgi apparatus supernatant (centrifuged 15 min at $5,000g$ to remove Golgi apparatus). This supernatant is then diluted 1:5 with a medium containing 37.5 mM Tris maleate, pH 6.5, 0.35 M sucrose, 1% Dextran, and 5 mM $MgCl_2$ to give a final sucrose concentration of ca. 0.35 M. At this point, the diluted fraction is centrifuged at $10,000g$ for 10 min to remove residual mitochondria.

The diluted supernatant from the differential centrifugation is then top-loaded onto a discontinuous sucrose gradient consisting of 2.0 M sucrose 1.5 M sucrose and 1.3 M sucrose in a volume/volume ratio of 3:4:4 (for a 40-ml centrifuge tube, 6 ml, 8 ml, 8 ml). The remainder of the tube is filled with the diluted Golgi apparatus supernatant and centrifugation is for 8–$9 \times 10^6 g$ min (for example, 90,000 X g for 90 min). Rough ER fragments are recovered from the 1.5/2.0 M sucrose interface. Free ribosomes and glycogen enter the 2.0 M sucrose layer and will form a pellet with longer centrifugation times. Smooth "microsomes" collect at the other interfaces (Fig. 3.4). The 1.3 M sucrose/homogenate interface is especially rich in plasma membrane and Golgi apparatus fragments and is usually discarded. Endoplasmic reticulum fragments collecting at the 1/3/1.5 M sucrose interface (ER-II) are largely rough ER and regions of smooth

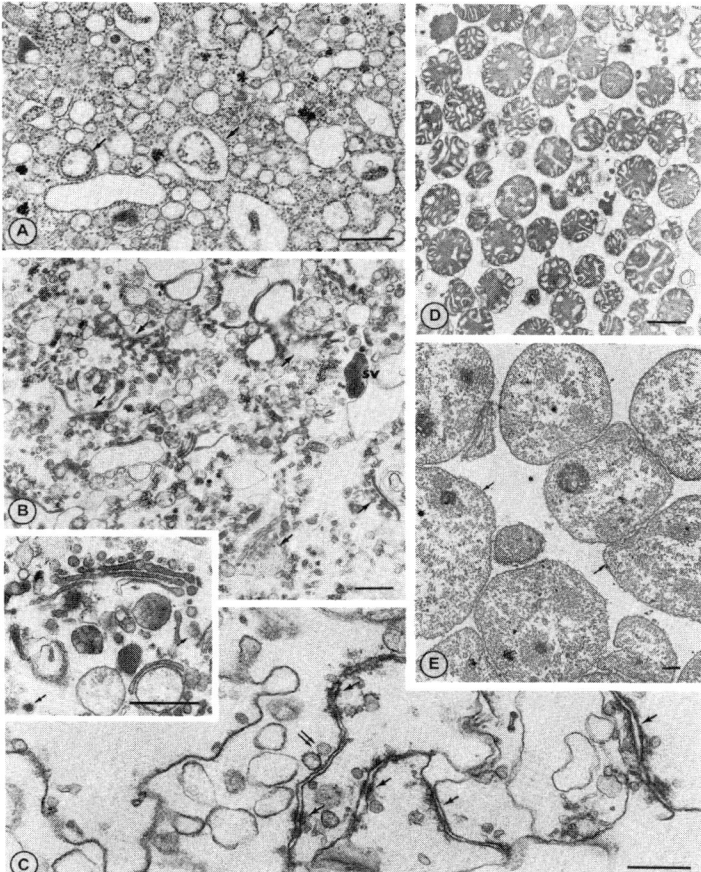

Fig. 3.3. Golgi apparatus and reference fractions isolated from single homogenates of mouse liver. (A) Endoplasmic reticulum characterized by vesicles with attached ribosomes (*arrows*). Bar=0.5 μm. (B) Golgi apparatus fraction composed of stacked cisternal saccules (*arrows*) and secretory vesicles filled with lipoprotein particles (sv). Small 50–100-nm clathrin-coated vesicles were present as well (*inset, small arrows*). Bar=0.5 μm. (C) Plasma membrane fraction with numerous junctional complexes (*arrows*), membrane sheets, and bile canaliculi. Bar=0.5 μm. (D) Mitochondrial fraction. Matrices were condensed and both the inner cisternae and outer membranes were intact. Bar=1 μm. (E) Nuclei fraction. The nuclear envelope was present (*arrow*) and the nuclear matrix showed areas of both condensed and dispersed chromatin. Reproduced from Croze and Morré, 1984 with permission from the American Physiological Society. Bar=1 μm.

ER continuous with rough ER. Smooth ER also enters the ER-II fraction; this is the fraction that contains the bulk of the enzyme activities induced, for example, in livers of animals where smooth ER proliferation has been induced by pheno-barbital or other drugs (Fig. 3.5). The fractions are collected from the gradients using a Pasteur pipette, diluted 1:1 with buffer or distilled water and collected by centrifugation at 90,000 X *g* (26,000 rpm, Spinco SW 27 rotor or equivalent) for 20 min.

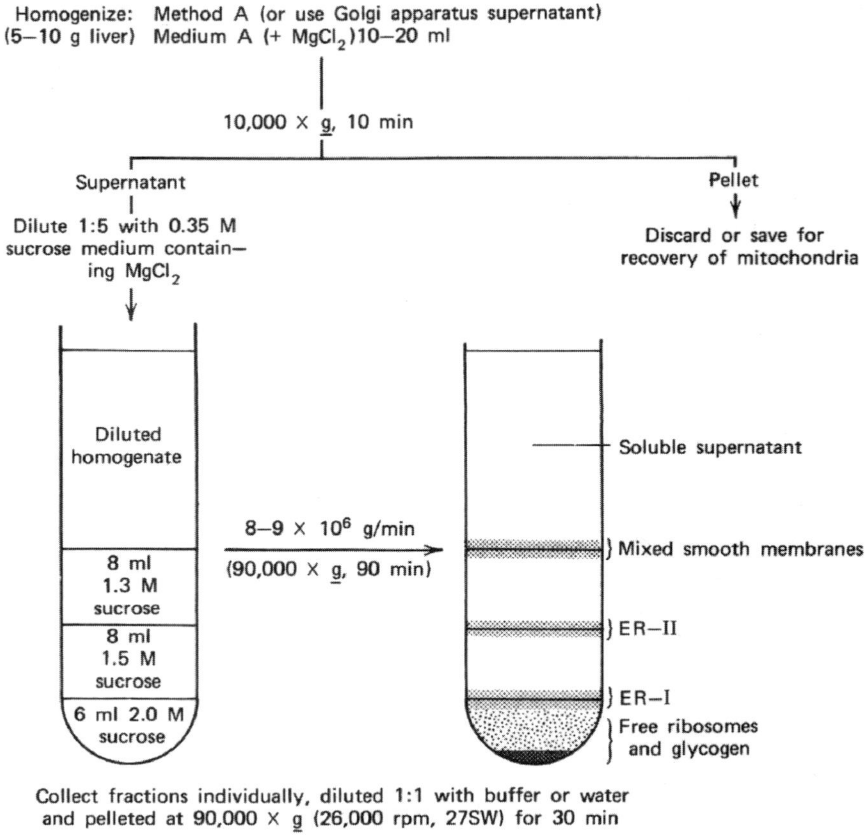

Homogenize: Method A (or use Golgi apparatus supernatant)
(5—10 g liver) Medium A (+ MgCl$_2$)10—20 ml

10,000 × g, 10 min

Supernatant   Pellet

Dilute 1:5 with 0.35 M   Discard or save for
sucrose medium contain—   recovery of mitochondria
ing MgCl$_2$

Diluted
homogenate   Soluble supernatant

8—9 × 10^6 g/min   } Mixed smooth membranes
8 ml
1.3 M
sucrose   (90,000 × g, 90 min)
8 ml   } ER—II
1.5 M
sucrose
6 ml 2.0 M   } ER—I
sucrose   Free ribosomes
                                                                        and glycogen

Collect fractions individually, diluted 1:1 with buffer or water
and pelleted at 90,000 × g (26,000 rpm, 27SW) for 30 min

Fig. 3.4. Sucrose gradient procedure for preparing ER from the Golgi apparatus supernatant. Reproduced from Morré, 1973 with permission from John Wiley & Sons, Inc.

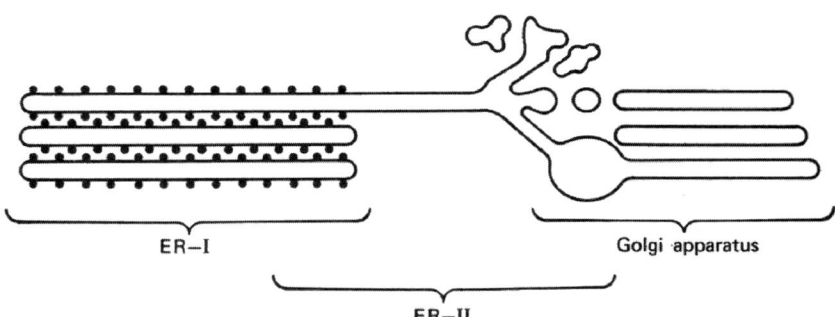

ER—I   Golgi apparatus

ER—II

Fig. 3.5. Diagrammatic interpretation of the origin of ER-I, ER-II, and Golgi apparatus fractions.

1. Yield: 10–15 mg protein/10 g fresh weight of liver. The ratio of ER-1 to ER-II is about 1.

2. Recovery: The total ER of rat liver is approximately 380 mg/10 g of liver so that the overall recovery (ER-I + ER-II) is on the order of only about 10%. The bulk of the remainder is found in the low-speed pellets. This can be recovered by resuspension of the pellets in the 0.35 M sucrose medium, rehomogenization and a repeat of the purification steps. Usually, the amount of rough ER is not limiting in comparison studies with Golgi apparatus and the procedure using the Golgi apparatus supernatant provides more than sufficient material.

3. Purity: ER-I (rough ER) is predominantly vesicles derived from rough ER with attached ribosomes and uncontaminated by plasma membrane or Golgi apparatus. Glycogen is the principal contaminant with small amounts of mitochondria occasionally entering the fraction depending on the efficacy of the initial dilution/centrifugation step to remove mitochondria. Total membranous contaminants rarely exceed 5%.

(b) Plasma membrane: Plasma membranes are readily purified by the technique of aqueous two-phase partition (Fig. 3.6) from crude mixtures or from the low-speed debris fraction underlying the Golgi apparatus-rich fraction obtained by differential centrifugation. Homogenates prepared either in dilute 1 mM bicarbonate or isotonic sucrose are first centrifuged to concentrate the plasma membrane vesicles. The concentrated membranes

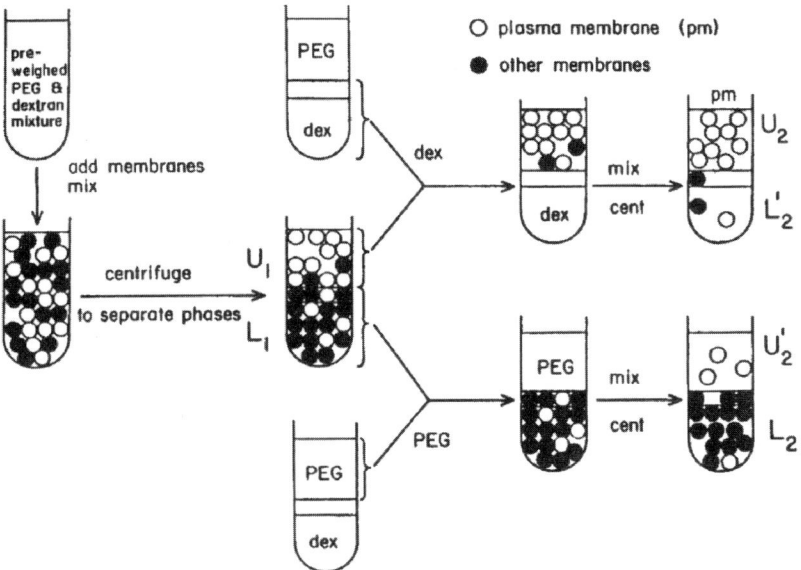

Fig. 3.6. Diagram depicting the purification of plasma membranes by aqueous two-phase partition. Reproduced from Morré and Morré, 1989 with permission from Informa Healthcare.

are then combined with a mixture of dextran and polyethylene glycol that will of itself spontaneously separate into a polyethylene glycol-rich upper phase containing the plasma membranes and a dextran-rich lower phase containing all other membranes. The yield of plasma membranes is 20% or more of those present in homogenates and the fraction purity is 90% or greater. The composition of a 16 g two phase system (6.4%) suitable for isolation of plasma membrane is given in Table 3.2. Details are given by Morré and Morré (1989).

(c) Nuclei, mitochondria, lysosomes, and microbodies: Nuclei are usually purified by centrifugation of the pellet from the isolation of plasma membrane to a plasma membrane gradient (Fig. 3.3E) yields an acceptable nuclei fraction and greatly increases the yield of plasma membrane from the overall procedure. Mitochondria are easily purified by repeated washing in a simple medium containing 0.25 M sucrose (Fig. 3.7), 0.01 M Tris HCl, pH 7.4 and 0.1 mM ethylenediaminetetraacetic acid (EDTA). After a 10 min, 650 X g centrifugation to remove debris, the supernatant is centrifuged at 4,000 X g for 10 min to collect the crude mitochondrial pellet. The supernatant is then centrifuged for 15 min at 8,000 X g. This pellet is resuspended along with the 5,000 g pellet and the centrifugation steps repeated until a homogenous brown and tightly packed pellet results from the 5,000g centrifugation. Details are given by Morré (1973).

To obtain purified fractions of minor cell components such as lysosomes and, peroxisomes which together account for only a few percent of the total protein of the liver cell, more specialized procedures are required (Wattiaux et al., 1963; Baudhuin et al., 1964; Piqueras et al., 1994). Glucose-6-phosphate dehydrogenase and several glycolytic enzymes such as lactate dehydrogenase are accepted as markers for the "soluble" cytosol.

In the assay or isolated fractions for marker activities, a second parameter to estimate membrane amount (protein, phospholipid phosphorous, or both) is required. This will permit calculation of relative specific activity or amount (ratio of fraction specific activity or amount to total homogenate specific activity

Table 3.2. Composition of a 16-g (6.4%) two-phase system.

Reagent	Amount
20% Dextran T-500 (Pharmacia)[a]	5.12 g
40% Polyethylene glycol 3350 (PEG) (Union Carbide)	2.56 g
0.2 M Potassium phosphate buffer, pH 7.2	0.4 mL
1 M Sucrose (not essential)	1.6 mL
Sufficient distilled water to bring the weight of the mixture to 14 g	
Crude plasma membranes (ca. 250 mg) resuspended in 1 mM	
bicarbonate or distilled water	2 g

[a] It is practical to prepare stock solutions of the phase system and store them frozen as aliquots. The stock solutions of dextran and polyethylene glycol are prepared on a weight basis, while sucrose, buffer, and salt stock solutions are prepared on a molar basis

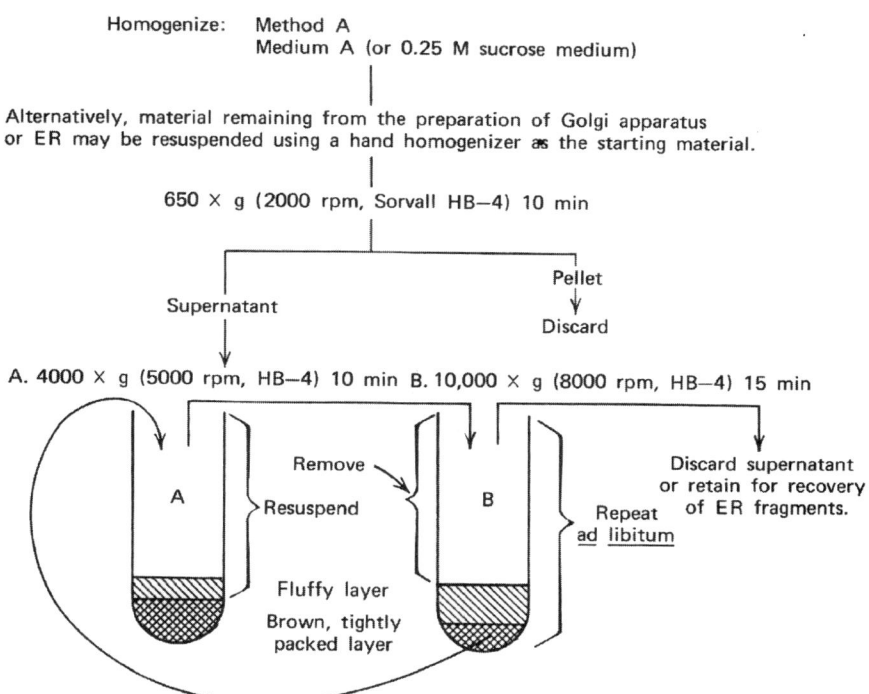

Homogenize: Method A
Medium A (or 0.25 M sucrose medium)

Alternatively, material remaining from the preparation of Golgi apparatus or ER may be resuspended using a hand homogenizer as the starting material.

650 × g (2000 rpm, Sorvall HB–4) 10 min

Supernatant

Pellet
Discard

A. 4000 × g (5000 rpm, HB–4) 10 min B. 10,000 × g (8000 rpm, HB–4) 15 min

A

Remove
Resuspend

B

Repeat
ad libitum

Discard supernatant
or retain for recovery
of ER fragments.

Fluffy layer
Brown, tightly
packed layer

Tube A should eventually contain a homogeneous pellet of brown tightly packed material = mitochondria. The fluffy layer concentrated in tube B may be used as a source of crude lysosomes or microbodies or discarded.

Fig. 3.7. Isolation of mitochondria. Reproduced from Morré, 1973 with permission from John Wiley & Sons, Inc.

or amount) providing that an assay of the starting material (commonly the total homogenate) is also carried out. It is helpful, as well, if all derived fractions are assayed and a balance sheet constructed which should add up to nearly 100% recovery both for protein or lipid phosphorous and the marker activity (Morré et al., 1979a).

A combined procedure where Golgi apparatus as well as reference fractions are obtained from a single homogenate and applicable to smaller amounts of starting material has been described for mouse liver (Croze and Morré, 1984).

3.1.3. Isolation of Golgi Apparatus Fractions from Plant Cells

For best results, low-shear homogenization that is provided by using a mechanized razor blade chopper (Morré, 1971) is recommended. Hand chopping, while used in the first successful isolations from onion stem (Morré and Mollenhauer, 1964), is effective but time consuming. The Polytron homogenizer, useful for liver and other animal tissues, gives a rotary scissors action and operated at 3–4,000 rpm for about 2 min also has yielded useful preparations.

The ratio of tissue to medium should be as high as 1:1 (w/v) and no lower than 2:1 (w/v). The composition of the medium is critical. It usually consists of an osmotically active solute (0.5 M sucrose), a buffer (0.01 M sodium phosphate, pH 6.5) and dextran (1%) and other agents to protect against degradative changes.

Differential centrifugation (differential pelleting) is used to remove contaminating cell components and to concentrate Golgi apparatus prior to gradient centrifugation. Normally, homogenates are centrifuged for 75,000g min to remove the bulk of the contaminating mitochondria, plastids, nuclei, cell walls and debris. This step may also remove up to 50% of the Golgi apparatus or more if the g min are increased to remove all of the mitochondria. The supernatant from the first differential centrifugation may then be applied directly to a gradient or a second centrifugation of 300,000g min (30 min at 10,000g) may be used to concentrate the Golgi apparatus (Fig. 3.8). The important consideration here, especially with plants, is to avoid a tightly-packed pellet. Centrifugation on a sucrose "cushion" or "minigradient" circumvents this difficulty.

In sucrose gradients, plant Golgi apparatus equilibrate at a density of about 1.12 g/ml similar to their animal counterparts. Based on extensive studies with linear gradients, the step gradient outlined in Fig. 3.8 was developed to correspond to a natural tendency for plant fractions to band into four major regions of the gradient. Step gradients offer advantages for preparative cell fractionation in being easily prepared and as a reproducible source of membranes in useful yield. If only Golgi apparatus fractions are to be prepared, the gradient procedure can be simplified further, concluding with flotation over a single sucrose layer. This has the advantage that the Golgi apparatus stacks never enter the sucrose gradient but remain in a cytoplasm-rich supernatant.

All steps should be completed expeditiously with under 2 h elapsing between homogenization and fixation of the fraction for morphological examination. The morphology of Golgi apparatus isolated in the presence of 0.1% final concentration of glutaraldehyde (Fig 3.9) is characteristic of that in vivo and has been used to provide an accurate basis for estimating fraction composition and relative enrichment.

Marker enzymes for plant Golgi apparatus remain problematic. There is a low K_m glucan synthetase activity associated with Golgi apparatus but a high K_m form of this activity is found associated with the plasma membrane (Ray et al., 1969; Van Der Woude et al., 1974). Even at low UDP-glucose concentration of 1.5 micromolar, much of the total homogenate activity is due to plasma membrane and not Golgi apparatus. Contamination of plant Golgi apparatus fractions by plasma membranes may be based on morphometric analysis of fractions following specific staining of the plasma membrane-derived vesicles by phosphotungstic acid at low pH (Roland et al., 1972).

Based on the above morphological and marker criteria, Golgi apparatus fractions from plants, prepared in the absence of chemical fixatives, are estimated to be about 50% Golgi apparatus-derived material. The remainder consists of a complex mixture of endoplasmic reticulum, plasma membrane, tonoplast and

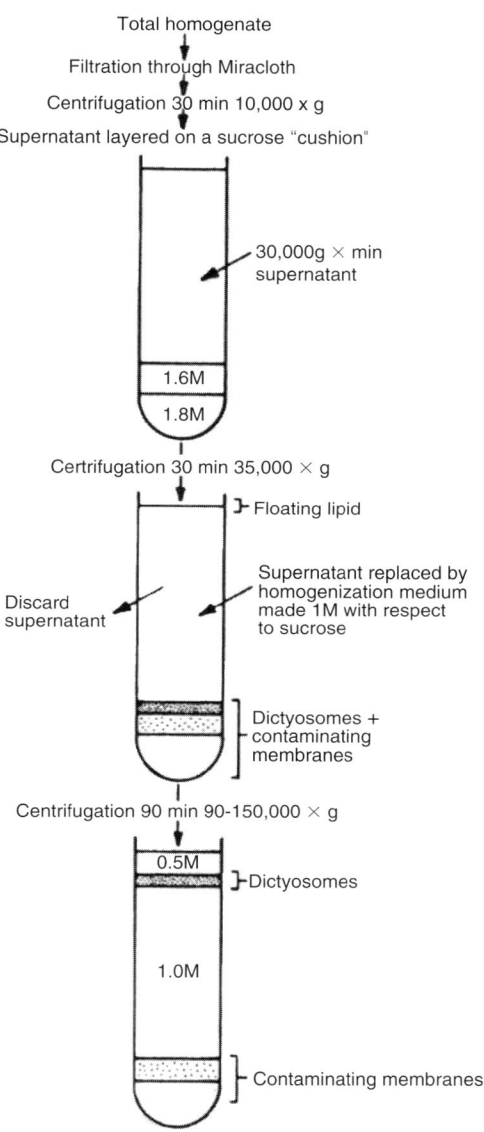

Fig. 3.8. A simple discontinuous sucrose gradient procedure for isolation of plant Golgi apparatus. Reproduced from Morré and Buckhout, 1979 with permission from Horwood Publishing.

plastid membranes. Mitochondria are usually relatively minor components in the fractions. With glutaraldehyde stabilization, it is possible to produce fractions of 85% dictyosome-derived material or greater. Major contaminants include plastids and endoplasmic reticulum vesicles.

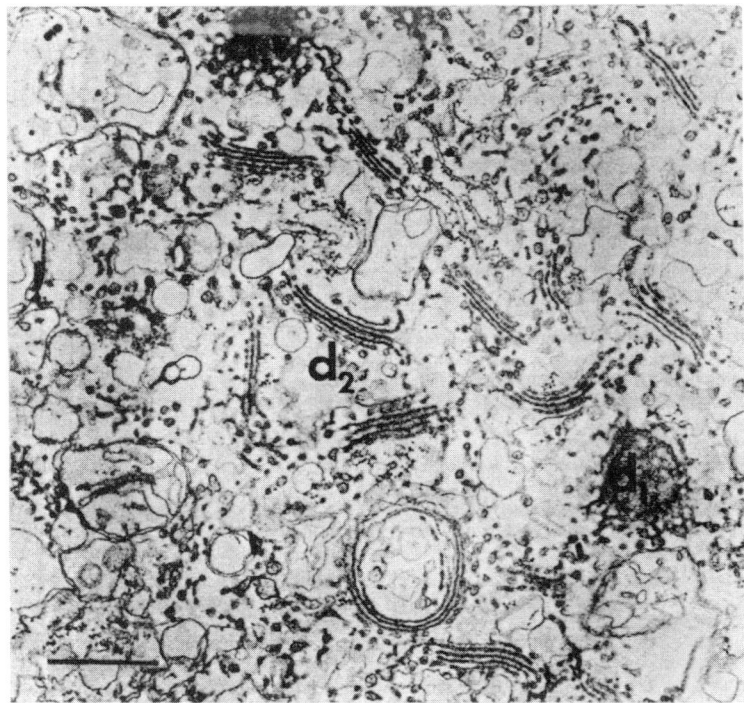

Fig. 3.9. Electron micrograph of a Golgi apparatus fraction from onion stem prepared from a homogenate to which 2.5% glutaraldehyde was added to preserve intact stacks. Stacks or dictyosomes (d_1) in face view and cross section (d_2) are illustrated. From a study with W. P. Cunningham cf.; Reproduced from J. Cell Biol., 1966, 28, 169–179. Copyright 1966 The Rockefeller University Press. Scale bar = 0.5 μm.

3.1.4. Isolation of Golgi Apparatus from Mammalian Cells Grown in Culture

Cells collected by centrifugation for 6 min at 1500g and 4°C are resuspended in 0.2 mM EDTA in 1 mM NaHCO$_3$ in an approximate ratio of 1 ml/10^8 cells and incubated on ice for 10 to 30 min to swell the cells. Homogenization is with a Polytron homogenizer (Kinematica, Lucerne, Switzerland, distributed by Brinkman Instruments, Westbury, NY, USA) for 40 s at 10,000 rpm using a 10ST-probe and 7-ml aliquots. Light microscopy may be used to monitor cell breakage. About 90% cell breakage without breakage of nuclei is usually sufficient.

The homogenates are centrifuged at 4°C for 10 min at 175g to remove nuclei and at 20,000g for 30 min to prepare a plasma membrane-enriched microsome fraction. The supernatant was discarded and the pellets were resuspended in 0.2 M potassium phosphate buffer (pH 7.2) in a ratio of approximately 1 ml per pellet from 5 × 10^8 cells. The resuspended membranes then were loaded onto the two-phase system with a polymer mixture containing 6.6% Dextran T500 (Pharmacia Biotech, Alameda, CA, USA), 6.6% (w/w) poly(ethylene glycol) 3350 (Fisher

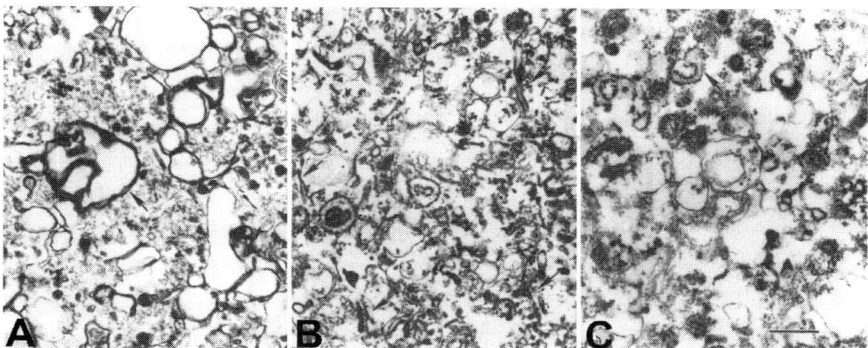

Fig. 3.10. Electron micrographs of cell fractions from BAL-17 cells isolated by combined aqueous two-phase partition and differential centrifugation. (A) Plasma membrane-enriched upper phase (*arrows*). (B) Golgi apparatus (*arrows*) isolated by centrifugation followed by aqueous two-phase partition. (C) Endoplasmic reticulum (*arrows*) from plasma membrane-depleted lower phase. Reproduced from Morré and Morré, with permission from Elsevier. Scale bar = 0.2 μm.

Scientific, Pittsburg, PA, USA) and 0.2 M potassium phosphate, pH 7.2. The weight of the system is brought to 14 g with distilled water. Resuspended microsomes (2 g) are added to the two-phase system to a final weight of 16 g. The tubes are inverted vigorously for 40 times in the cold (4°C). The phases are separated by centrifugation at 1150g in a Beckman JS-12.1 (Beckman Instruments) rotor for 5 min at 4°C. The upper phase containing primarily plasma membranes is usually >85% as determined by electron microscopy and assay of marker enzymes. The yield is about 20 mg plasma membrane protein from 10^{10} HeLa cells.

For isolation of Golgi apparatus, the lower phase is diluted 10-fold with 1 mM sodium bicarbonate and centrifuged for 30 min at 25,000g. The supernatant, containing the Golgi apparatus is diluted 2-fold with 1 mM sodium bicarbonate and centrifuged for 45 min at 31,700g. The supernatant after isolation of Golgi apparatus is then used as a source of endoplasmic reticulum by further centrifugation at 80,000g for 30 min. The electron microscope morphology of the fractions is illustrated in Fig. 3.10 for BAL-17 cells.

3.2. Subfractionations of Golgi Apparatus based on Density

The Golgi apparatus is clearly a heterogeneous cell component consisting of structurally, and presumably functionally, specialized but interassociated cisternae, tubules, and vesicles represented by a progressive change across the stack or along the polarity axis known collectively as membrane differentiation (Chapter 4). Thus, not only is each cisterna expected to differ from the other in the stack but within each cisterna are found saccules, tubules and vesicles of potentially unique composition and functional state.

To increase our understanding of biochemical and functional gradients within the Golgi apparatus much work has been devoted to procedures for Golgi apparatus subfractionation. Most often, intact Golgi apparatus have served as the starting material. The procedures begin by unstacking the apparatus followed by collection of each successive saccule in a separate fraction. Finally, it would be instructive to further subfractionate each cisternal fraction into vesicles, saccules and peripheral tubules. For a stack with 5 cisternae, a total of 15 or more subfractions might be necessary to fully represent the two orthogonal polarity gradients that characterize the Golgi apparatus.

While this may seen a formidable task at first consideration, considerable progress has been recorded in developing methodology for each of the necessary steps. Simple procedures have already been described for intact Golgi apparatus in high yield and fraction purity from rat liver. Unstacking of intact Golgi apparatus can be carried out yielding fractions enriched in cisternae and/or secretory vesicles. A subfractionation procedure for rat liver that yields both saccules and peripheral tubules as well as secretory vesicles has also been described (Ovtracht et al., 1973). Finally, it is possible to resolve cis and trans Golgi apparatus elements as well as individual cisterna from different positions within the stack by preparative free-flow electrophoresis (Morré et al., 1983b).

Ehrenreich et al. (1973; see also Bergeron et al., 1973) described a procedure for rat liver yielding three Golgi apparatus subfractions designated GF_1, GF_2, and GF_3. GF_1 consists principally of secretory vesicles. GF_2 and GF_3 are mixed fractions containing cisternal vesicles. Merritt and Morré (1973) further recognized the existence of subpopulations of secretory vesicles derived from different cisternae within the stacked cisternae of Golgi apparatus of rat liver (Figs. 3.11 and 3.12). These properties are described in Table 3.3 to contrast the two main populations – immature vesicles from the forming face and mature vesicles from the secretory or mature face. Thus, secretory vesicles dissociated

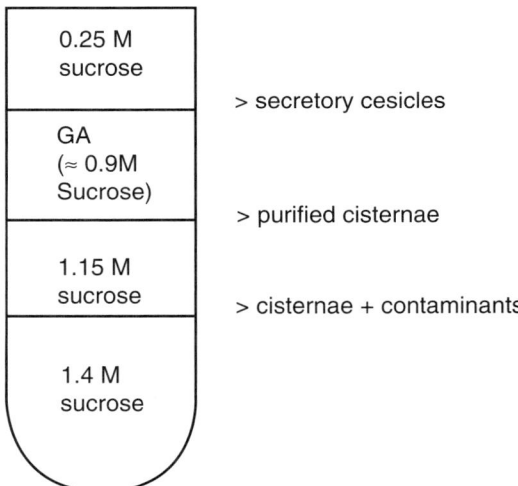

Fig. 3.11.   Sucrose gradient procedure for purification of secretory vesicles and cisternal fractions from rat liver Golgi apparatus (GA). Reproduced from Merritt and Morré, 1973 with permission from Elsevier.

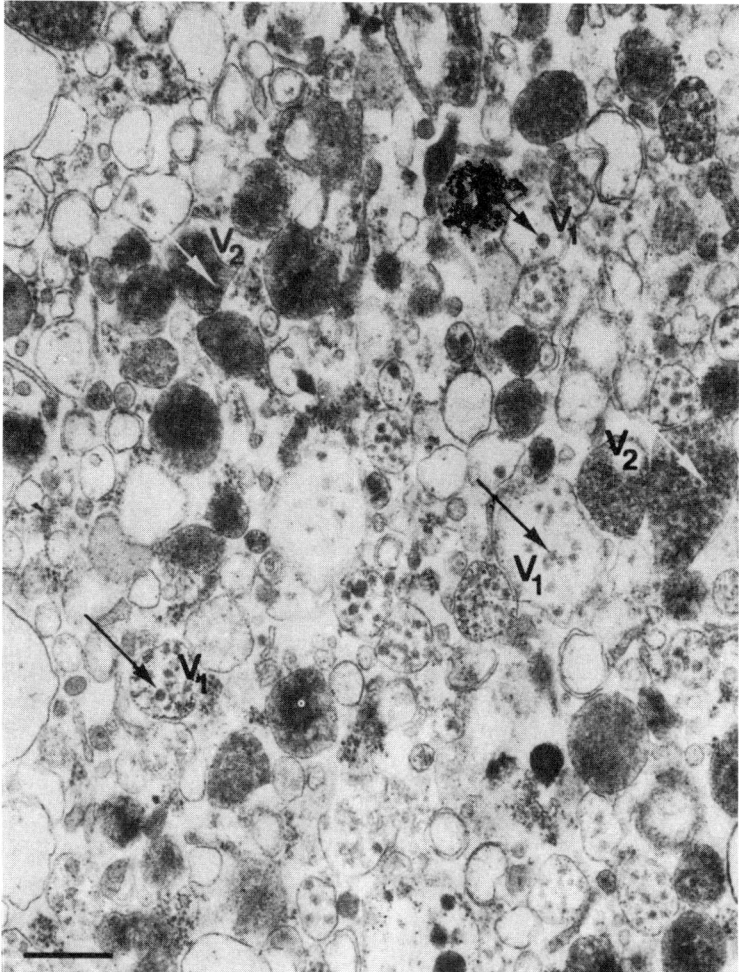

Fig. 3.12.   Secretory vesicle fraction isolated from a Golgi apparatus fraction prepared from rat liver. The range of secretory vesicle types characteristic of the intact tissue are illustrated. Immature secretory vesicles from the forming face have larger lipoprotein particles (*black arrows*) and an electron-transparent matrix which impart a swollen appearance to the vesicles (V_1). Mature secretory vesicles from the secreting face of the Golgi apparatus contain smaller lipoprotein particles (*white arrows*) within a darkly staining vesicle matrix (V_2). Other secretory vesicles are characterized by an appearance intermediate between these two extremes. Reprinted from Merritt and Morré, 1973 with permission from Elsevier. Scale bar = 0.5 μm.

from Golgi apparatus of rat liver consist in themselves of heterogeneous mixtures of immature and mature vesicles as well as intermediate vesicles derived from intermediate (intercalary) cisternae. There is no basis for the assumption, for example, that GF_1 is enriched for trans (mature) Golgi apparatus elements or that GF_3 is enriched in cis (forming) Golgi elements. In fact, subfractionation of rat

Table 3.3.   Characteristics of immature and mature secretory vesicles in rat liver.

Property	Immature secretory vesicles	Mature secretory vesicles
Matrix after osmium fixation[a]	Light	Dark
Appearance after osmium fixation[a]	Swollen	Compact
Average diameter of lipoprotein particles[a,b] of vesicle contents	60 nm	45 nm
Cytochemical fixation[c]-resistant NADH-FeCN oxidoreductase	Negative	Positive
Protein/phospholipid[d]	0.57	0.56
Triglyceride/phospholipid[d]	0.30	0.24
Cholesterol/phospholipid[d]	0.02	0.03
Cholesterol ester/phospholipid[d]	0.02	0.03

[a] Merritt and Morré (1973)
[b] Twaddle et al. (1981)
[c] Morré et al. (1978b); Morré and Vigil (1979)
[d] Hess et al. (1979)

liver Golgi apparatus elements depends almost entirely on relative numbers of lipoprotein particles trapped within membranous elements and less on membrane properties associated with the polarity axis. However, this property is advantageous in that it permits separation of vesicles from cisternae and peripheral tubules, one of the necessary requisites for Golgi apparatus subfractionation. Secretory vesicles derived from Golgi apparatus were first isolated from a plant source (Van Der Woude et al., 1971) using filtration based on size. A protocol for isolation of secretory vesicles from mammary gland is given by Keenan et al. (1979).

There appear to be insufficient differences in the lipid/protein ratio across the Golgi apparatus stack to permit decisive fractionation by conventional gradient separations based on density. An alternative approach is the use of imposed markers. Such markers to be useful must not only label a particular Golgi apparatus face or specific cisterna but also impart density differences to permit subsequent gradient separations. An example is a glutaraldehyde-resistant NADH-ferricyanide reductase of the mammalian plasma membrane. Within the Golgi apparatus, this activity is expressed only on the mature or secretory face (Morré and Vigil, 1979). The procedure is basically a cytochemical one whereby the different membranes can be marked for subsequent identification by electron microscopy and, most important, related to the situation in situ. However, at the same time, the reactive membranes are rendered more electron-dense than unreactive membranes through the deposition of cuprous ferricyanide (Hatchett's brown). Using this procedure, it has been possible to resolve mature and immature secretory vesicles of hepatocyte Golgi apparatus (Hess et al., 1979).

The use of deposition of lead phosphate (Leskes et al., 1971a, b) in combination with beta-glycerol phosphate as substrate at low pH affords an opportunity to isolate and identify lysosome-forming elements. Antibodies coupled to latex beads or other solid supports of electron dense materials (e.g., Bretz et al.,

1980) for glycosyltransferase-rich regions have also been utilized. Limitations are imposed primarily by the specificity of the label, the steepness of the activity gradient across the Golgi apparatus and the fidelity of in vitro applications to fractions of cytochemical procedures established in vivo for tissues.

3.3. Golgi Apparatus Subfractionation by Free-Flow Electrophoresis

Separation of cis and trans Golgi apparatus elements and a general sub-fractionation of Golgi apparatus cisternae across its polarity axis is afforded by the technique of free-flow electrophoresis (Fig. 3.13). In this procedure, the mixture to be separated is introduced as a fine jet into a separation buffer moving across the field lines of an electric field. Membranes bearing different electrical charges migrate different distances across the separation chamber as the basis for the separations. An expected gradient of the acidic glyco-conjugates (sialic acid-containing glycoproteins and glycolipids) across the polarity axis provides one basis for the separation (Fig. 3.14). Such a gradient had been indicated previously from cytochemical studies (Rambourg et al., 1969) and from the increasing

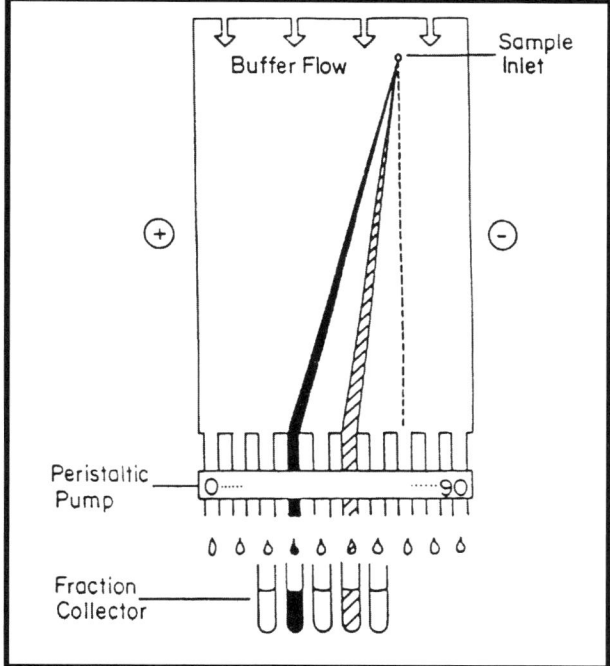

Fig. 3.13. Principles of preparative continuous free-flow electrophoresis used to subfractionate Golgi apparatus from cis to trans across their polarity axes (Hannig and Heidrich, 1977; Heidrich, 1981).

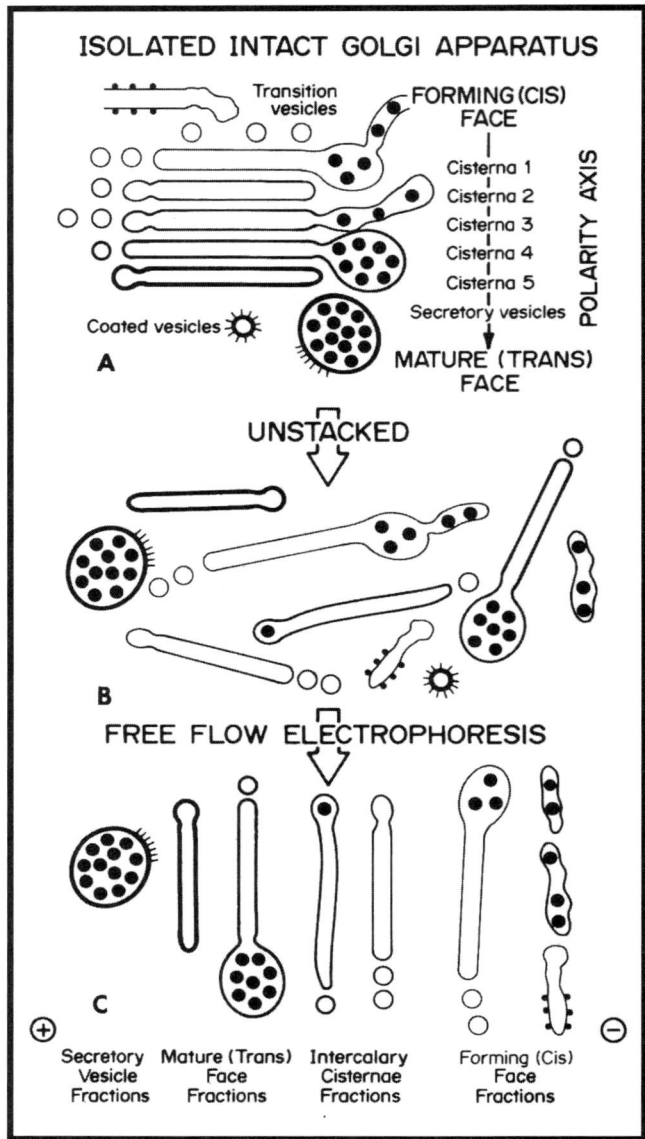

Fig. 3.14. Diagrammatic representation of Golgi apparatus subfractionation by preparative free-flow electrophoresis. Details are given in Morré et al. (1984a).

sialic acid content comparing endoplasmic reticulum, Golgi apparatus and plasma membrane. Additionally, the enzymes for terminal sialic acid additions are found in the Golgi apparatus.

   In the approach used, intact Golgi apparatus were isolated as the starting material. These intact Golgi apparatus were then unstacked enzymatically and mechanically. The unstacking procedure and conditions for free-flow electrophoresis are given by Morré et al. (1984a).

As a morphological marker for the cisternae from the mature (trans) face of the Golgi apparatus, the fixation-resistant deposition of cupric ferrocyanide (Hatchett's brown) (Morré and Vigil, 1979; Goldenberg et al., 1979) was used (Fig. 3.15). This activity was associated with those fractions nearest the anode, i.e., to the left of the midpoint of the separation. It was low or absent from fractions nearest the point of injection, i.e., to the right of the midpoint of the separation.

To mark the immature (cis) face, osmication was according to the procedure of Friend and Murray (1965). Pellets were fixed for 1 hour in 1% osmium tetroxide in 0.01 M sodium phosphate, pH 7.2, rinsed with water, and then treated with unbuffered 2% osmium tetroxide at 40° for 4 h. Under these conditions membranes and vesicles to the right of the separation center were reactive while vesicles to the left of center were unreactive.

To mark intercalary cisternae (cisternae of the midregion of the stack), the enzyme NADPase was used (Navas et al., 1986). As shown originally in rat incisor ameloblasts (Smith, 1980), this activity marks the central cisternae of the stack while the cis- and trans-most cisternae are unmarked (Fig. 7.9). This activity, when

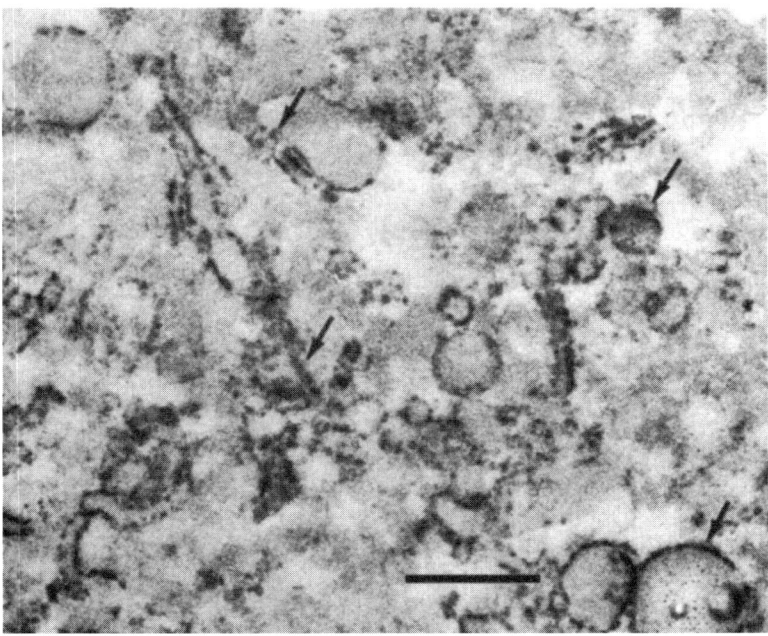

Fig. 3.15. Mature (trans) face electrophoretic fraction marked by a glutaraldehyde-resistant NADH-ferricyanide reductase activity demonstrated by the presence of electron-dense deposits of Hatchett's brown associated with the cisternal membranes (arrows). This cytochemical activity is restricted to cisternae of the distal or trans pole of the rat liver Golgi apparatus. Reproduced from Morré et al., 1983b, with permission from Wissenschaftliche Verlagsgesellschaft mbH. Scale bar = 0.5 μm.

applied to the free-flow electrophoresis fractions was restricted to the center of the separation.

The separation of cis and trans elements shown by morphological and cytochemical procedures is confirmed from the distributions of various marker constituents. Two glycosyltransferases of terminal sugar additions, sialyl and galactosyl transferase, show a gradient of increasing activity toward the anode. In contrast, nucleoside diphosphate phosphatase (UDP as substrate) and NADH-cytochrome c reductase, both enzymatic activities concentrated in endoplasmic reticulum of rat liver and putative cis face markers, show the opposite distribution. NADH-cytochrome c reductase shows two peaks of activity, one around fraction 39 and another at fraction 32. The latter may correspond to the NADH-ferricyanide reductase activity resistant to cytochemical localization used as a morphological marker for trans Golgi apparatus elements of rat liver.

While charged surface molecules are a primary determinant of electrophoretic mobility, diffusion potential, negative outside, has been shown experimentally to also contribute (Morré et al., 1994a). A diffusion potential,

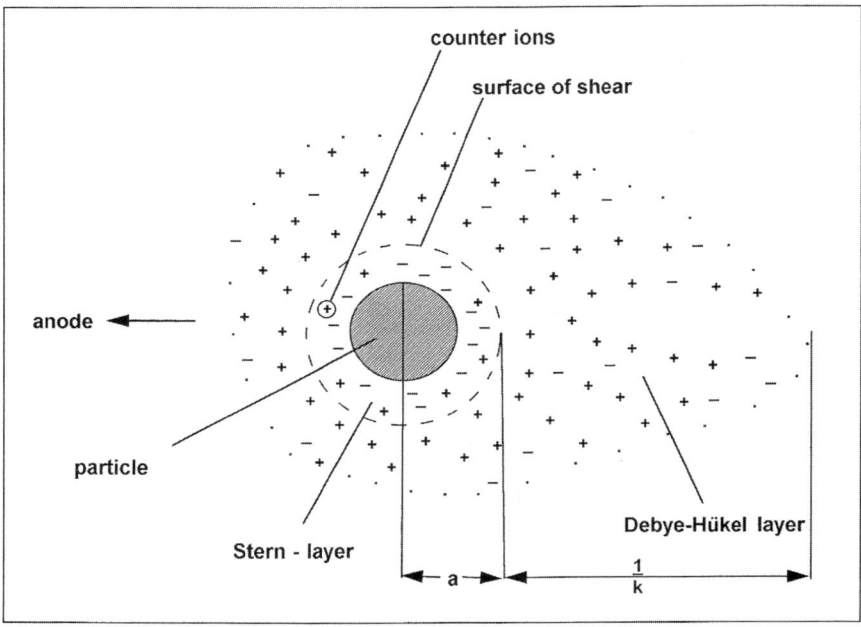

Fig. 3.16. Theoretical basis for free-flow electrophoretic separations based on diffusion potential. The figure illustrates how the electric double layer of a particle in a solution of electrolytes is thought to be influenced by an electric field (Hannig and Heidrich, 1990). The inner region includes absorbed ions and a diffuse region in which ions redistribute according to the electrical force exerted on the particle. The ζ potential, the so-called actual electrical charge, arises from a "surface of shear" at the border with the inner hydrate – Stern layer to then influence electrophoretic mobility. It is the electrical charge at this surface of sheer that may be modified by diffusion potential to alter electrophoretic mobility. Reprinted from Morré Lawrence, Safranski, Hammond and Morré, 1994a, and Hannig and Heidrich, 1990, copyright 1990 with permission from Elsevier.

however, will not of itself alter the electrophoretic mobility of a particle in buffered solution. Electrophoretic mobility is universally considered to be determined by a mobile ion cloud around the particle, the so-called Debye–Hückel layer (Hannig and Heidrich, 1990; Fig. 3.16). The Debye–Hückel layer is the outer region of an electric double layer of ions surrounding bioparticles. In an electric field, the layer is deformed such that the outer ion cloud is shifted in a direction opposite to that of particle movement to expose underlying charged groups contributing to diffusion potential to allow their contribution as well to electrophoretic mobility.

Thus, the process of membrane differentiation across the polarity axis deduced first from analyses of total Golgi apparatus fractions compared to endoplasmic reticulum and plasma membrane and from morphological considerations is reflected in the Golgi apparatus subfractions prepared by free-flow electrophoresis.

3.4. Summary

Subcellular fractionation of cells and tissues provides an important source of biochemical information concerning the structure, identification, characterization, biosynthesis, turnover and function of Golgi apparatus. Procedures are described that yield purified preparations of Golgi apparatus and appropriate reference fractions such as endoplasmic reticulum, plasma membrane, mitochondria and nuclei from liver tissue as well as from plant and mammalian cells grown in culture. Advantageously, the procedures can be carried out in a few hours using normal laboratory equipment. What are presented for the most part are methods developed over nearly three decades in the authors' laboratories and used routinely not only to prepare Golgi apparatus but a full range of reference fractions as well. Also presented are specialized protocols to subfractionate Golgi apparatus orthogonal to the cis to trans polarity axis to isolate secretory vesicle tubule and cisternal fractions and to subfractionate Golgi apparatus across the cis to trans polarity axis into immature, intermediate (intercalary) and mature cisternae, the latter being achieved by preparative continuous free-flow electrophoresis.

4

Tubules

Tubules, 30 nm in diameter, are a conspicuous feature of all Golgi apparatus (Fig. 4.1). Their function, however, has remained elusive. Tubules may extend for considerable distances from the stack. The total surface area of the tubules and associated fenestrae is equivalent to or greater than that of the flattened portions of cisternae. Thus the tubules appear to contribute significantly to Golgi apparatus mass.

The tubules are continuous with the peripheries of the stacked cisternae (Figs. 4.1 to 4.4). Cisternae from the exterior faces of the stack consist almost entirely of tubules with much reduced central plate-like structures (Fig. 4.3). Cisternae near the middle of the stacks often begin to anastomose close to their edges and then blend gradually into the resultant fenestrae (Fig. 4.2). When fenestrae are absent, tubules may emerge directly from the peripheries of the flattened saccules (Fig. 4.3). These features are illustrated in the model and diagram of Fig. 4.5.

Tubules interconnect closely adjacent stacks of the Golgi apparatus (Figs. 2.11 and 4.4) and connect secretory vesicles to the fenestrated cisternal peripheries (Figs. 4.6 and 2.12). One hypothesis is that tubules comprise a more static component of the Golgi apparatus in contrast to the stacked cisternal plates which may continuously turnover. Peripheral tubules, often with coated buds attached, may represent the means whereby adjacent Golgi apparatus stacks exchange carbohydrate-processing enzymes or where resident Golgi apparatus proteins are introduced into and out of the stack during membrane flow-differentiation (Chapter 6).

Electron micrographs of cells where stacks are closely adjacent reveal the tubules that interconnect the cisternae of one Golgi apparatus stack with cisternae in another Golgi apparatus stack (Figs. 4.4 and 2.12). Such connections most often maintain a positional equivalency to the cisterna to which they are attached. That is, tubules are seen most often to interconnect adjacent cisternae at the same level within the stack. Tubules probably also interconnect widely separated stacks, but this has been more difficult to demonstrate because the tubules tend to branch and change directions within the plane of a thin section. Therefore, most estimates of tubule connections between stacks rely on fortuitous sections or on specialized procedures such as isolated and negatively stained preparations, precipitate stains, thick sections, or 3-D reconstructions. Unfortunately, such approaches are sometimes misleading because of the harsh preparative procedures

D. James Morré and Hilton H. Mollenhauer, *The Golgi Apparatus.*
© Springer 2009

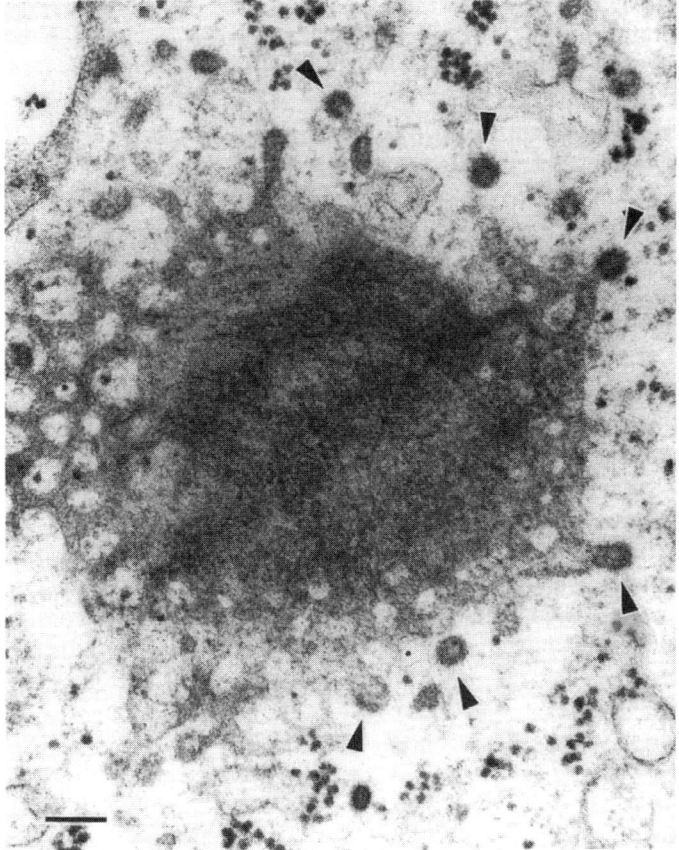

Fig. 4.1.   Face view of a Golgi apparatus cisterna from a daffodil flower fixed in glutaraldehyde and osmium tetroxide, at an angle nearly tangential to the plane of the cisternae to show the system of peripheral tubules, some with coated buds (*arrowheads*). Reproduced from Mollenhauer and Morré, 1998 with permission from Springer Science+Business Media. Scale bar = 0.1 μm.

necessary or because of the loss of resolution that occurs when viewing preparations such as thick sections where structures may overlap. Small flat saccules, termed junctional cisternae, may act as intermediate connecting points for tubules arising from different locations within the cell. These are regularly observed in negatively stained isolated cisternae (Fig. 4.7). Connections via tubules between cisternae within the same stack usually are not observed (Clermont et al., 1994).

Tubules are also clearly a major feature of the TGN (Geuze and Morré, 1991). When compared to other Golgi apparatus tubules, those of the post-Golgi apparatus structures are generally more variable in size with somewhat greater

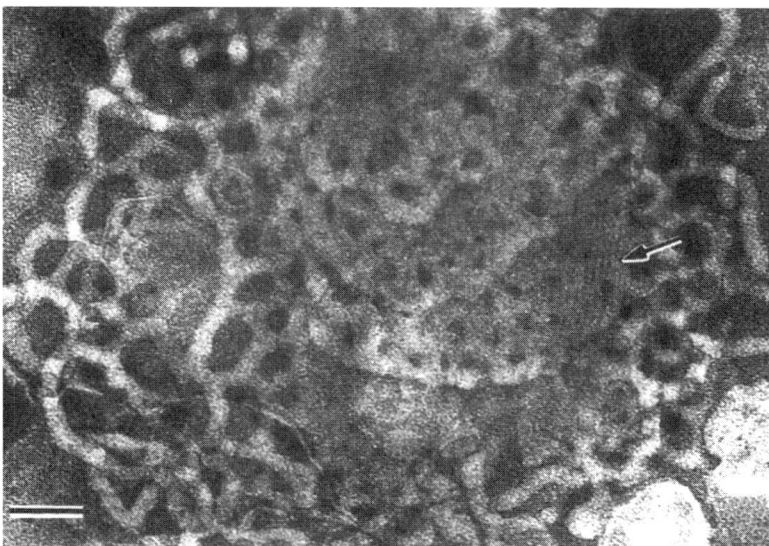

Fig. 4.2. A plant (radish root) Golgi apparatus stack isolated without glutaraldehyde stabilization. Unstacking had begun so that portions of all component cisternae were partially revealed in the same positional relationships as in the original stack. The cis pole is toward the upper edge of the micrograph and the trans pole is toward the bottom edge. The peripheries of the mid-cisternae are mostly fenestrated whereas the peripheries of the trans cisternae are mostly tubular. *Arrow* indicates a mat of intercisternal filaments. Reproduced from J. Cell Biol., 1966, 29, 373–376. Copyright 1966 The Rockefeller University Press.

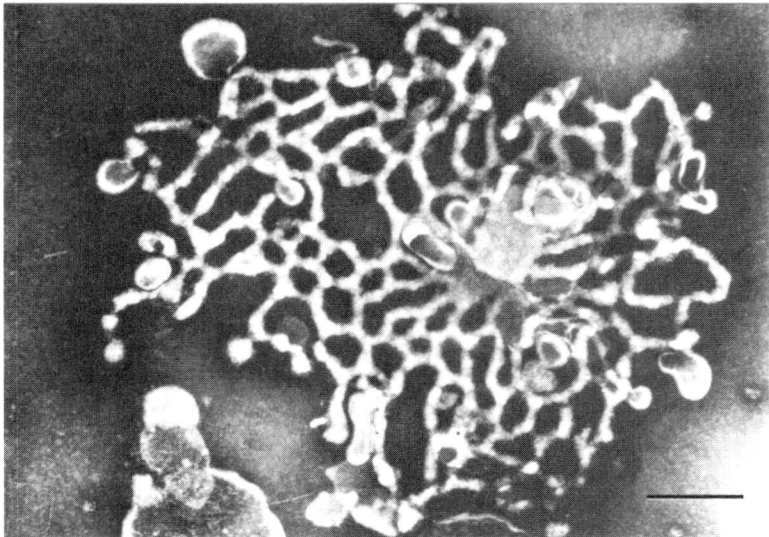

Fig. 4.3. Single cisterna from an exterior face of a Golgi apparatus stack isolated from onion (*Allium capa*) stem and stained on the electron microscope grid with phosphotungstic acid. Reproduced from Cunningham et al., 1966, Journal of Cell Biology, 28:169–179. Copyright 1966 The Rockefeller University Press. Scale bar = 0.2 µm.

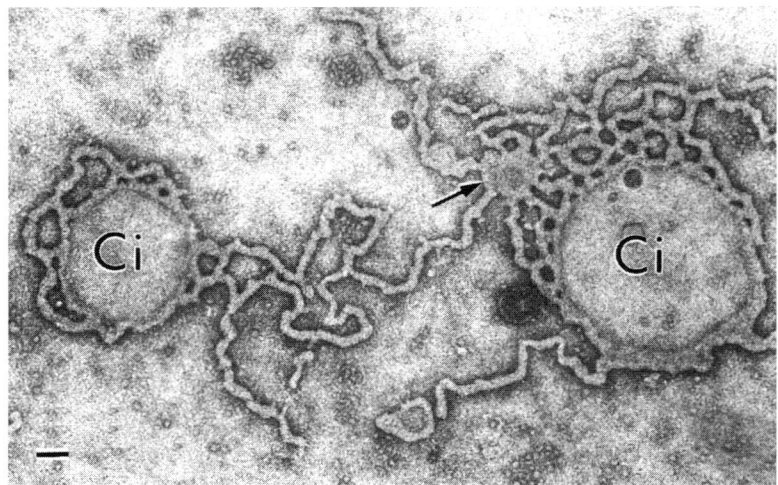

Fig. 4.4. Two bean (*Phaseolus vulgaris*) root tip Golgi apparatus cisternae (Ci) negatively stained with phosphotungstic acid. The cisternae are connected via peripheral tubules through a small subcisterna or junctional complex (*arrow*). Reproduced from Mollenhauer and Morré, 1998 with permission from Springer Science+Business Media. Scale bar = 0.1 μm.

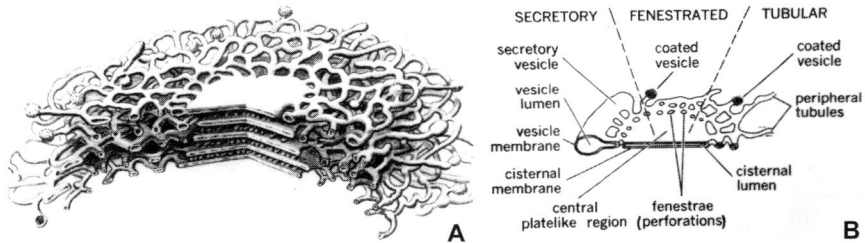

Fig. 4.5. An early Golgi apparatus model showing a single stack composed of four cisternae with emphasis on the tubular cisternal peripheries. Cisternal maturation is depicted from *top* to *bottom*. Reproduced from Mollenhauer and Morré, 1966a. Reprinted, with permission, from the Annual Review of Plant Physiology, Volume 17 © 1966 by Annual Reviews.

diameters and thick membranes. TGN tubules and associated vesicles frequently exhibit clathrin coats.

4.1. Function of Golgi Apparatus Peripheral Tubules in Delivery of Cargo from Endoplasmic Reticulum to the Golgi Apparatus

4.1.1. Liver Parenchyma and Intestinal Absorptive Cells

The Golgi apparatus of intestinal absorptive cells and from liver parenchyma (hepatocytes) share many characteristics in common. These Golgi apparatus are engaged in the segregation, modification, and secretion of lipoproteins to the

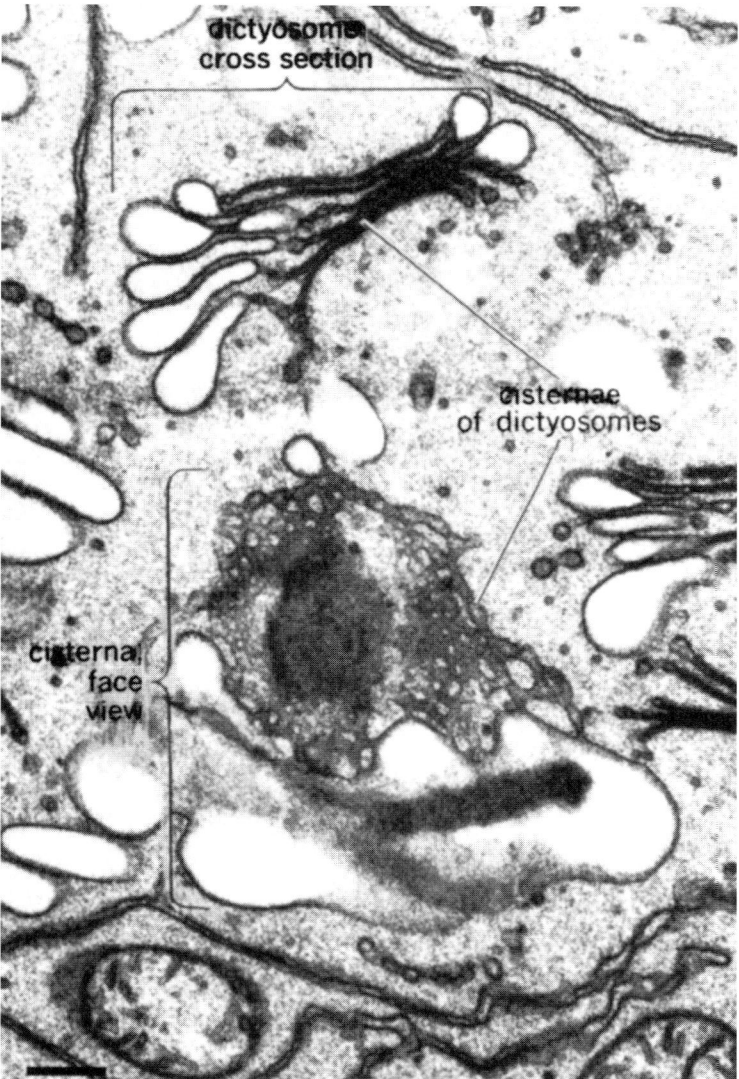

Fig. 4.6. Portion of the Golgi apparatus of a secretory cell from the root cap of maize. The stack on the *bottom* was sectioned tangentially and illustrates multiple tubule connections to attached forming secretory vesicles. Reproduced from J. Cell Biol., 1966, 29, 373–376. Copyright 1966 The Rockefeller University Press. Scale bar = 0.5 μm.

circulation (to the capillary space for liver and to the lymphatics for intestine). The lipoproteins are packaged in the form of discrete particles (low density and very low-density lipoprotein particles for liver and very low density lipoprotein particles and chylomicra for intestine). The particles vary in size, buoyant density,

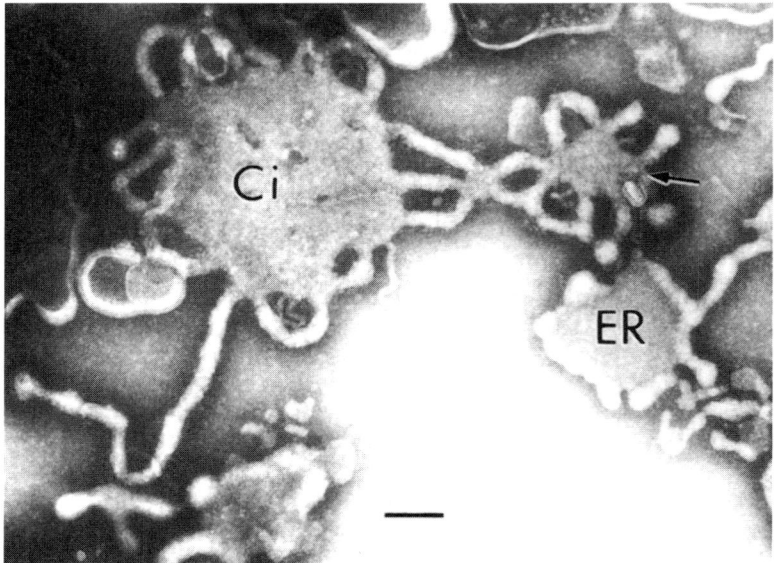

Fig. 4.7. Plant Golgi apparatus cisterna (cauliflower) negatively stained with phosphotungstic acid to illustrate junctional cisternae (*arrow*). Ci, Cisterna; ER, endoplasmic reticulum. Reproduced from Mollenhauer and Morré, 1998 with permission from Springer Science + Business Media. Scale bar = 0.2 μm.

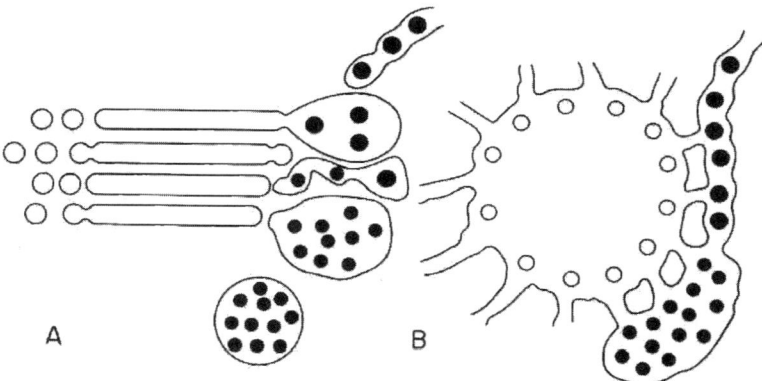

Fig. 4.8. Diagram illustrating the peripheral route followed by lipoprotein particles in the liver from site of their synthesis in endoplasmic reticulum to the forming secretory vesicles during which the central plate-like portions of the cisternae are bypassed.

and lipid and protein composition. Of the three types of particles, the protein and cholesterol content is greatest in low density lipoproteins and least in chylomicra. Triglycerides form the bulk of the lipid. Abnormal production and/or metabolism of these particles contributes to cardiovascular disease. Under normal conditions, these particles are a major source of circulating triglyceride for delivery to cells of the body.

The lipoprotein particles are elaborated first in smooth portions of the endoplasmic reticulum and then pass via direct connections (boulevard périphérique) to the secretory vesicles (Figs. 4.8 and 2.10). Lipoprotein particles rarely, if ever, enter the saccular portions of the Golgi apparatus. In the Golgi apparatus, particle diameter is reduced presumably due to lipase action and the loss of triglyceride (Hess et al., 1979). Terminal carbohydrates are added to oligosaccharide chains. Additionally, in liver, very low-density lipoprotein particles appear to be secreted directly via endoplasmic reticulum-derived vesicles (so-called Golgi apparatus bypass) (Morré, 1981). Particles reaching the circulation by this route appear not to be as extensively processed as those released via the Golgi apparatus route.

Liver may be estimated to secrete more than 50 different types of proteins, lipoproteins, and glycoproteins to the circulation. Included among these are the low and very low density lipoproteins already mentioned, serum albumin, plasma fibronectin, lysosomal enzymes, fibrinogen, and various clotting factors. Most, if not all, of these proteins are packaged and processed via the Golgi apparatus and follow a peripheral route to enter the vesicles from sites of synthesis on endoplasmic reticulum. To what extent sorting takes place, beyond that which segregates lysosomal enzymes from secreted proteins, is not known.

As a means to test the hypothesis that rough and smooth endoplasmic reticulum membranes were functionally continuous with Golgi apparatus and secretory vesicles in liver as required by a peripheral route of direct cargo delivery, studies were undertaken to determine if the newly synthesized enzymes stimulated by phenobarbital administration could be detected in Golgi apparatus. The microsomal oxidative O-demethylation of p-nitroanisole to p-nitrophenol (O-demethylase) was utilized as a marker activity to monitor transfer. A rapid increase in O-demethylase activity was detected in rough endoplasmic reticulum over a period of about 10 min (Fig. 4.9 and Table 4.1). This response paralleled that reported for the drug-induced synthesis of NADPH-cytochrome c oxidoreductase. The increase in O-demethylase activity in rough endoplasmic reticulum was followed after about 10 min by an increase in activity in an endoplasmic reticulum fraction which contained the smooth portions of the endoplasmic reticulum plus regions of rough endoplasmic reticulum continuous with smooth endoplasmic reticulum. Increased O-demethylase in Golgi apparatus fractions lagged 10–15 min behind that in the smooth endoplasmic reticulum-containing fractions (Table 4.1). The sequential appearance of enzyme activity in rough endoplasmic reticulum (rough ER) , smooth endoplasmic reticulum (smooth ER), and Golgi apparatus was seen in the lag times, the half times of drug-induced increase (rough ER < smooth ER < Golgi apparatus), in the time at which maximum activity was attained, and in the decay kinetics. The times required for the appearance of the enzyme in the smooth endoplasmic reticulum (ER-II) and its appearance in the Golgi apparatus were much longer than those required to traverse the Golgi apparatus stack. Rather the values were interpreted as times for a membrane protein to move to the Golgi apparatus-attached secretory vesicles from rough ER via the peripheral system of Golgi apparatus tubules.

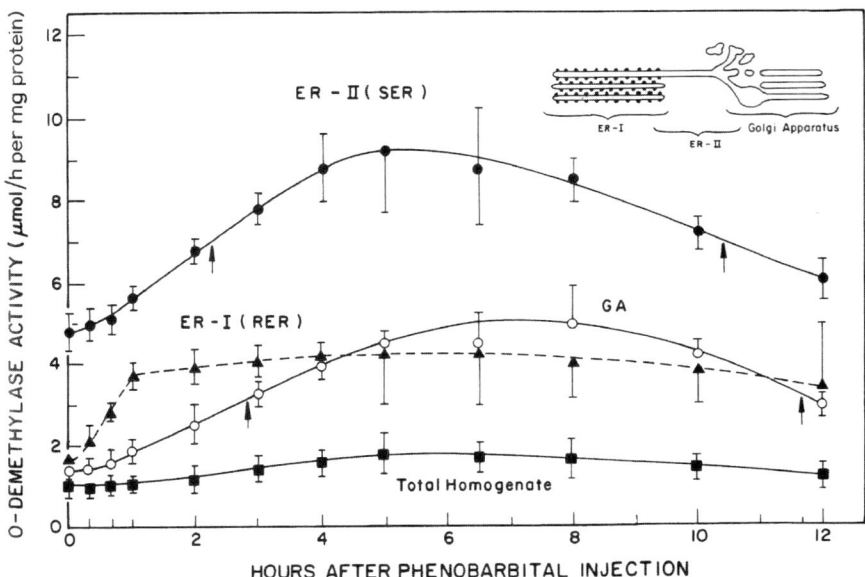

HOURS AFTER PHENOBARBITAL INJECTION

Fig. 4.9. Modulations in specific enzyme activity for the oxidative *O*-demethylation of *p*-nitroanisole by rat liver fractions following a single injection of phenobarbital. A diagrammatic interpretation of the origin of the cell fractions is shown in the *upper right*. Specific activity rose rapidly in a fraction derived from rough endoplasmic reticulum (RER = ER-I). This was followed by a more prolonged increase in specific activity of a fraction containing smooth elements of endoplasmic reticulum (SER) and rough–smooth transition regions (= ER-II). Lastly, the specific activity of the enzyme was observed to increase in Golgi apparatus (GA) fractions. No *O*-demethylase activity could be detected in plasma membrane fractions either before or after phenobarbital injection (data not shown). *Arrows* indicate the halftimes of drug-induced increase (*left*) and decay (*right*). Details are given in Table 4.1. Reproduced from Morré et al., 1974a with permission from Lippincott Williams & Wilkins.

Table 4.1. Time constants for the appearance and disappearance of drug-induced *O*-demethylase in rough (ER-I) and smooth (ER-II) endoplasmic reticulum and Golgi apparatus (GA) fractions from rat liver following a single injection of phenobarbital.

Time constant	Fraction	Time after phenobarbital injection (min)
$t_{1/2}$ of appearance	ER-I	40
	ER-II	168
	GA	192
Lag time by extrapolation (t_o)	ER-I	9
	ER-II	18
	GA	27
$t_{1/2}$ of decay	ER-I	n.d.
	ER-II	620
	GA	700

Source: From Morré et al. (1974a)
n.d., Not determined

4.1.2. Peripheral Tubule Function in Acinar Cells of Pancreas and Parotid Gland and Chromaffin Cells of the Adrenal Medulla

Since rough endoplasmic reticulum and smooth endoplasmic reticulum form a peripheral continuum with Golgi apparatus cisternae and secretory vesicles in liver, a similar pathway may be followed in zymogen secreting cells for example. Proteins secreted by pancreas and parotid gland and catecholamines secreted by chromaffin cells of the adrenal medulla arise in structures at or near the Golgi apparatus which are termed condensing vacuoles. Condensing vacuoles, once formed, often remain in the vicinity of the Golgi apparatus where they assume a spherical shape as they accumulate the dilute secretory products and develop into mature granules over periods of sometimes several hours (Fig. 2.20). The condensing vacuoles do not increase in diameter. Rather, condensing vacuoles appear to concentrate the secretory product as evidenced by changes in electron opacity of the granule contents. It is not known if the condensing vacuoles are attached to the Golgi apparatus during much of this filling period but most protein transport to the vacuoles from the endoplasmic reticulum appears to occur via a peripheral route most likely involving tubules (see below).

The proteins of the zymogen granule contents are synthesized on polyribosomes attached to the rough endoplasmic reticulum. They enter the cisternal space of the rough endoplasmic reticulum. When [^{3}H]leucine is administered to animals or pancreatic slices, autoradiographic grains are seldom concentrated over the stacked cisternae of the Golgi apparatus but appear to leave the endoplasmic reticulum and enter the condensing vacuoles via a route peripheral to the cisternal stacks (Jamieson and Palade, 1967a, b; Fig. 8.3). The central saccules of the Golgi apparatus appear to be bypassed. It is hypothesized that this bypass is accomplished by small transition vesicles which bud off the endoplasmic reticulum, migrate to the granule, and fuse with it to deliver both secretory proteins and membranes. This would lead to delivery of excess amounts of membrane to the granules and an increase in diameter during filling which has not been observed.

An alternative mechanism would be entry of secretory product via direct tubular connections. Such a mechanism would allow vesicles to be formed from the Golgi apparatus and then be filled with product in a manner independent of further Golgi apparatus participation or of a shuttle mechanism. Moreover, direct movement of secretory material from rough endoplasmic reticulum to condensing vacuoles does not necessarily argue for an alternative pathway for membrane flow which completely bypasses the Golgi apparatus cisternae. Despite the possibility that condensing vacuoles are filled independently of the Golgi apparatus, new condensing vacuoles appear to be formed from Golgi apparatus membrane at the ends of Golgi apparatus cisternae.

4.2. Golgi Apparatus Buds – Vesicles or Coated Ends of Tubules?

A major consideration in evaluating the potential importance of peripheral tubules to Golgi apparatus function is whether or not free vesicles as an alternative to tubules are involved in transport of membrane constituents and/or cargo from one cisternae to another across the stack. There is no doubt that vesicles exert such roles at both the entry and exit faces. However, convincing evidence for a similar role for vesicles in transfer between intercalary cisternae has not been forthcoming.

The supposition that there was vesicular transfer across the stack as an alternative to cisternal maturation derived initially from a cell-free assay developed by Rothman and colleagues (1984) to study inter-Golgi apparatus transport. Revealed was a phenomenon whereby membrane glycoproteins from one cell type were processed by enzymes located in Golgi apparatus from another cell type. To explain these observations, Rothman and colleagues postulated that vesicles bud from one Golgi apparatus stack and migrate to and fuse with cisternae of other Golgi apparatus stacks in the cell-free system. An extension of this hypothesis was that these same or similar vesicles were involved in the trafficking of membrane material from one cisterna to the next in the same Golgi apparatus stack (Dunphy and Rothman, 1985).

The morphological entity associated with intercompartment Golgi apparatus transfer was a coated bud first described by Orci et al. (1986; see, however, Mollenhauer et al., 1976) using tannic acid-containing fixatives. Treatment of transport reactions with GTP-γ-S resulted in accumulation of these distinctive coated buds (Melancon et al., 1987).

Bud formation was induced by GTP-γ-S (Malhotra et al., 1988) and the coat proteins were identified as a complex of COPI coatomers, including β-COP. Brefelin A appeared to exert its disruptive action at the Golgi apparatus by promoting the dissociation of β-COP from the Golgi apparatus buds.

The coated buds of the Golgi apparatus nearly always are attached to the ends of the cisternal tubules (Morré and Keenan, 1994) which of themselves are an integral structural aspect of the Golgi apparatus cisternae (Fig. 4.10). The

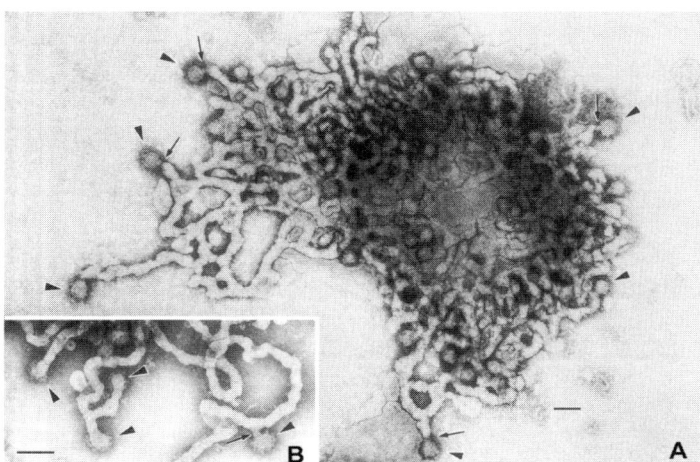

Fig. 4.10. Golgi apparatus isolated from bean root, stabilized with glutaraldehyde, and negatively stained with phosphotungstic acid using the procedures described by Cunningham et al. (1966). Coated buds are attached to some of the tubules (*arrowheads*). The buds may appear at ends of tubules or may be attached to the tubules by small neck-like appendages (*arrow* in B). Reproduced from Mollenhauer and Morré, 1998 with permission from Springer Science+Business Media. Scale bar = 0.1 μm.

tubules exhibit a variety of lengths and may be attached at any level within the stack or even interconnect adjacent stacks.

Extensive morphological analyses of Golgi apparatus zones of rodent liver, cultured mammalian cells, and various plant species have not revealed free coated vesicles associated at or near medial Golgi apparatus cisternae in the location of the buds. The majority of the buds appear to be attached to tubules. Free vesicles occur at the Golgi apparatus network (clathrin-coated vesicles) but not in between. Most of the so-called small vesicle profiles associated with median Golgi apparatus elements are cross sections through the many peripheral tubules that characterize that region of the Golgi apparatus (Figs. 2.5 to 2.8). This is evidenced by the electron translucent centers of the profiles, which suggests that they are cross sections of tubules and not vesicles. If they were vesicles, the top and bottom of each vesicle would be within the plane of the thickness of an ordinary thin section. The vesicle centers would appear filled rather than hollow.

It is widely assumed that the mechanism of transfer involved a vesicular intermediate that formed, detached and migrated as a free vesicle and fused. However, this assumption, while consistent with transfer kinetics and characteristics if it existed as a free vesicle, has never been verified in studies that might be excepted to trap the vesicle intermediate.

The coated buds of intermediate Golgi apparatus compartments do not appear to dissociate from isolated Golgi apparatus even under conditions optimized for inter Golgi apparatus transfer. This is further exemplified by the observation that, in order to isolate the buds, it was necessary to employ harsh chaotropic agents to first dissolve away the tubular connections (Malhotra et al., 1989). This contrasts with transition vesicles, coated vesicles or secretory vesicles which, once formed, do not require harsh extraction methods for release and subsequent isolation (Fig. 10.2). Also noteworthy is that inter compartment Golgi apparatus transfer has never been established for a single Golgi apparatus stack and has been studied only between adjacent stacks in heterologous transfer systems. Neither has inter Golgi apparatus transfer been observed with Golgi apparatus fully immobilized on nitrocellulose, for example (Chapter 10).

It may be that coated ends of tubules are involved in the communication between adjacent Golgi apparatus stacks or that the Golgi apparatus buds selectively stained with tannic acid can function as vehicles for transfer of resident Golgi apparatus in or out of the cisternae within a single stack during membrane differentiation. Deng et al. (1992) studied intermixing of resident Golgi apparatus membrane proteins in rat–hamster polykaryons. Their results suggest that a relatively slow intermixing of resident Golgi apparatus proteins does occur between cisternal stacks and that direct physical continuity between Golgi apparatus elements is required for the intermixing to take place.

Mironov et al. (2004) reported that breakdown of the tubular system by dicumarol, markedly slowed intra-Golgi traffic of VSV-G transport from the endoplasmic reticulum to the medial Golgi, and inhibited the diffusional mobility of both galactosyltransferase and VSV-G tagged with green fluorescent protein. However, it did not alter transport from the trans-Golgi network to the cell surface, Golgi-to-endoplasmic reticulum traffic of ERGIC58, COPI-dependent Golgi vesiculation by AIF_4 or ADP-ribosylation factor and β-COP coat protein binding (COPI) to Golgi membranes. These authors concluded that dicumarol induced the selective breakdown of the tubular components of the Golgi complex and inhibited intra-Golgi transport. Their findings support that lateral diffusion between adjacent stacks has a role of Golgi apparatus tubules in protein transport through the Golgi complex.

If the coated ends of Golgi apparatus tubules, i.e., buds, remain attached to tubules and serve to facilitate transfer among contiguous Golgi apparatus stacks, this would explain why the basic assay procedure of Rothman et al. (1984), where transfer among heterologous Golgi apparatus stacks is measured, operates much more efficiently than transfer among immobilized Golgi apparatus stacks.

While Golgi apparatus cisternae in the same stack appear not to be connected either by tubules or by shuttling vesicles, vesicles do appear to contribute to cargo transport in some but not all instances (Pelham, 2001). Recent findings do emphasize that segregation of cargo and enzymes does occur within contiguous stretches of Golgi membrane. The flat cisternal membranes are accessible to both enzymes and possibly certain cargo whereas tubules derived from them, represent a system for delivery of cargo directly to secretory vesicles. Tubules may also serve as sites where resident Golgi proteins may be delivered during membrane maturation (Morin-Ganet et al., 2000).

Transfer among Golgi stacks via peripheral tubules could represent a major retrograde transport mechanism of membrane trafficking (Weidman, 1995). It would allow lateral segregation as a major sorting mechanism without a need for a complicated system of autograde and retrograde transport vesicles. Consistent with this view would be the greatly enhanced number of Golgi apparatus tubules seen in electron micrographs following brefeldin A treatment (Ulmer and Palade, 1989). One action of brefeldin A is to interfere with the binding of β-COPs, and if β-COPs are important for inter Golgi apparatus transfer, then it would follow that inter Golgi apparatus retrograde transfer of resident Golgi apparatus proteins also would fail. Consistent with these observations would be addition and removal of resident Golgi apparatus proteins via β-COP-coated tubules from the saccule portions of Golgi apparatus cisternae as required during flow differentiation of Golgi apparatus membranes within the context of the membrane maturation model of Chapter 6 (see also Chapter 11).

4.2.1. Isolation of Tubule-Enriched Fractions

The first reported isolation of Golgi apparatus tubules (Ovtracht et al., 1973) began with intact isolated Golgi apparatus from rat liver (Morré, 1971). Lysosomal extracts or crude Taka-diastase preparations were used to unstack the Golgi apparatus to cleave earlier added stabilizing detran polymers followed by disruption by means of repeated excursions through a fine-bore Pasteur pipette. The unstacked and separated Golgi apparatus components were then subjected to discontinuous sucrose density gradient centrifugation and were analyzed by electron microscope morphometry. Fraction II, collected just about and just below the 0.9/1.2 M sucrose interface, was enriched in the lipoprotein-filled tubules of the rat liver Golgi apparatus. Attempts to separate the peripheral tubules not containing lipoprotein particles from fractions just containing the cisternal plates has, thus far, seemed to require the use of chaotropic agents to dissolve the plates and release the tubules (Malhotra et al., 1989). We have utilized such procedures successfully in preliminary experiments to prepare tubule-enriched fractions both from rat liver and from Golgi apparatus isolated from rat liver (Fig. 4.11).

There is little information as to what biochemical activities or metabolic functions might be associated with Golgi apparatus tubules. Golgi apparatus contain high concentrations of ubiquinone (Nyquist et al., 1970; Zambrano et al., 1975; Kalen et al., 1987) and of ubiquinone biosynthetic enzymes, for example, the enzymes that transfer solanesol pyrophosphate to 4-hydroxybenzoate (nonaprenyl-4-hydroxybenzoate transferase) (Teclebrhan et al., 1995). The latter authors have suggested that ubiquinone is synthesized sequentially in the endoplasmic reticulum–Golgi apparatus system and is, thereafter, translocated to other cellular membranes.

Dallner (1978) described an unusual microsomal fraction which could not be aggregated by divalent cations. This is a property shared with Golgi apparatus tubule fractions such as those isolated from Golgi apparatus rat liver by chaotropic disruption (Fig. 4.11; D. J. Morré, unpublished results). The unusual microsome fraction, designated by Dallner and colleagues as SII microsomes, have been assumed to represent some compartment of the Golgi apparatus and are enriched in both enzymes of ubiquinone and sterol biosynthesis (Teclebrhan et al., 1995). Nonaprenyl-4-hydroxybenzoate transferase which transfers the solanesol side chain to the precursor ring during ubiquinone biosynthesis is concentrated in this fraction. Following condensation, a number of additional reactions take place, which convert the 4-hydroxybenzoate moiety to a methoxy- and methyl-substituted benzoquinone ring. These reactions also have been suggested to take place in the Golgi apparatus (Teclebrhan et al., 1995). A major role for Golgi apparatus tubules in the biosynthesis and transport of isoprenoid compounds including, but not restricted to, ubiquinone and cholesterol would not be unexpected based upon these findings.

Tubules may also account for cholesterol or quinone transport and targeting, for example, the selective transfer of ubiquinone to mitochondria and of plastoquinone to the chloroplast in plants where both ubiquinone and plastoquinone appear to be synthesized in the endoplasmic reticulum–Golgi apparatus system (Swiezewaska et al., 1993; Osowska-Rogers et al., 1994).

Measurements of the flux of membrane constituents through the Golgi apparatus, inhibitor studies, and direct light and electron microscopic observations in favorable cell types, reveal a pattern of events suggestive of cisternal turnover (Morré, 1987; Mollenhauer and Morré, 1991). In actively secreting cells, the plate-like portions of Golgi apparatus cisternae appear to be formed from transition vesicles at the cis face of the Golgi apparatus and discharged as secretory vesicles plus a cisternal remnant at the trans face at a rate of one every 3–4 min (Morré, 1987). With a total transit time through the Golgi apparatus of 15–20 min and with new cisternae being formed as existing cisternae are lost, turnover of tubules may not occur to the same extent as that of plate-like portions of cisternae.

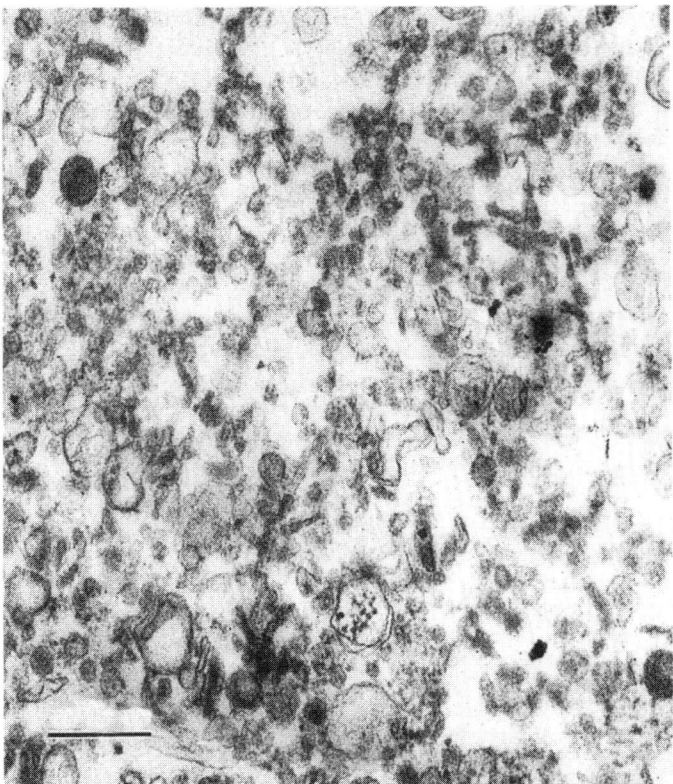

Fig. 4.11. Purified fraction of Golgi apparatus tubules isolated by a procedure involving use of chaotropic agents to disrupt low-density lipoprotein-depleted Golgi apparatus fractions isolated from rat liver into tubule- and plate-like cisternal fractions. This fraction, like the SII microsome fraction of Dallner (1978), could not be aggregated by divalent cations and was found to be enriched in nonaprenyl-4-hydroxybenzoate transferase. Reproduced from Mollenhauer and Morré, 1998 with permission from Springer Science+Business Media. Scale bar = 0.5 μm.

4.3. Summary

Tubules, 30 nm in diameter, are universally associated with Golgi apparatus peripheries. The system of tubules may extend to many micrometers through the cytoplasm and may exceed the central saccules in total membrane surface. Tubules function to interconnect individual Golgi apparatus stacks, in the delivery of certain types of cargo to secretory vesicles and condensing vacuoles, in the attachment of secretory vesicles to cisternae and in activities of the trans Golgi apparatus network. Additionally, Golgi apparatus tubules may contribute to the biosynthetic activities of the Golgi apparatus through the formation of ubiquinone and related isoprenoid substances. Tubule-associated COPI-coated buds add new dimensions to tubule function in the inter-Golgi apparatus transfer, addition and removal of specific Golgi apparatus enzymes and membrane constituents.

5

Endomembrane Biogenesis

Biological membranes which serve as interfaces between different cellular regions to achieve compartmentation, vary widely in their chemical composition. Gas vacuole membranes of blue-green algae are almost entirely protein, whereas artificial membranes may be constructed entirely from phospholipids. Some membranes contain glycosylated proteins or lipids, while others do not. Striking variations in protein and lipid composition occur among species often to the extent that the compositions of different organelles within a species have greater similarities than the same organelle from different species. Plasma membranes, while serving the same essential functions in all cells, exhibit wide variations in composition according to cell type and, with regard to specific receptors, considerable variation in composition among species. Variation may be encountered even within a single cell type and with stage of development or in response to environmental or cellular effectors. Thus, membrane biogenesis emerges as a process of considerable complexity and diversity. The outcome is an interface or compartment that is important to the regulation or segregation of cellular metabolism. Even within a single morphologically identifiable cell compartment such as endoplasmic reticulum or Golgi apparatus, not all regions morphologically defining that cell compartment are alike.

Membrane constituents are first assembled from simpler precursors by well-established biochemical pathways (Fig. 5.1). Whether it is the assembly of amino acids into membrane proteins or incorporation of head groups into phospholipids, the final stages of biosynthesis frequently are catalyzed by enzymes closely associated with a membrane. Few, if any, membrane constituents or terminal biosynthetic enzymes of membrane constituents occur free in the cytoplasm in significant concentrations (Table 5.1).

Within the endomembrane system, plasma membranes, for the most part, are devoid of membrane biosynthetic capabilities (Table 5.1). Possible exceptions include enzymes leading to the synthesis of diglyceride, CDP-diglyceride, and phosphatidylglycerol, terminal sugar transfers in some cell types, and various exchange reactions involving fatty acids and phospholipid head groups. The bulk of the membrane biosynthetic capacity of membrane constituents of the plasma membrane occurs in the Golgi apparatus and endoplasmic reticulum. The Golgi apparatus has a limited capacity to form most phospholipids and possibly to acquire proteins translated on free polyribosomes located within the Golgi apparatus zone as well but does serve as a major contributor to the synthesis of glycolipids of the plasma membrane and in the glycosylation of plasma membrane proteins.

D. James Morré and Hilton H. Mollenhauer, *The Golgi Apparatus.*
© Springer 2009

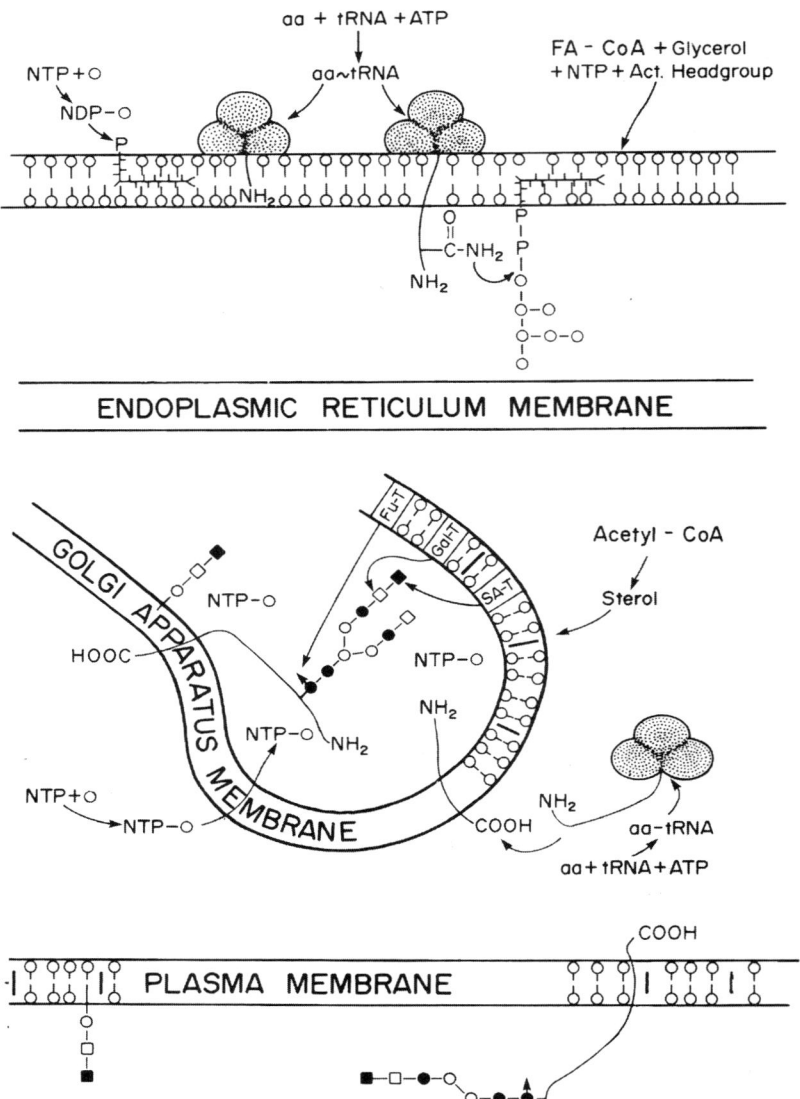

Fig. 5.1. Simplified representation of major biochemical pathways of membrane biogenesis. Modified from Schachter and Roseman, 1980; Struck and Lennart, 1980. Reproduced from Morré, 1987 with permission from Elsevier.

5.1. Role of Endoplasmic Reticulum in Membrane Biogenesis

In in vivo experiments with rats, isotopically labeled amino acids are rapidly incorporated without discernable lag into membranes of rough endoplasmic reticulum. Major components of membrane lipids are synthesized by enzyme

Table 5.1. Comparative lipid compositions of nuclear envelope, endoplasmic reticulum, Golgi apparatus, and plasma membrane fractions from rat liver[a].

	Grams per 100 g total membrane			
Constituent	Nuclear envelope	Endoplasmic reticulum	Golgi apparatus	Plasma membrane
Total lipid	30	30	35	42
Neutral lipid	4	4	6	11
Sterols	2	2	3	7
Cholesterol	1.8	1.2	2	>6
Cholesterol esters	0.2	0.8	1.0	<1
Triglycerides	2	2	3	3–4
Glycolipids	0.025	0.05	0.2	0.6
Phospholipids	26	27	29	30
Phosphatidylcholine	16	16	15	14
Phosphatidylethanolamine	6	7	7	7
Phosphatidylserine	2	1	1	1
Phosphatidylinositol	2	2	2.5	2
Sphigomyelin	<1	1	3.5	6

Source: From Morré and Ovtracht (1977)
[a]Based on data from Benedetti and Emmelot (1968); Kleinig (1970); Yunghans et al. (1970); Keenan and Morré (1970); Morré et al. (1971b, Table 2); Keenan et al. (1972b); Emmelot et al. (1974); Franke and Kartenbeck (1976). The values represent averages and calculations from several experiments and determinations and must be considered only approximate. Minor components such as fat-soluble vitamins (Nyquist et al., 1970, 1971a, b) and minor phospholipids (see Emmelot et al., 1974) were not included in the calculations

systems that terminate more or less exclusively in endoplasmic reticulum. Glycoproteins and glycolipids are synthesized in a step-wise manner but, for glycoproteins at least, evidence indicates that polypeptides of protein portions synthesized on polysomes migrate to or are inserted into new membranes such as those of the endoplasmic reticulum where initial glycosylation takes place followed by transfer to the Golgi apparatus where glycosylation is completed.

5.2. Biosynthetic Capabilities of Golgi Apparatus Relevant to Membrane Biogenesis

Certain glycosyltransferases and sulfotransferases have been localized in Golgi apparatus fractions. These provide clear examples of the potential role of the Golgi apparatus in membrane biogenesis. Other membrane biosynthetic activities are either absent from Golgi apparatus when examined critically with isolated cell fractions or are shared with endoplasmic reticulum and/or nuclear envelope.

5.3. Biosynthesis of Membrane Lipids

The same five major phospholipids are present in all major classes of membranes within the endomembrane system of most mammalian cells, although in plants, spingomyelin is absent and phosphatidylserine is a minor component (Table 5.1). While the phospholipid and fatty acid compositions are similar among endomembrane types, they are not identical and the phospholipid to protein ratios vary as do the relative amounts of individual phospholipids. The phospholipid composition is greatest in plasma membrane and Golgi apparatus and least in endoplasmic reticulum and nuclear envelope. Thus, some opportunity for phospholipid biosynthesis and/or addition of phospholipids by Golgi apparatus is provided. The major differences in lipid distribution for endomembranes of rat liver are an increasing proportion of sphingomyelin between endoplasmic reticulum and plasma membrane (nuclear envelope = endoplasmic reticulum < Golgi apparatus < plasma membrane) and a compensating decrease in phosphatidyl choline comparing the same membranes in the same order (Fig. 5.2). With mammary gland of the rat and bovine, in addition, an increasing proportion of phosphatidylethanolamine and a decreasing proportion of phosphatidylinositol were observed when endoplasmic reticulum, Golgi apparatus, and plasma membrane were compared (Keenan et al., 1972a). Determinations for endomembranes of pancreas, mammary gland, plant stems, and other tissues showed similar trends.

Translated on a protein basis, the phosphatidylcholine content of endomembranes remains essentially constant, whereas there is a net increase in the contents of the serine, inositol, and ethanolamine phosphatides going from endoplasmic reticulum to Golgi apparatus to plasma membrane (Fig. 5.2). There is also a very large net increase in sphingomyelin content and a parallel increase in the content of cholesterol for these membranes from mammals (Yunghans et al., 1970). The fatty acid composition of the various phospholipids also shows a pattern of change, with those from the Golgi apparatus frequently being intermediate between those from endoplasmic reticulum and plasma membrane. For example, the proportion of unsaturated fatty acids increases (endoplasmic reticulum < Golgi apparatus < plasma membrane) while the contents of 16:0 fatty acids of phosphatidylcholine decrease (endoplasmic reticulum > Golgi apparatus > plasma membrane). These findings demonstrate that different endomembranes have characteristic fatty acid compositions but not unique fatty acids. They also indicate that selective incorporation and/or exchange of these phospholipids or their acyl moieties may take place within the Golgi apparatus (Keenan and Morré, 1970).

There are several mechanisms by which the accretion of lipids into membranes are known to occur. Synthesis of phospholipids, especially phosphatidylcholine, is a characteristic reaction of the endoplasmic reticulum. However, Golgi apparatus also have a modest capacity to synthesize phospholipids (Jelsema and Morré, 1978; Table 5.2). Alternative explanations to account for the characteristic lipid composition of the Golgi apparatus include transfer of regions of endoplasmic reticulum enriched in the required lipids via a flow mechanism and/or transfer of specific lipids by exchange involving carrier proteins of the soluble cytoplasm.

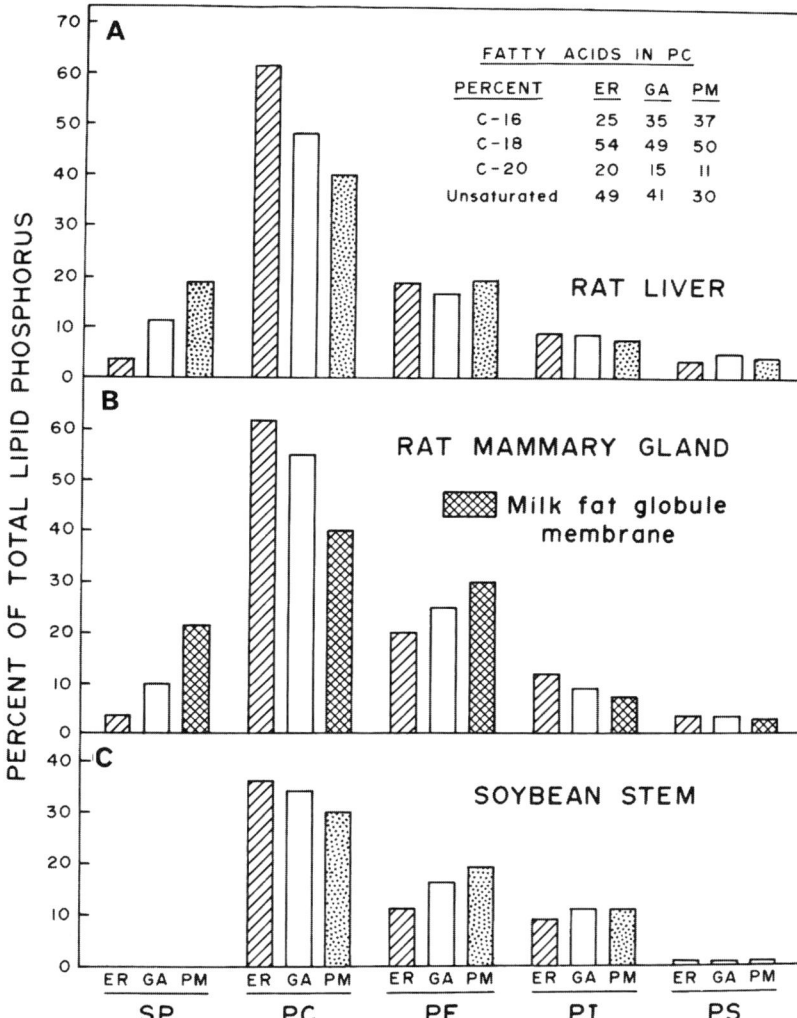

Fig. 5.2. Phospholipid composition of animal and plant endomembranes comparing rat liver (A), fat mammary glands (B), elongating hypocotyls of etiolated soybeans (*Glycine max*) (C). SP, Sphingomyelin; PC, phosphatidylcholine; PE, phosphatidylethanolamine; PI, phosphatidylinositol; PS, phosphatidylserine. Reproduced from Morré, 1987 with permission from Elsevier.

Data of Table 5.2 show that Golgi apparatus of rat liver have nearly a full complement of glycerophospholipid biosynthetic enzymes (Fig. 5.3). Enzymes of CDP-glyceride and phosphatidylglycerol biosynthesis are present in endoplasmic reticulum (about 70% of the total) (Table 5.2). However, the specific activities of the Golgi apparatus were 85% those of the endoplasmic reticulum (Table 5.2). Activities for the synthesis of phosphatidylglycerol in Golgi apparatus were 40% those of endoplasmic reticulum.

The synthesis of sphingomyelin occurs primarily via transfer of choline from CDP-choline to ceramide. Golgi apparatus as well as endoplasmic reticulum

Table 5.2.   Distribution of the terminal enzymes of biosynthesis of glycerophosphatides among purified cell fractions of rat liver.

Cell fraction	Phosphoglyceride biosynthetic enzymes[a]						
	PC	PE	CDP-DG	PI	PI (C)	PS (E)	PG
Total homogenate	56	34	51	180	250	79	185
Endoplasmic reticulum	232	142	164	820	1,020	151	472
Golgi apparatus	86	58	140	230	410	40	322
Plasma membrane	2	3	36	6	4	53	88
Supernatant	3	2	4	17	1	42	3
Mitochondria							

Source: From Jelsema and Morré (1978)

[a]Units of specific activity are nanomoles product formed per hour per milligram protein. The identity of products was verified by thin-layer chromatography. Cell fractions were prepared according to Morré (1973). The soluble supernatant is the microsome-free supernatant after centrifugation for 2.5 h at 100,000g. Assays for terminal glycerophosphatide biosynthesizing enzymes were as follows: PC, phosphatidylcholine; CDP, choline 1,2-diglyceride cholinephosphotransferase; PE, phosphatidylethanolamine; CDP, ethanolamine 1,2-diglyceride ethanolaminephosphotransferase; CDP-DG, CDP-diglyceride cytidyltransferase; PI (C), phosphatidylinositol, complete synthesis, Williamson and Morré (1976); PS (E), phosphatidylserine, exchange reaction only; PG, phosphatidylglycerol, α-glycerophosphotransferase.

fractions from rat liver contain ceramide and the requisite transferase enzymes. The acyltransferase catalyzing the conversion of sphingosine to ceramide is largely an endoplasmic reticulum enzyme in rat liver although the activity also is present in the Golgi apparatus. It is a reasonable expectation, therefore, that both endoplasmic reticulum and Golgi apparatus possess the biosynthetic capacity to form sphingomyelin and the relative abundance of this phospholipid (plasma membrane > Golgi apparatus > endoplasmic reticulum) can be understood on the basis of direct transfer of the ceramide precursors via ceramide transfer protein (CERT) (Handa et al., 2003) from biosynthetic sites in the endoplasmic reticulum directly to the trans Golgi apparatus possibly involving contact sites between the two structures, rather than via a flow mechanism across the stacked cisternae. Synthesis of sphingomyelin for delivery to the plasma membrane is then completed almost exclusively in the trans Golgi apparatus or trans Golgi apparatus network where the enzyme sphingomyelin synthease is located (Allan and Obradors, 1999).

In addition to containing enzyme systems for de novo biosynthesis of phospholipids and triglycerides, Golgi apparatus also contain various enzymes catalyzing acylation, deacylation and exchange reactions potentially capable of effecting glycerol lipid remodeling and turnover. Some of these enzymes exhibit positional specificity with regard to various fatty acids, others do not.

Because of the potential for phospholipid synthesis in the Golgi apparatus and the possibility of extensive redistribution of intact phospholipids among cellular membranes by exchange as well as various remodeling reactions possible, the analysis of the kinetics of synthesis and transfer of labeled phospholipids and triglycerides among various endomembrane compartments is expected to be relatively more complex than those of individual membrane proteins or glycoconjugate molecules. These expectations are borne out by the data available (Morré et al., 1979b).

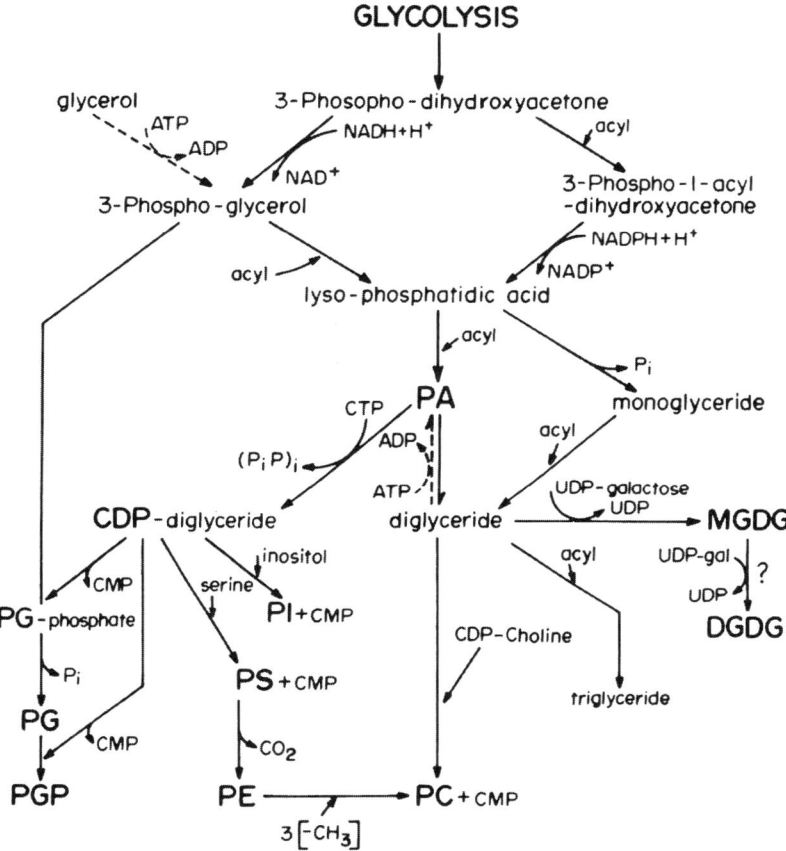

Fig. 5.3. Biosynthetic pathways for glycerolipids. Various base-exchange, acylation–deacylation, and degradative reactions are not shown.

5.4. Biosynthesis of Membrane Sterols

Considerably less is known to help account for an increasing sterol content comparing endoplasmic reticulum, Golgi apparatus and plasma membrane in a variety of cell types (Table 5.1). Sterol biosynthesis is partly a function of internal cytomembranes and those biosynthetic enzymes that are membrane associated enter the microsome fraction. Surprisingly, the specific cellular locations of these enzymes is not known except that the enzymes are consistently associated with smooth membranes (i.e., smooth endoplasmic reticulum and/or tubular portions of the Golgi apparatus; see Chapter 4 for localization of steroid biosynthetic enzymes in Golgi apparatus tubules).

Sphingolipid- and sterol-enriched microdomains (lipid rafts) occur in the exoplasmic leaflet of membranes of both the Golgi apparatus and plasma membranes (Chapter 7; Hansen et al., 2000). In the biosynthetic exocytic pathway, raft association with the trans-Golgi apparatus network has suggested delivery of pre-formed rafts to the cell surface via Golgi apparatus-derived transport vesicles.

5.5. Biosynthesis of Membrane Proteins

Membrane proteins may be divided into two broad categories based on conformation and location in the membrane. Extrinsic or peripheral membrane proteins are associated with the surface of the membrane by ionic interactions and can be extracted by salt solutions or by metal chelating agents. Intrinsic or integral proteins, on the other hand, are associated by hydrophobic reactions and with lipids. They can be removed only by chaotropic and other agents that disrupt hydrophobic interactions. Generally, intrinsic proteins are isolated in association with lipids and some, such as glucose-6-phosphatase and 5'-nucleotidase require an association with lipids for optimum activity. Membrane proteins are defined as those proteins associated with the membrane per se, either extrinsic or intrinsic as distinct from secretory proteins or content (cargo) proteins found free within the luminal interiors or matrix of membrane enclosed compartments.

Protein synthesis is the exclusive domain of polyribosomes and associated messenger RNA. Polyribosomes exist in at least two distinct subpopulations in eukaryotes; those free in the cytoplasm and those associated with membranes. The latter, plus the membranes to which they attach, form principally the rough-surfaced endoplasmic reticulum and the outer membrane of the nuclear envelope.

A generally accepted thesis is that the attachment of polyribosomes to membranes facilitates, or even may be obligatory, for the secretion of cargo proteins destined for export (see also Chapter 8). Now widely accepted is the concept that proteins to be retained with cells and tissues, so-called "house-keeping proteins" are synthesized predominantly, if not exclusively, on free polyribosomes. However, in contrast, synthesis of membrane proteins is clearly an activity shared by both free and bound polyribosomes. Synthesis of membrane proteinase on membrane-bound polyribosomes provides an especially attractive mechanism since it allows for direct insertion of the relatively hydrophobic regions directly into the membrane and obviates the need for migration through the hydrophilic cytoplasm. Many membrane proteins are synthesized and inserted by this route. However, some intrinsic membrane proteins, e.g., NADPH cytochrome c reductase of endoplasmic reticulum and several mitochondrial proteins as examples, are synthesized exclusively or predominantly on free polyribosomes with subsequent migration to and insertion into the membrane.

Elder and Morré (1976a) compared the abilities of total rough endoplasmic reticulum, polyribosomes released from rough endoplasmic reticulum, and free polyribosomes from rat liver, to incorporate amino acids into intrinsic membrane proteins using in vitro assay procedures with insertion into membrane vesicles and resistance to extraction with 1.5 M KCl and 0.1% deoxycholate as the end point. Polyribosomes bound to endoplasmic reticulum were shown to have the greatest capacity to synthesize radioactive products that were either copurified with intrinsic proteins of endoplasmic reticulum or were precipitated by antisera raised against intrinsic proteins of endoplasmic reticulum. It was estimated that at least two-thirds were synthesized on polyribosomes bound to membranes of the endoplasmic reticulum per se while the remainder appeared to be synthesized on free polyribosomes. Subsequent investigations of individual membrane proteins have been consistent with such a distribution.

5.5.1. The Origins of Golgi Apparatus Proteins

As noted in Chapter 4, when ribosomal, extrinsic and secretory proteins are excluded from analysis by high salt plus detergent treatment to remove them, Golgi apparatus and plasma membrane possess a number of protein bands in common with endoplasmic reticulum. Kinetic analyses demonstrate that such proteins are synthesized first on polyribosomes attached to endoplasmic reticulum and then transferred via a flow mechanism to the Golgi apparatus and subsequently to the plasma membrane as well. However, there are a few proteins intrinsic to the membrane which may be unique to the Golgi apparatus or shared by Golgi apparatus and plasma membrane that are absent from the endoplasmic reticulum (Franke and Kartenbeck, 1976).

Franke et al. (1972) and Franke and Scheer (1972) reported that stacked cisternae of plant Golgi apparatus had as a consistent feature of their morphology, a single polyribosome or group of several polyribosomes associated with their immature or forming faces of the cisternal stacks. In *Euglena gracilis*, the association of polyribosomes with cisternal stacks was particularly striking. For each stack, a single row of polyribosomes occupied a clearly defined zone limited on one side by the immature Golgi apparatus cisternae and on the other side by conventional rough endoplasmic reticulum (Mollenhauer and Morré, 1974). Golgi apparatus associated polyribosomes have subsequently been reported for rat liver and a variety of other animal tissues and species (Figs. 5.4 and 5.5; Table 5.3). Golgi apparatus associated

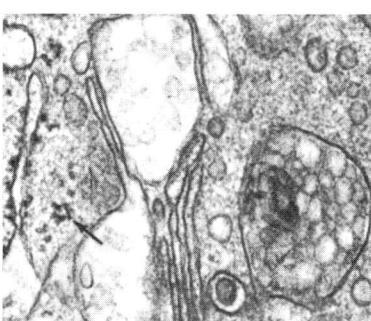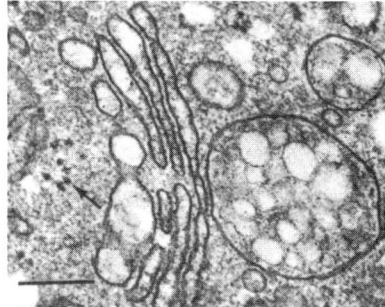

Fig. 5.4. Portions of rat hepatoma cells showing polyribosomes (*arrows*) adjacent to dictyosomes of the Golgi apparatus. The cisternal stacks are surrounded in the cell by a differentiated region of cytoplasm or zone of exclusion in which several Golgi apparatus-associated polysomes are embedded, usually in the near proximity of the forming face. The polysomes are not directly associated with membranes and represent a class of free polyribosomes. Such Golgi apparatus-associated polyribosomes have been shown to be active in the synthesis of proteins in vitro and add a new dimension, that of protein synthesis, to the biosynthetic capacity of the Golgi apparatus complex. Reprinted from Morré, 1977b, Cell Surface Reviews, Morré, p. 17, Copyright 1977b with permission from Elsevier. Scale bar = 0.2 μm.

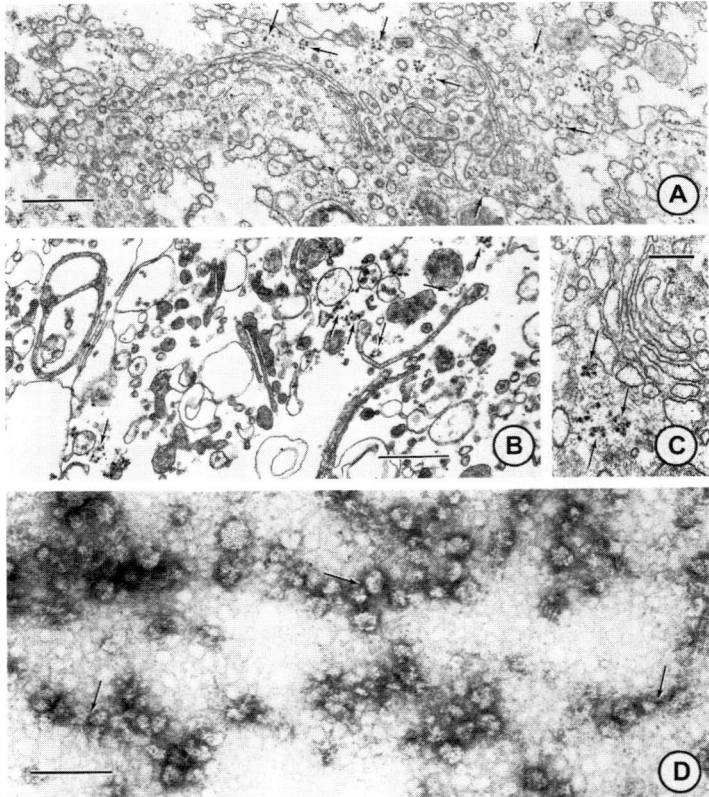

Fig. 5.5. Electron micrograph of isolated Golgi apparatus and Golgi apparatus-associated polyribosomes (*arrows*). (A) Golgi apparatus of rat liver fixed in situ showing associated polysomes. Scale bar = 0.5 μm. (B) Isolated Golgi apparatus and Golgi apparatus-associated polyribosomes (*arrows*). Scale bar = 0.5 μm. (C) Higher magnification of Golgi apparatus-associated polysomes (*arrows*) in situ showing matrix material of the Golgi apparatus zone. Scale bar = 0.2 μm. (D) Negative stain (2.5% uranyl acetate) of polysomes (*arrows*) isolated from a Golgi apparatus fraction by treatment with 0.25 m KCl. The polysomes appear to be surrounded by a fibrillar material and exhibited anomalous behavior on sucrose gradients as if still embedded in the matrix material of the Golgi apparatus zone. Reproduced from Elder and Morré, 1976a, with permission from the authors. Scale bar = 0.1 μm.

polyribosomes are not membrane-associated in the conventional sense as with rough endoplasmic reticulum. They are a class of free polyribosome located in the specialized region of cytoplasm surrounding the Golgi apparatus referred to as the zone of exclusion (Chapter 2).

Originally, it was suggested that Golgi apparatus-associated polyribosomes might be responsible for the synthesis of a few proteins or enzymes specifically localized in the Golgi apparatus, such as one or more of the Golgi apparatus-specific glycosyltransferases (Mollenhauer and Morré, 1974). That suggestion, however, has never been subjected to rigorous experimental verification.

When translated in the presence of Golgi apparatus membranes, the translation products were inserted immediately into the Golgi apparatus membranes and

Table 5.3. Quantitation of Golgi apparatus–endoplasmic reticulum–polyribosome associations in carrot cells grown in suspension culture[a].

Parameter	Units	Number ± SD[a]
Endoplasmic reticulum	Per 100 profiles	
In register with forming face[b]		31 ± 8
Adjacent to Golgi apparatus periphery		27 ± 17
In register with mature face		5 ± 5
Nuclear envelope		
In register with forming face		6 ± 5
Stacks exhibiting no obvious		
Association with either endoplasmic reticulum or nuclear envelope		40 ± 14
Transition vesicles (nap-like coats)	Per dictyosomes profile	
Within Golgi apparatus zone of exclusion		1.5 ± 0.25
Stacks with transition vesicles (nap-like coats)	Per 100 profiles	87 ± 3
Polyribosomes within Golgi apparatus zone of exclusion	Per dictyosome profile	
At forming face		2.8 ± 1.0
At Golgi apparatus periphery		2.5 ± 0.6
At mature face		2.7 ± 0.7
Spiny-coated vesicles within Golgi apparatus zone of exclusion	Per dictyosome profile	0.8 ± 0.2
Stacks with spiny-coated vesicles	Per 100 profiles	54 ± 5
Cisternae per stack	Number	4.6 ± 0.3

Source: From Morré et al. (1984b)
[a]Results are from analyses of 100 stacks with standard deviations among averages of 25 stacks from four determinations
[b]Including those permeating the forming face regions from the Golgi apparatus periphery

became resistant to extraction with 1.5 M KCl and 0.1% deoxycholate, a characteristic of intrinsic membrane proteins. The majority of these peptide bands when analyzed electrophoretically had mobilities similar to proteins in plasma membranes which were absent from endoplasmic reticulum but were relatively minor components of Golgi apparatus (Elder and Morré, 1976a).

A capacity for synthesis of intrinsic membrane proteins at or near the Golgi apparatus is consistent with in vivo studies of the kinetics of incorporation of labeled amino acids into membrane proteins (Ray et al., 1968; Franke et al., 1971b; Morré et al., 1974a; Croze and Morré, 1981). Pulse-chase experiments with [³H]leucine show that incorporation of label into the Golgi apparatus fraction indicate two phases of incorporation, one delayed and one rapid (cf. Chapter 6). The rapid phase occurs with the same kinetics as incorporation into rough endoplasmic reticulum. To explain these findings, input into Golgi apparatus from two sources has been assumed. Direct insertion of proteins, possibly including some newly synthesized by the Golgi apparatus polyribosomes has been evoked to explain the rapid incorporation phase. Transfer via a flow mechanism of membrane proteins from endoplasmic reticulum has been suggested to explain the delayed phase. As a result, membrane proteins of Golgi apparatus attain higher specific radioactivities of protein label than do those of endoplasmic reticulum.

The phenomenon of Golgi apparatus-associated polyribosomes offers a mechanism whereby the protein composition of Golgi apparatus membranes might be altered to facilitate membrane differentiation. Input at or near the Golgi apparatus of certain membrane proteins absent from endoplasmic reticulum would be important to account adequately for the functional polarity of the Golgi apparatus. Yet the relative amounts and limited numbers of proteins synthesized by the Golgi apparatus-associated free polyribosomes limits this biosynthetic role to one of modification. Most of the intrinsic proteins of the Golgi apparatus still appear to be synthesized at or near the rough endoplasmic reticulum.

5.6. Glycosylation of Membrane Glycoproteins

A central role of the Golgi apparatus in glycosylation of both membrane and secretory glycoproteins is well established. Results of cell fractionation experiments confirm initial findings from autoradiography that certain sugar groups, for example, galactose, fucose, sialic acid, distal *N*-acetylglycosamine and mannose, are added preferentially at the Golgi apparatus. Others, such as the internal glucosamine, mannose and glucose of complex asparagine-linked glycoproteins, for example, are incorporated earlier at the endoplasmic reticulum perhaps soon after peptide synthesis or even while the peptide chain is still attached to the ribosome (Fig. 6.1; Walter and Lingappa, 1986). For the most part, the sugar transfers catalyzed by endoplasmic reticulum involve lipid intermediates of the dolichol pathway whereas Golgi apparatus transferases utilize sugar nucleotides directly.

5.7. Glycosylation of Membrane Glycolipids

Glycolipids, especially the sialic acid-containing glycosphingolipids known as gangliosides, are enriched in plasma membranes of most, if not all, mammalian cells. The sugars making up the carbohydrate portions are added sequentially from appropriate sugar nucleotide donors in reactions catalyzed by membrane-located glycosyltransferases largely associated with the Golgi apparatus (Fig. 5.6; Table 5.4).

5.8. Formation of Sugar Nucleotides and Other Active Intermediates of Glycosylation Reactions

The glycosyltransferases of Golgi apparatus that function in glycosylation of membrane glycolipids and glycoproteins transfer glycoses from appropriate nucleotide sugars to acceptors which are usually incomplete carbohydrate side chains of glycoproteins or glycolipids, according to the following overall scheme:

Nucleotide – glycose + acceptor → glycose – acceptor + nucleotide

The enzymes are usually specific for both acceptor and donor. In ER, the acceptor may be a lipid intermediate including various oligosaccharide–polyisoprenol

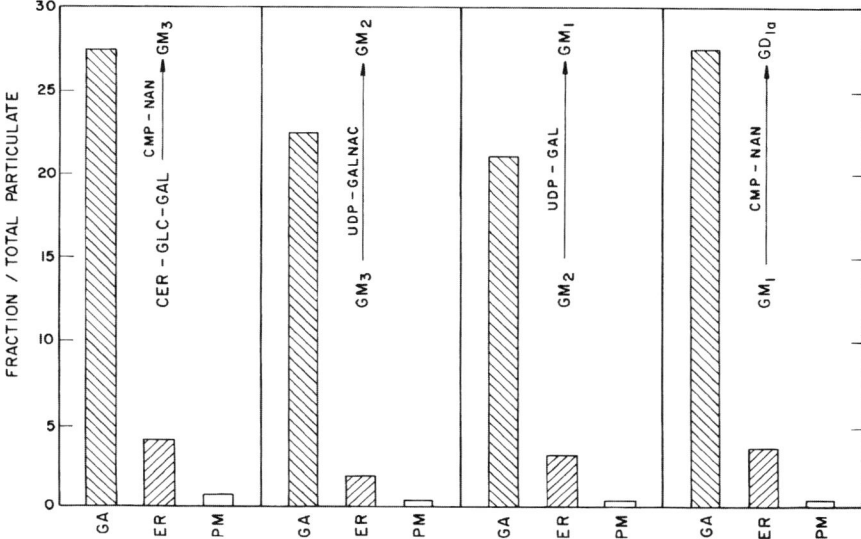

Fig. 5.6. Distribution of several sugar–nucleotides:glycolipid transferase activities among purified endomembrane fractions from rat liver.

Table 5.4. Comparative carbohydrate composition of nuclear envelope, endoplasmic reticulum, Golgi apparatus, and plasma membrane fractions from rat liver[a].

Constituent (units)	Nuclear envelope	Endoplasmic reticulum	Golgi apparatus	Plasma membrane
Total carbohydrate (% dry weight)	<1	1	2	4 (2–7)
Constituent sugars (μg carbohydrate/mg protein)				
Sialic acid	Trace	2.0	13.6	18.0
Hexosamine[b]	5.0	5.5	6.5	19.0
Fucose	Trace	0.5	3.0	3.0
Galactose	2.0	2.5	7.0	14.0
Glucose	2.0	0.8	0.3	Trace
Mannose	9.5	9.0	8.4	13.0
Total sialic acid (nmoles/mg protein)	2	2–5	20	30–50
Ganglioside sialic acid (nmoles/mg protein)	<0.3	<0.3	2	5
Protein sialic acid (nmoles/mg protein)	<2	2–5	18	30–45

Source: From Morré and Ovtracht (1977)
[a]Based on data from Dod and Gray (1968); Keenen et al. (1972b); Emmelot et al. (1974); Kasper (1974); Franke and Kartenbeck (1976); T. W. Keenan and D. J. Morré (unpublished); unpublished data of Dr. J. Stadler, Institute of Experimental Pathology, German Cancer Research Center, Heidelberg. The values represent averages and calculations from several experiments and determinations and must be considered only approximate
[b]Predominantly *N*-acetylglucosamine

derivatives but within the Golgi apparatus transfers are most often directly to the growing glycoprotein or glycolipid oligosaccharide chain.

The relevant glycosyltransferases of the Golgi apparatus are oriented predominantly at the inner or environmental surface of the vesicle membranes so that disruption of the vesicles with detergents or other means facilitates access to the substrate. Significantly, this coincides with the asymmetric substitutions of the membranes with carbohydrate groups of glycolipids and glycoproteins. That these carbohydrate groups are exposed on the inner face of internal membrane has been confirmed by the inaccessibility of intact vesicles to lactoperoxidase labeling and concanavalin A binding. Thus, the conclusion is often reached that glycosyltransferase enzymes of the Golgi apparatus function by the addition of sugars within the lumina of individual cisternae, rather than at the Golgi apparatus–cytoplasm interface.

This spatial arrangement of glycosyltransferases poses a problem in logistics. Charged and hydrophilic sugar nucleotide substrates must traverse the Golgi apparatus membrane in situ from the cytosol where they are synthesized to the Golgi apparatus cisternal lumens where they are utilized. Sugar nucleotide traverse of the Golgi apparatus membranes is facilitated by sugar nucleotide-specific transporters, hydrophobic protein dimers with 6 to 10 transmembrane domains (Hirschberg et al., 1998).

5.9. Sulfation Reactions

Cytochemical and autoradiographic studies first suggested that Golgi apparatus sites were important sites of polysaccharide sulfation (Lane et al., 1964). In different mammalian types, sulfate was detected initially in Golgi apparatus using electron microscope autoradiography to localize radioactivity in cells exposed for very short periods to ^{35}S sulfate. Subsequently, sulfotransferase activities were localized in Golgi apparatus fractions from kidney (Fleischer and Zambrano, 1974).

5.10. Distribution of Glycosyltransferases Across the Polarity Axis of the Golgi Apparatus

There have been numerous autoradiographic and cytochemical studies that have mapped the cis to trans distribution of glycosyltransferase activities within Golgi apparatus of various cell types (Roth, 1997). However, relatively few studies have provided direct biochemical demonstrations of such distributions and gradients of activities. One method, applicable to such studies, was to use the technique of preparative free-flow electrophoresis described in Chapter 3 to resolve cisternae of rat liver Golgi apparatus after unstacking into fractions enriched in cisternae from the cis, median, or trans faces of the Golgi apparatus. Comparisons with fractions from unstacked Golgi apparatus from rat hepatoma revealed potentially important differences in enzyme distributions.

Findings are illustrated in Chapter 12 (Figs. 12.3 to 12.6). Sialyltransferases, either with endogenous acceptors or with asialofetuin as acceptor, were located, as expected for liver, in the most electronegative fractions representing the trans-most cisternae. Also located in the trans half of the Golgi apparatus were the galactosyltransferases. *N*-acetylglucosaminyltransferases were more cis located. A bimodal distribution with both cis and trans locations was exhibited by fucosyl-transferases. The findings for liver contrasted with those for rat hepatoma where, surprisingly, both sialyl and galactosyltransferases exhibited a predominantly cis location whereas *N*-acetylglucoaminyl and fucosyltransferases gave bimodal distributions with both cis and trans localizations (Chapter 12; Hartel-Schenk et al., 1991).

With liver, the cis to trans order of activities paralleled the order of terminal sugar additions in N-linked glycoprotein biogenesis. For hepatomas, the order most closely paralleled the order of terminal sugar additions for O-linked glycoprotein biogenesis.

5.11. Summary

The synthesis, assembly, processing, and directed transport of membrane constituents is important to growing or dividing cells and for membrane renewal or replacement in differentiating and/or nongrowing cells. The bulk of most eukaryotic membranes can be accounted for by proteins and lipids, either simple or complex (i.e., glycosylated). Polar lipids and sterols are synthesized predominantly by endoplasmic reticulum. Synthesis of membrane proteins is on cytoplasmic polyribosomes (either free or membrane-bound). Diversity and the emergence of unique compositional patterns is due largely to specific mechanisms of assembly or directed trafficking of membrane constituents to ensure proper co- or posttranslational insertion and processing. But for many constituents, the site of insertion and/or processing is not at the final destination. The majority of constituents of plasma membranes appear to be first synthesized at the rough endoplasmic reticulum followed by sequential vesicular transport to the Golgi apparatus, maturation including glycosylation and sulfation at the Golgi apparatus, and finally, delivery to the plasma membrane per se via Golgi apparatus derived secretory vesicles. The contribution of the Golgi apparatus to membrane biogenesis seems primarily to be one of modification and sorting although important roles in the biosynthesis of phospholipids, protein insertion and deletion, and quinone and/or sterol biosynthesis (Chapter 4) are indicated as well.

6

Function in the Flow-Differentiation of Membranes

Membrane differentiation within the cell's endomembrane system may be defined as the progressive change in the appearance, composition, organization, and/or functional specialization observed within a single membrane or among different but ontogenetically related membranes or systems of membranes. Local differentiation is visualized by various cytochemical procedures or by measurements of membrane thickness or distributions of intramembranous particles measured from electron micrographs. Functional specialization of different but ontogenetically related membranes is reflected in various biochemical and physical parameters determined from analyses of isolated and purified cell fractions or from cytochemistry. Differentiation occurs along established export pathways and traffic routes within the cell and, by definition, proceeds from precursor-like to product-like. In the sense of secretion or plasma membrane biogenesis, differentiation proceeds from endoplasmic reticulum like to plasma membrane-like. While observations providing evidence for membrane differentiation are based primarily on static analyses (morphological observations, analyses of isolated cell fractions), they do provide an important conceptual framework within which dynamic events may occur.

For membrane differentiation to contribute significantly as a major mechanism of membrane biogenesis whereby one kind of precursor membrane is converted into another kind of product membrane, e.g., derivation of Golgi apparatus from endoplasmic reticulum or of plasma membrane from Golgi apparatus, transfer of membrane from one compartment to another must accompany, precede or follow the process of differentiation. This dynamic component has been termed membrane flow. Without a flow mechanism to account for the transport of membrane constituents from one region of the cell to another, processes of membrane differentiation could account for the origin of new membrane types only in situations where there was no intermixing of membrane compartments. Working in concert, the combined processes of flow-differentiation offer a mechanism of membrane biogenesis consistent with both ultrastructural and biochemical information. With the availability of highly purified cell fractions and various correlative investigative approaches, the hypothesis has been verified experimentally in several test systems. It is perhaps the most universal function of the Golgi apparatus when a wide range of species and cell types is compared; more universal even than secretion. Products elaborated by the Golgi apparatus for secretion are packaged in membrane-enclosed transport vesicles.

D. James Morré and Hilton H. Mollenhauer, *The Golgi Apparatus*.
© Springer 2009

These vesicles have membranes that are plasma membrane-like and capable of fusing with the plasma membrane. They are products of the flow-differentiation of membranes. A major premise of the membrane flow-differentiation hypothesis of Golgi apparatus function is that Golgi apparatus do not participate in secretion without carrying out membrane flow-differentiation but membrane flow-differentiation may proceed in the absence of delivery of cargo.

6.1. Morphological Evidence for Membrane Differentiation within the Golgi Apparatus

Morphological differences among different classes of internal cellular membranes were first noted in 1956 by F. Sjöstrand (1956, 1963, 1968) and confirmed for both plants (Ledbetter, 1962; Ueda, 1966) and animals (Yamamoto, 1963). In these early investigations, membranes of endoplasmic reticulum were found to be thinner and to have a less pronounced dark-light-dark pattern than plasma membrane. The appearance of membranes of the Golgi apparatus was intermediate between those of endoplasmic reticulum and the plasma membrane (Mollenhauer and Morré, 1978c; Fig. 6.1; Tables 6.1 and 6.2).

It was subsequently shown that membranes of the Golgi apparatus exhibited a progression of change from thin to thick across the cisternae of each stack. This was first reported by Grove et al. (1968) for the fungus *Pythium ultimum* based on observations of changes in both membrane thickness and in staining intensity (Fig. 6.2). Membranes of cisternae at one pole of the stack (the pole proximal to endoplasmic reticulum or nuclear envelope; forming or cis face) appeared to be similar to endoplasmic reticulum and the nuclear envelope. Membranes of cisternae at the opposite face were similar to the plasma membrane. Membranes of intercalary cisternae were intermediate in appearance, so that each successive cisternae was more like the plasma membrane from the forming to the maturing face (that is, denser, thicker, and have a clearer dark-light-dark pattern). Membrane differentiation within Golgi apparatus stacks has been observed in other organisms (Hicks, 1966; Ledbetter, 1962; Porter et al., 1967; Sakai and Shigenaka, 1967; Ueda, 1966; Yamamoto, 1983). However, the pattern in *Pythium*, since observed in other plant and animal cells, was so striking as to be interpreted as morphological evidence that a major Golgi apparatus function was to transform membranes received from endoplasmic reticulum or the nuclear envelope into a membrane type morphologically indistinguishable from plasma membrane.

6.1.1. Measurements of Membrane Thickness

In general, membrane thickness progresses from thin to thick along the nuclear envelope–endoplasmic reticulum–Golgi apparatus–secretory vesicle–plasma membrane pathway. Staining intensity, however, can vary. Endoplasmic reticulum may be lightly stained, the unusual observation, or vice versa, depending on conditions. Yet, regardless of specific staining pattern, membranes of the forming face of the Golgi apparatus always resemble endoplasmic reticulum, while those

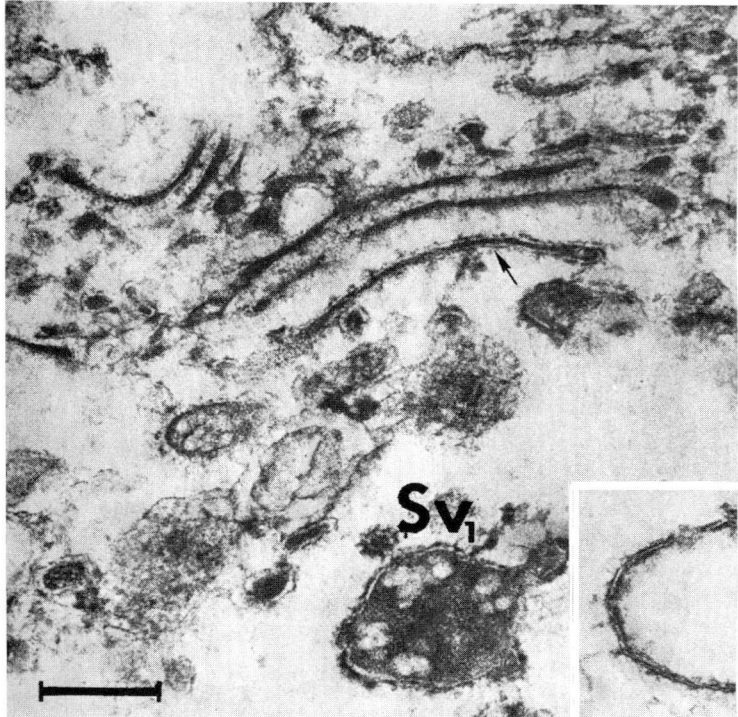

Fig. 6.1. Thin section of an isolated stack of Golgi apparatus cisternae from rat liver fixed for 20 h with 6% glutaraldehyde and section stained with uranyl acetate and lead citrate with omission of osmium tetroxide post-fixation to accentuate membrane differentiation and increase in membrane thickness from endoplasmic reticulum (ER)-like (top) to plasma membrane-like (arrow and *insert*) across the stacked cisternae. Reproduced from Morré et al., 1971a with permission from Raven Press. Scale bar = 0.2 μm.

Table 6.1. Endomembrane differentiation based on membrane dimensions in cells of onion (*Allium cepa*) stem[a].

Endomembrane component	Average membrane thickness (nm)
Outer membrane of nuclear envelope or rough endoplasmic reticulum	4–5
Outer mitochondrial and plastid membrane (peroxisome membrane)	4–5.5
Golgi apparatus cisternae (central plate or saccule)	5–7
"Free" secretory vesicles	8–9.5
Plasma membrane	9.5

Source: From Mollenhauer and Morré (1994)
Glutaraldehyde–osmium tetroxide fixation (measurements are approximate but taken from the same electron micrographs)
[a] The thickness of the tonoplast (membrane of the central vacuole) could not be determined accurately because of dense deposits on one or both membrane surfaces but was estimated to be in the range of 6.0–7.5 nm

Table 6.2.   Differentiation of endomembranes of rat liver.

Parameter	Endoplasmic reticulum	Golgi apparatus	Plasma membrane
Membrane thickness (osmium tetroxide fixation) (Å)[a]	65	65–85	85
Spacing between phosphate head groups in lipid bilayer (Å)[b]	40	45	50
Lipid content (%)[c]	30	35	40
Protein content (%)[d]	70	65	60

[a] Morre et al. (1971a); D. J. Morré, unpublished results
[b] Morre et al. (1974a)
[c] Average of all cisternae in the stack
[d] Keenan and Morré (1970); Yunghans et al. (1970)

at the maturing face (or membranes of secretory vesicles) always resemble the plasma membrane. If fixation conditions are employed that cause the endoplasmic reticulum to swell, these same conditions will result in swelling of the Golgi apparatus forming (cis) face. If osmium impregnation methods load the luminal spaces of endoplasmic reticulum with electron-dense deposits, similar deposits will be seen in the forming (cis) face cisternae of the Golgi apparatus. Thus, membrane staining and/or morphological characteristics seem to reflect intrinsic properties of the membrane and seem to provide reliable criteria for membrane differentiation.

6.1.2. Organization of Membrane Constituents

The differentiation of membranes within the Golgi apparatus is evident, also, from the organization of constituents within membranes. Low-angle, x-ray diffraction analysis reveals an increasing center-to-center spacing between the phospholipid polar groups in the bilayer comparing endoplasmic reticulum and Golgi apparatus (Fig. 6.3; Table 6.2). The average spacing for Golgi apparatus membranes yields a value intermediate between endoplasmic reticulum and plasma membrane.

Membrane differentiation is frequently revealed in terms of the numbers and distributions of intramembranous particles obtained in freeze-etching analyses. In plant cells, progressive increases in the number of particles, comparing endoplasmic reticulum, the Golgi apparatus, secretory vesicles, and the plasma membrane have been observed. Increases were apparent from analyses of both fracture faces (Vian, 1974). The final density and arrangement of particles characteristic of the plasma membrane were achieved in secretory vesicles prior to their fusion with the plasma membrane. A reverse situation has been found in mammalian cells where plasma membranes may have fewer particles and relatively fewer particles are associated with membranes of certain secretory granules. Interpretation of findings involving the 75 Å intramembranous particles is complicated by

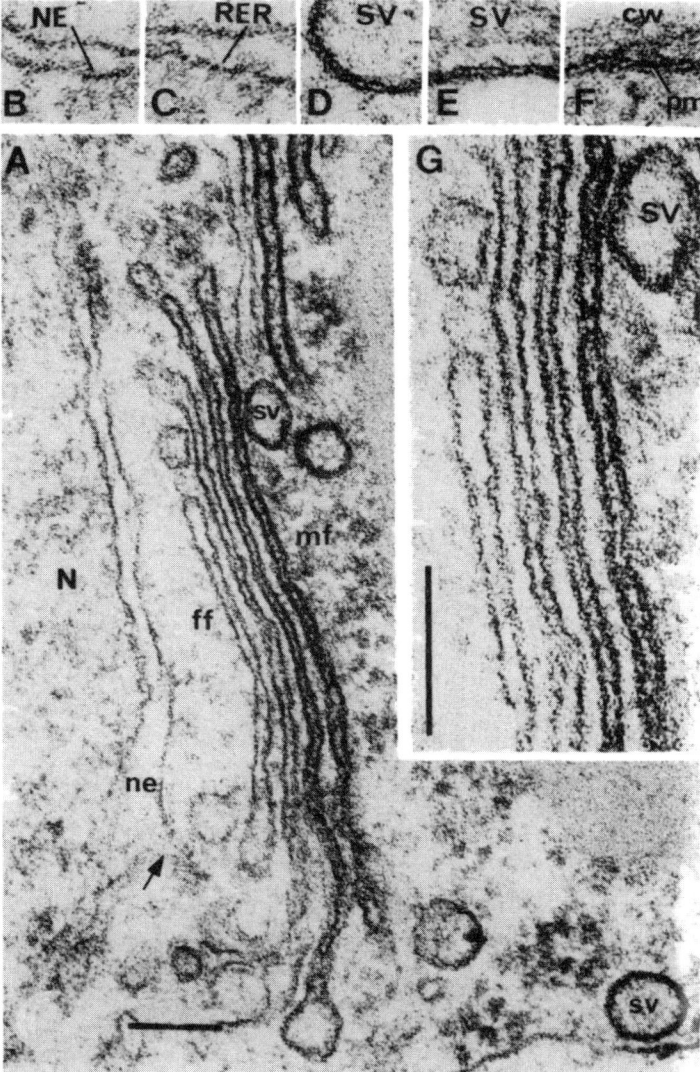

Fig. 6.2. Endomembrane differentiation in the fungus *Pythium ultimum* as revealed by a progressive increase in membrane thickness and intensity of staining from endoplasmic reticulum-like or nuclear envelope-like at the forming (cis) face (ff) to plasma membrane-like at the maturing face (mf). (A) A cisternal stack adjacent to the nucleus (N). When stacks are so positioned, blebs (*arrow*) and transition vesicles from the nuclear envelope (NE) are found in the space between the forming face of the stacked cisternae and the nuclear envelope, sv, secretory vesicle. (B) Membranes of the nuclear envelope (NE). (C) Membranes of rough endoplasmic reticulum (RER). (D and E) Membranes of secretory vesicles (SV). (F) Plasma membrane (pm) adjacent to the cell wall (cw). (G) Enlargement of a portion of the cisternal stack in (A). Reproduced from Grove et al., 1968 with permission from the AAAS. Scale bar = 0.1 μm.

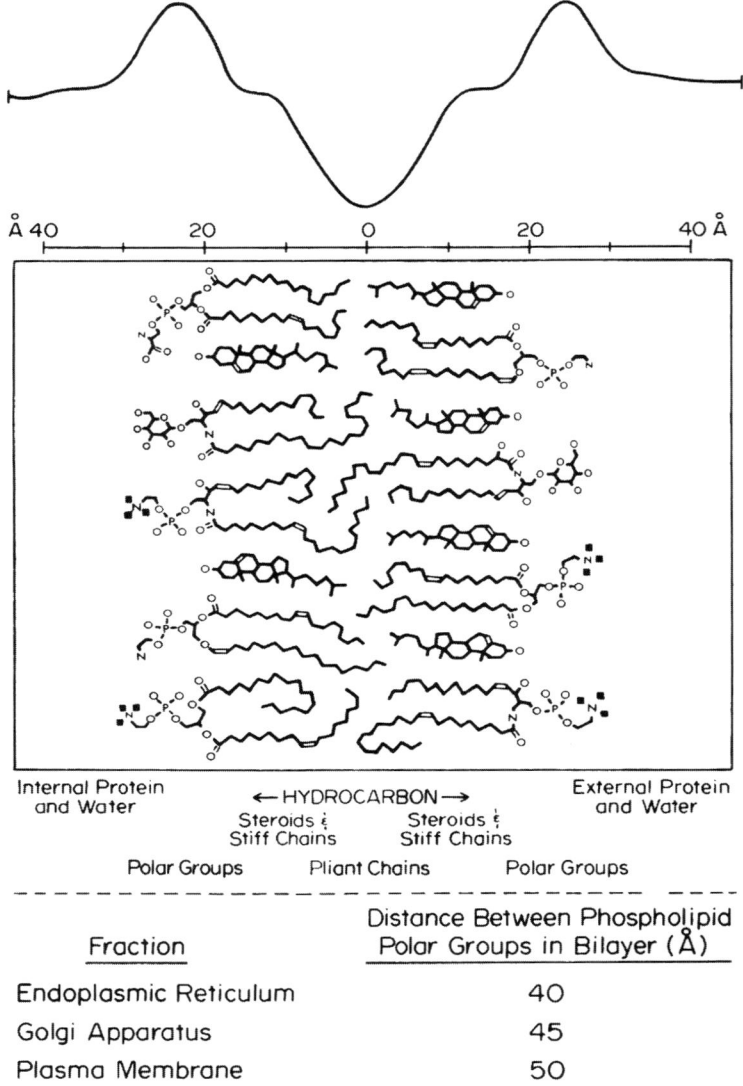

Fraction	Distance Between Phospholipid Polar Groups in Bilayer (Å)
Endoplasmic Reticulum	40
Golgi Apparatus	45
Plasma Membrane	50

Fig. 6.3. Schematic illustration of membrane structure as interpreted from low-angle x-ray analysis. The values in the table are from rat liver fractions with assistance from Alain Blaurock at the Cardiovascular Research Institute, San Francisco Medical Center, California. Reproduced from Morré et al., 1974a with permission from Lippincott Williams & Wilkins.

possibilities that their distribution may be subject to rapid change in response to a variety of environmental and developmental signals and stimuli. Also, their identification as specific enzymes or antigens has not been conclusive (assumed) nor is it certain that they are always proteins or glycoproteins inserted into the lipid layer.

6.1.3. Evidence from Cytochemistry

The most precise methods for morphological demonstration of membrane differentiation came from cytochemistry. Here a relatively clear visualization of gradients within cell components or systems of components is possible that exceeds that normally obtained with autoradiography or conventional biochemical assays. For example, the use of cytochemical methods to detect carbohydrate components over membranes of the Golgi apparatus in cells that do not normally secrete large quantities of mucins shows a progressive elaboration of cell coat material across the stacked cisternae of the Golgi apparatus (Rambourg, 1971; Fig. 6.4). Acidic groups, e.g., sialic acid, are apparently added near the maturing face suggestive of a step-wise assembly of coat materials by membranes of the Golgi apparatus. At the plasma membrane, coat materials are most evident in association with the outer (external) membrane leaflet. Within the cell as demonstrated for Golgi apparatus cisternae and secretory vesicles, the carbohydrate-rich material is oriented toward the luminal surfaces. Because of the manner in which the vesicles fuse with the plasma membrane, the inner surfaces of the cisternae or vesicles (those adorned with the saccharide coats within the cell) are equivalent to the external surfaces of the plasma membrane.

In plant cells, a phosphotungstic acid stain at low pH has been employed as a selective stain for the plasma membrane (Roland et al., 1972). This stain, under somewhat different conditions, is used to detect glycoproteins in animal

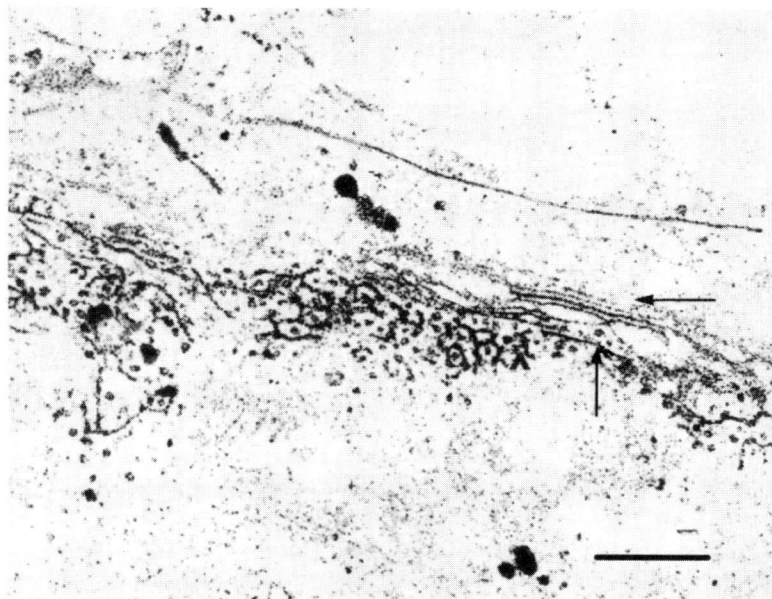

Fig. 6.4. The Golgi apparatus region of a columnar cell of the duodenal epithelium of the rat impregnated with phosphotungstic acid–chronic acid–silver to demonstrate progressive differentiation of saccule membranes from the cis (immature) to trans (mature) face. Reproduced from Rambourg et al., 1969, Journal of Cell Biology, 40:395–414, Copyright 1969, The Rockefeller University Press.

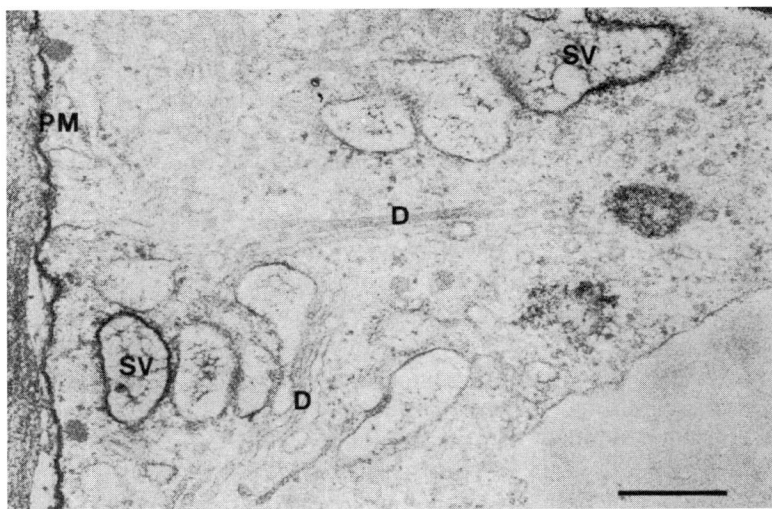

Fig. 6.5.   Golgi apparatus of a secretory cell of the outer root cap of maize. Phosphotungstic acid at low pH staining of mature secretory vesicles (SV) and plasma membrane (PM) develops as a cis to trans polarity gradient across the cisternal stacks or dictyosomes (D). Reproduced from Morré and Mollenhauer, 1983 with permission from Wissenschaftliche Verlagsgellschaft mbH. Scale bar = 0.1 μm.

cells (Marionozzi, 1967). Through the use of this stain, it has been shown that membranes of secretory vesicles derived from the Golgi apparatus acquire progressively the cytochemical characteristics of plasma membranes (Vian and Roland, 1972; Fig. 6.5). By means of this staining procedure, it has been possible to establish that membranes of secretory vesicles develop characteristics of the plasma membrane in advance of their fusion with the plasma membrane and that the transformation is progressive from one pole of the cisternal stack to another. It should be emphasized that in this use of phosphotungstic acid at low pH, some component of the membrane, perhaps a glycolipid (Yunghans et al., 1978), is stained rather than a component of the surface coat.

    When applied to plants, other carbohydrate-detecting procedures yield results similar to those obtained for animal cells. Since the coat is elaborated progressively, it provides an important index of membrane differentiation. Especially evident is the marked polarity of the Golgi apparatus where the increase in staining intensity observed from the forming to the maturing face of the stacked cisternae provides considerable indirect evidence for the dynamic model of Golgi apparatus function developed later in this chapter.

    Enzyme cytochemistry has also been employed to advantage in the study of Golgi apparatus to provide an important link between morphology and biochemistry (Appendix Table 3). Despite the realization that rarely can cytochemical differences in enzyme activity be equated, a priori, to real distributions of enzyme activity in situ, a more precise localization of enzymatic activities is given than is possible by conventional biochemical analyses of isolated cell fractions. Major

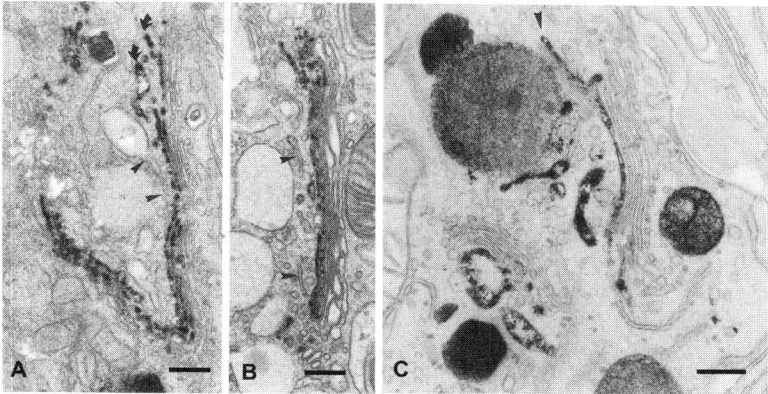

Fig. 6.6. Morphology and enzyme cytochemical localization of phosphatase activities in the Golgi apparatus of serous secretory cells of rat tracheal epithelium. (A) Inosine diphosphatase. (B) Thiamine pyrophosphatase. (C) Acid phosphatase. Reproduced from Pavelka and Ellinger, 1991 with permission from John Wiley and Sons, Inc. Scale bar = 0.5 μm.

uncertainties in cytochemistry derive from differences that arise due to fixation and/or interference by the electron-dense-detecting reagents that must be used to enhance visualization.

An important contribution of enzyme cytochemistry to the understanding of the Golgi apparatus is the demonstration of gradients of enzyme activity across the stacked cisternae. Such gradients would be predicted from the biochemical analyses (Chapter 4). This is especially evident with the membrane-bound enzyme activities amenable to cytochemical localization such as the nucleoside phosphatases (Fig. 6.6). In general, enzyme activities concentrated in the nuclear envelope and/or endoplasmic reticulum show a gradient of decreasing cytochemical activity from the forming to the maturing face while enzyme activities concentrated in the plasma membrane show the opposite gradient (increasing toward the maturing face of the Golgi apparatus or associated with mature secretory vesicles). Thiamine pyrophosphatase, an enzyme concentrated in the Golgi apparatus, shows a more variable pattern of distribution, sometimes with no gradient at all, depending on the cell type and stage of development. In some examples, however, the enzyme marks only one or, at most, a few cisternae, usually at the mature face of the apparatus.

6.2. Biochemical Evidence for Membrane Differentiation within the Golgi Apparatus

The biochemical basis for morphological and functional differences of membranes that seems to arise within the stacked cisternae of the Golgi apparatus have been sought from studies where purified fraction, especially those from rat liver (Chapter 3), of endoplasmic reticulum, Golgi apparatus, Golgi apparatus-derived secretory vesicles, and plasma membranes have been compared. Not only are such findings expected to reveal the chemical nature of the events contributing to membrane differentiation but important clues to the mechanisms are also provided.

Equally important in the context of this chapter, the findings implicate the Golgi apparatus as a principal locus where membrane differentiation takes place. As emphasized repeatedly, one gains the impression from morphological observations that endoplasmic reticulum membranes are transformed within the Golgi apparatus to a type of membrane that is at least plasma membrane-like if not plasma membrane. This function of the Golgi apparatus in membrane transformation should be and is expressed by a chemical composition that is intermediate between the two types of membrane (endoplasmic reticulum and plasma membrane) between which it serves as the morphological bridge.

This does not mean that there are not some biochemical constituents or arrangements of constituents that are unique to or concentrated in each of the three morphologically distinct types of membranes under comparison. Such constituents and arrangements of constituents exist and contribute operationally to the concept of marker enzymes. Marker enzymes are essential to cell fractionation studies (Chapter 3) and may contribute to the obvious morphological and functional aspects that serve to identify particular types of membrane in comparison to others.

6.2.1. Survey of Biochemical Constituents Common to All Endomembranes

Biological membranes, endomembranes of hepatocytes being no exception, are composed primarily of lipids and proteins with lesser amounts of saccharide (carbohydrate) residues and other constituents. Some proteins and most, if not all, lipids are widely distributed among different endomembranes within a given species, tissue, or cell type although the proportions may differ.

6.2.1.1. Lipids

As emphasized in Chapter 4, all endomembranes contain the same major classes of phospholipids, neutral lipids, and sterols as well as the same spectrum of fatty acyl side chains associated within these lipids. Yet, absolute quantities and percentages of each of the different lipids, especially among phospholipids, reveal important differences and trends. In general, membranes of endoplasmic reticulum contain the greatest proportion of phosphatidylcholine (lecithin). The proportion of phosphatidylcholine is less in Golgi apparatus and least in the plasma membrane. The decrease in phosphatidylcholine is compensated for by an increase in sphingomyelin. This phospholipid is low in endoplasmic reticulum, greatest in the plasma membrane, and intermediate in the Golgi apparatus. A similar pattern is exhibited by cholesterol, cholesterol esters, and triglycerides of the membranes. Other phospholipids remain relatively constant in rat liver comparing the different membranes but, in the mammary gland, a pattern of change is exhibited with an increasing phosphatidylethanolamine content compensated for by a decrease in phosphatidylinositol when endoplasmic reticulum, Golgi apparatus, and plasma membranes are compared.

Although data is more limited for other species, plant membranes which lack sphingomyelin and are very low in serine show similar trends to animals with those phospholipid species they contain (Fig. 5.2). Similar results are also available, although in less detail, from pancreas (Meldolesi et al., 1971).

6.2.1.2. Proteins

Early analyses of the overall protein composition of different endomembranes were based on disk gel electrophoresis patterns. While endoplasmic reticulum, Golgi apparatus, and plasma membrane fractions exhibited unique patterns of band distribution and intensity, they also possessed a number of bands in common especially when the intrinsic membrane fraction freed of all ribosomal, extrinsic and secretory proteins, were analyzed. Such studies were subject to the valid criticism that protein bands displayed on a gel are normally unidentified, uncharacterized and incompletely resolved. Similar appearing bands may contain completely different proteins whereas the same peptide chain, due to some post-translational modification, may migrate differently depending on its cell component of origin.

Therefore, it was important that findings from disk gel electrophoresis be supplemented by information from the laborious task of isolating and characterizing individual protein species from each of the cell components in question. This was done for the limited number of proteins where based on immunological, precursor–product or detailed kinetic analyses, endoplasmic reticulum, Golgi apparatus, and plasma membrane were deduced to contain the same or at least very similar proteins.

Current detailed and quantitative proteomic analyses of the secretory pathway have resolved >1400 by now mostly known or otherwise characterized proteins. Findings have yielded distributions consistent with the membrane maturation or flow-differentiation model as originally proposed (see Morre and Mollenhauer, 2007) to explain endomembrane trafficking (Gilchrist et al., 2006).

6.2.1.3. Glycoproteins

Much of what can be said for proteins applies equally well to glycoproteins although since proteins become glycosylated in a step-wise manner as posttranslational events often involving enzymes localized in the Golgi apparatus (Chapter 3), the expectation would be that endoplasmic reticulum, Golgi apparatus and plasma membrane would exhibit substantial differences in the spectrum of individual glycoproteins they display. For the most part, glycoproteins of endoplasmic reticulum would be expected to be incomplete with respect to their saccharide chains or at least be poor in terms of the terminal saccharides added at the Golgi apparatus. In contrast, plasma membrane as an end product of membrane differentiation would contain the greatest concentration of glycoproteins of highest complexity. For the most part, these expectations are borne out suggesting that the stepwise glycosylation sequence clearly established for secretory glycoproteins is equally applicable to membrane glycoproteins. Relative carbohydrate compositions of endomembranes ranges from 2% to 7% of the dry weight for plasma membrane (average 4%) to about 1% for endoplasmic reticulum and 2% for the

Golgi apparatus (Table 5.4). Of this, about 90% are contributed by glycoproteins and about 10% by glycolipids. The electrophoretic patterns of different glycoprotein molecular weight classes also bear out predictions based on biosynthetic pathways. In particular there appears to be an increase in high-molecular-weight glycoprotein constituents from rough endoplasmic reticulum to the Golgi apparatus to the plasma membrane (Elder and Morré, 1976b; Evans, 1970) and a compensating decrease in the proportion of lower molecular weight glycoprotein constituents. This increase in the amount and complexity of the carbohydrate portions of glycoproteins, especially in terms of additions of acidic terminal saccharides such as sialic acid contribute significantly to morphological patterns associated with membrane differentiation within the Golgi apparatus. As noted in the previous section, the overall increase in membrane glycoproteins associated with membrane differentiation has been detected by various cytochemical procedures and is readily visualized by electron microscopy.

6.2.1.4. Glycolipids

Cerebrosides, gangliosides and other glycolipids, although concentrated in plasma membranes, are also widely distributed among other endomembrane components (Keenan et al., 1972b). The amount present in Golgi apparatus fractions, although less than that found in plasma membranes, is somewhat greater than that found in endoplasmic reticulum. In rat liver homogenates less than 65% of the total gangliosides and also of neutral glycosphingolipids are in the plasma membrane. The remainder is distributed among rough endoplasmic reticulum, mitochondria, Golgi apparatus, nuclei and supernatant fractions. The plasma membrane and the supernatant fractions contain a complete spectrum of gangliosides similar to total liver homogenates. Mitochondria lack the gangliosides of the disialoganglioside pathway. Golgi apparatus contain little or no higher homologs, while rough endoplasmic reticulum contains those in Golgi apparatus plus G_{T1}. Ganglioside biosynthetic enzymes, on the other hand, are localized primarily, if not exclusively, in Golgi apparatus (Chapter 7). These observations in combination with metabolic labeling experiments discussed later suggest that during membrane differentiation gangliosides synthesized in the Golgi apparatus are transported not only to the cell surface as newly formed plasma membrane but to other cell organelles as well possibly by mechanisms involving "carriers" of the soluble cytoplasm.

6.2.2. Changes in Constituents Concentrated in a Particular Membrane Compartment

The concept that each morphologically distinguishable cell component or some constituents or combination of features that are unique lead to the eventual demise of the "unit membrane" theory where all membranes were considered to be a continuous extension of the nuclear envelope and of a similar, if not identical, composition. The antitheses of the "unit membrane" is the concept that all membrane compartments are different or at least characterized by *marker* constituents that are unique. It is now clear that absolute single location markers (i.e., cell constituents whose subcellular distributions are restricted to one given cell component or to domains within a given component) are relatively rare. In fact, single location markers may be limited to certain constituents of mitochondria (e.g., cardiolipin, cytochrome oxidase) and of plastids (ribulose biphosphate carboxylase; galactolipids) where secondary locations of these constituents are unknown. A much more common phenomenon is for such markers to have a primary location in one cell component that accounts for most of the total cell content but, also, to have one or more secondary locations where the amount may be small but sufficient to be of functional significance. The fact that few, if any, absolute markers exist among the major endomembrane components (nuclear envelope, endoplasmic reticulum, Golgi apparatus, secretory vesicles, plasma membrane) is one of the strongest lines of evidence favoring a flow-differentiation mechanism for their biogenesis.

6.2.2.1. Endoplasmic Reticulum Markers and/or Nuclear Envelope Markers

In those tissues where accurate comparisons are possible, enzymatic activities of the nuclear envelope and of the endoplasmic reticulum are qualitatively and, often quantitatively, similar (Appendix Table 3). As a result, there are operationally few markers for fragments of nuclear envelope. The microsomal oxidoreductases of drug detoxification of nuclear envelope are not induced by phenobarbital in contrast to those of endoplasmic reticulum (Kasper, 1971). They are, however, induced in response to other drugs. The basis for this difference in responsiveness of two very similar, if not identical, enzyme systems is not known.

Not only are most enzymatic activities shared by nuclear envelope and endoplasmic reticulum but with the Golgi apparatus as well. These activities include constituents of the "microsomal" electron transport system, nucleoside phosphatases, and glucose-6-phosphatase, the most widely accepted endoplasmic reticulum "marker" for rat liver. These same activities may also be present in residual amounts in plasma membrane but specific activities 0.1–0.3 those of the endoplasmic reticulum (Fig. 6.7).

Some microsomal electron transport activities shared by endoplasmic reticulum and Golgi apparatus are listed in Table 6.3. Present in the Golgi apparatus are both NADH and NADPH oxidoreductase, cytochrome b_5, and cytochrome P_{450} (also a b-type cytochrome most often associated with detoxification-related mixed function oxidation). Additionally, Golgi apparatus contain substantial quantities of non-heme iron in amounts approaching those of the endoplasmic reticulum.

6.2.2.2. Golgi Apparatus Markers

Certain glycosyltransferases and the enzyme thiamine pyrophosphatase are among the few proteins concentrated in the Golgi apparatus (Roth and Berger, 1982; Fig. 6.8). Even with these activities, they seem to be more widely distributed among other compartments. In liver and other tissues and organs, thiamine pyrophosphate may also be present in endoplasmic reticulum and sometimes also

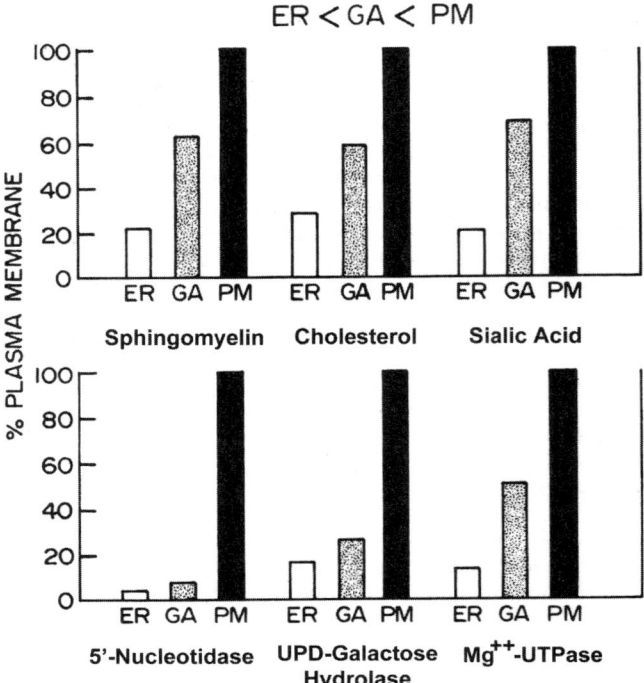

<figure>ER < GA < PM

% PLASMA MEMBRANE

Sphingomyelin   Cholesterol   Sialic Acid

5'-Nucleotidase   UPD-Galactose   Mg^{++}-UTPase
                   Hydrolase</figure>

Fig. 6.7. Biochemical evidence for endomembrane differentiation. Relative specific activities or amounts of constituents concentrated in membranes of endoplasmic reticulum (ER) comparing endoplasmic reticulum, Golgi apparatus (GA), and plasma membrane (PM) of rat liver. Other constituents showing a similar trend (ER > GA > PM) include NADPH-cytochrome c oxidoreductases, UDP-glucuronyltransferases, arylsulfatase c, glucose-6-phosphatase, nucleoside diphosphatase (UDP, GDP, or IDP as substrate), cytochrome P_{450}, and unsaturated fatty acids (studies of Keenan and Morré, 1970; Morré et al., 1971a, d; Yunghans et al., 1970) (from Morré, 1977b).

associated with plasma membranes. Glycosyltransferases have been ascribed to the plasma membrane of many cell types and low levels of specific activity, perhaps 5% of Golgi apparatus levels are associated with highly purified rough endoplasmic reticulum fractions. In recovery or balance sheet experiments, the glycosyltransferase activity associated with endoplasmic reticulum may account for as much as 20% of the total.

6.2.2.3. Plasma Membrane Markers

A number of enzymatic activities that characterize the plasma membrane also have been found associated with the Golgi apparatus (Appendix Table 3). These include 5'-nucleotidase, Mg^{2+}-ATPase, adenylate cyclase, alkaline phosphatase and γ-glutamyltranspeptidase from studies with rat liver. For the most part, a Golgi apparatus localization for these enzymes has been first indicated from biochemical measurements and subsequently confirmed by enzyme cytochemistry or vice versa. These enzymes may be present also in endoplasmic

Table 6.3. Membrane differentiation based on relative enzyme-specific activities of endomembrane fractions from rat liver.

Enzyme or constituent[a]	Relative specific activity[b]			References
	ER	GA	PM	
Glucose-6-phosphatase	10	1	0	Morré et al. (1969, 1974b); Cheetham et al. (1970); Fleischer et al. (1971)
UDP-glucuronyltransferase	10	1	0	Nyquist and Morré (1971)
NADPH-cytochrome c reductase	10	1	0	Morré et al. (1971d, 1974a, b, 1977b); Fleischer et al. (1971)
Cytochrome P-450	4	1	0	Fleischer et al. (1971); Morré et al. (1971d)
Terminal phosphoglyceride biosynthetic enzymes for PC, PE, and PI	4	1	0	Jelsema and Morré (1978); Morré (1977a, b);
O-Demethylase	2	1	0	Morré et al. (1974a)
Glycolipid glycosyltransferases (monosialoganglioside pathway)	1	10	0	Keenan et al. (1974)
Glycoprotein glycosyltransferases (exogenous acceptors)	1	10	0	Schachter et al. (1970); Schachter (1974)
NADH-cytochrome c ferricyanide reductase	9	2	1	Morré et al. (1971d, 1974b); Morré (1977a, b)
Cytochrome b_5	3	1	1	Fleischer et al. (1971); Morré et al. (1971d, 1974b); Morré (1977a, b)
NADH juglone reductase	2	1	1	Crane and Morré (1977)
Nucleoside diphosphatase (IPD, GDP, and TDP as substrates)	2	1	1	Cheetham et al. (1970); Cheetham and Morré (1970)
Thiamine pyrophosphatase	3	10	1	Cheetham et al. (1971)
UDP-galactose hydrolase	1	1.5	6	Morré et al. (1969)
Nucleoside diphosphatase (ADP, CDP, and TDP as substrates)	1	2	5	Cheetham et al. (1970); Cheetham and Morré (1970)
Mg^{2+}-ATPase	1	1	5	Morre et al. (1974b); Elder and Morré (1976a)
5′-Nucleotidase (5′-AMP as substrate)	1	2	40	Morré et al. (1969, 1974a, b); Cheetham et al. (1970)
Xanthine oxidase	0	1	8	Crane and Morré (1977)
Adenylate cyclase	2	1	4	Morré et al. (1974b); Yunghans and Morré (1978)
Na^+-, K^+-, Mg^{2+}-ATPase	1	0	7	Morré et al. (1974b); Elder and Morré (1976b)

[a] PC, Phosphatidylcholine; PE, phosphatidylethanolamine; PI, phosphatidylinositol
[b] ER, Endoplasmic reticulum (rough microsomes); GA, Golgi apparatus; PM, plasma membrane. For the majority of these studies, fraction purities exceeded 85%

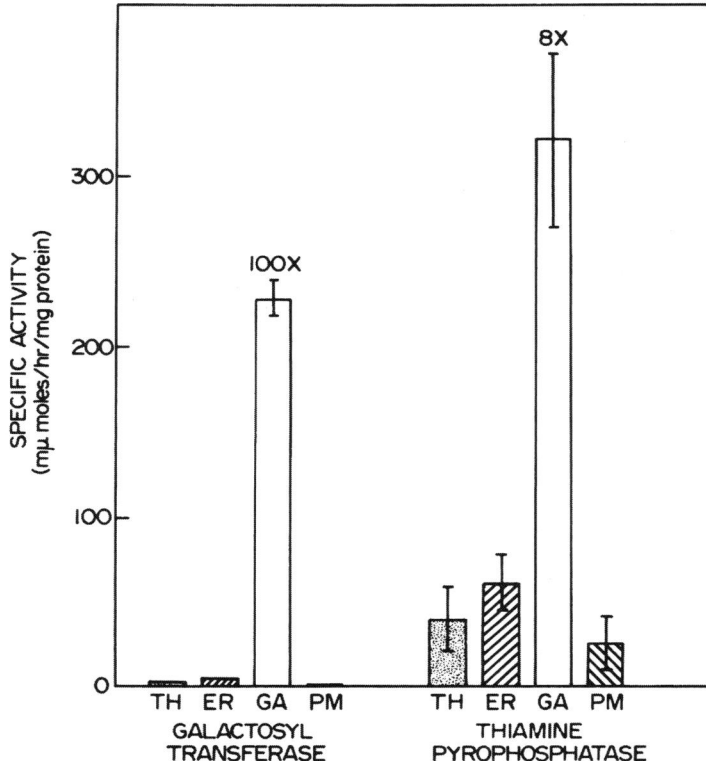

Fig. 6.8. Enzymes concentrated in Golgi apparatus. Specific activities of galactosyltransferase and thiamine pyrophosphatase comparing total homogenate (TH), endoplasmic reticulum (ER), Golgi apparatus (GA), and plasma membrane (PM) fractions from rat liver. Whereas Golgi apparatus fractions show a 100-fold enrichment relative to the total homogenate for galactosyltransferase, the comparable enrichment for thiamine pyrophosphatase is only 8-fold (Studies of Morré et al., 1969; Cheetham et al., 1971). Reproduced from Morré, 1977b with permission from the author.

reticulum. Only one enzyme, the ion-stimulated ATPase (Na^+-, K^+-stimulated Mg^{2+}-ATPase) is active only with plasma membrane. This is a result of a separation of biosynthesis and transport of the two subunits that comprise the active enzyme. Only in the plasma membrane do the two subunits first associate.

6.2.2.4. Patterns of Distribution of Enzymatic Markers

As illustrated in Fig. 6.9, the patterns of distribution of plasma membrane marker constituents is predominantly one of progressive increase comparing endoplasmic reticulum, Golgi apparatus and plasma membrane. Among more than 100 enzyme–substrate combinations compared, only the two exceptions were found. The monovalent ion-stimulated ATPase is active only in the plasma membrane. 5'-nucleotidase increases only gradually comparing endoplasmic reticulum and Golgi apparatus and then rises sharply for the plasma membrane.

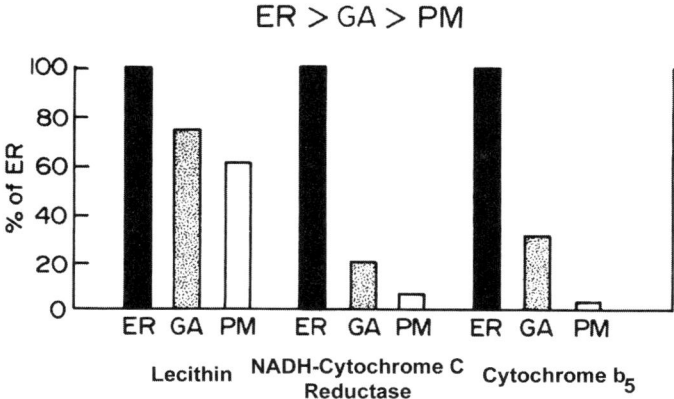

Fig. 6.9. Biochemical evidence for endomembrane differentiation. Relative specific activities or amounts of constituents concentrated in plasma membranes (PM) comparing endoplasmic reticulum (ER), Golgi apparatus (GA) and plasma membrane (PM) of rat liver. Other constituents showing a similar trend (ER<GA<PM) include Mg^{2+}-nucleoside triphosphatases (ATP, UTP, CTP, ITP, or TTP as substrate), alkaline phosphatase nucleoside diphosphatase (ADP, CDP, or tDP as substrate), cerebrosides, higher gangliosides, total glycoprotein carbohydrate, and saturated fatty acids (studies of Cheetham et al., 1970; Keenan and Morré, 1970; Keenan et al., 1972a; Morré et al., 1971a, 1974a; Yunghans et al., 1970). Reproduced from Morré, 1977b with permission from the author.

All other enzymes and chemical constituents show the pattern expected for a linear flow-differentiation of membranes where the relative specific activity (or amount per unit mass of membrane protein or phospholipid) present in Golgi apparatus is at a level intermediate between the two reference fractions, endoplasmic reticulum and plasma membrane.

For endoplasmic reticulum markers of rat liver, only ribosome attachment proteins such as the ribophorins appear to be totally absent from the Golgi apparatus. Of the enzymatic activities surveyed, glucose-6-phosphatase of rat liver emerges as one of the more endoplasmic reticulum specific. It is present in the immature elements of the Golgi apparatus, both by direct estimation in highly purified Golgi apparatus fractions and from enzyme cytochemistry but perhaps not in the plasma membrane. With all other activities and constituents a pattern of distribution opposite to that seen with plasma membrane markers is observed. It would seem as though these activities are those concentrated in endoplasmic reticulum and then either gradually lost, removed or diluted as membrane flow-differentiation progresses.

For Golgi apparatus markers, thiamine pyrophosphatase and glycosyltransferases of terminal sugar addition to glycoproteins and glycolipids, the pattern illustrated in Fig. 6.8 prevails. These activities which are few in number among known enzymes and identified proteins are consistent with a flow-differentiation mechanism only in so far as fulfilling the requirement that certain activities be concentrated in each major compartment of the pathway. At least with the glycosyltransferases, these enzymes themselves appear to be involved in the flow-differentiation mechanism where they function at the level of the

Golgi apparatus. With thiamine pyrophosphatase, its functional significance is unknown.

More important, it is readily possible to accommodate the Golgi apparatus markers into a flow-differentiation scheme because of the unique features associated with their biosynthesis, insertion into and removal from the membrane. They seem uniquely designed to function only at the Golgi apparatus whereas other enzymes present in all three major compartments seem designed to serve functions throughout the endomembrane system.

6.3. Immunological Manifestations of Endomembrane Differentiation

Distribution of membrane-associated antigens as well as lectin receptors are among the potentially most powerful tools for the study of flow-differentiation of membranes. Cell fractionation studies have revealed examples of similar or identical lectin receptors or antigens present in all major compartments. In evaluating such studies, it is important to realize that while determinants of the plasma membrane may be exposed to the external or environmental surface, similarly oriented determinants of endoplasmic reticulum or Golgi apparatus will be directed toward the membrane lumens. Luminally directed determinants will be inaccessible to antibodies or lectins presented from the outside so that determinations require the use of permeabilized vesicles. Thus, the cross-reactivity or apparent lack of cross-reactivity of internal and external membranes with respect to surface antigens is very much dependent on the experimental conditions of the assay. Care must be taken to assure access of antigen and antibody without significant impairment of reactivity.

Lectins have demonstrated specificity only for a particular type of sugar linkage, e.g., concanavalin A, the lectin from the jack bean, complexes with α-mannosides and any α-mannoside will compete with any other α-mannoside for the binding of this particular lectin (Table 6.4). As such, it is not surprising that the various lectins bind to each of the different endomembranes under comparison. Quantities bound may vary, however, and the usual pattern follows that of the overall carbohydrate composition of the fractions (Table 5.4). Endoplasmic reticulum binds less lectin than Golgi apparatus which binds less lectin than plasma membrane. An additional advantage of both lectins and antibodies is that they can be combined with electron dense markers for electron microscope localization. The localization of α-mannosides in Golgi apparatus as demonstrated using *Lens culinaris* lectin A linked to peroxidase is illustrated in Fig. 6.10.

Comparisons of antigens have dealt, in the past, mostly with cell surface antigens. Cell surface antigens, by definition, carry determinants expressed or accessible (see above) at the exterior of the cells. Implicit, but not necessarily correct, is a predominant or exclusive subcellular location of these antigens at the exterior of the plasma membrane. Antibodies offer advantages over lectins, for example, in being directed to specific molecules or specific parts of molecules.

Table 6.4. Plant lectins for glycoconjugate labeling.

Canavalia ensiformis	ConA	α-Man[a] > α-Glc > GlcNAc
Lens culinaris	LCA[b]	α-Man > α-Glc > GlcNAc (+Fuc)
Pisum sativum	PSA	α-Man > α-Glc = GlcNAc (+Fuc)
Triticum vulgare	WGA	GlcNAc > NANA
Helix pomatia	HPA	α-GalNac > α-GlcNAc >> Gal
Vicia villosa	VVA	α-GalNac > α-Gal
Griffonia simplicifolia I-A$_4$	GSI-A$_4$	α-GalNac > α-Gal
Maclura pomifera	MPA	α-GalNac > α-Gal
Ricinus communis I	RCAI	β-Gal > α-Gal >> GalNAc
Erythrina cristagalli	ECA	Gal-β1,4GlcNAc > α-GalNAc
Griffonia simplicifolia I-B$_4$	GSI-B$_4$	α-Gal >> α-GalNAc
Ulex europeus I	UEA I	αL-Fuc
Limax flavus	LFA	NANA

Source: From Pavelka and Ellinger (1991)
[a] Man, mannose; Glc, glucose; GlcNAc, *N*-acetylglucosamine; GalNAc, *N*-acetylgalactosamine; Gal, galactose; Fuc, fucose; NANA, sialic acid
[b] A = agglutinin

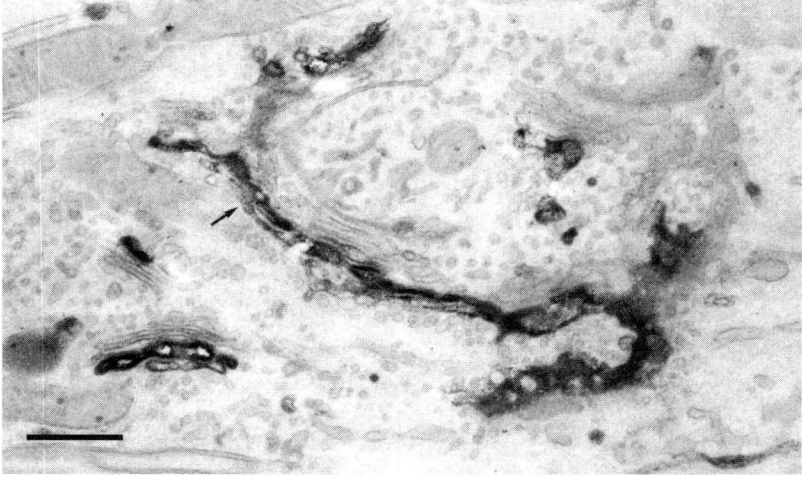

Fig. 6.10. Small intestinal absorptive cell of the rat illustrating localization of α-mannosides using horseradish peroxidase conjugated *Lens culnvaris* lectin + free *N*-acetylglucosamine to enhance lectin specificity. Label is restricted to cisternae at the cis side of the Golgi apparatus (*arrow*) and is absent from the endoplasmic reticulum. Reproduced from Pavelka and Ellinger, 1991 with permission from John Wiley and Sons, Inc. Scale bar = 1 μm.

Like lectins, they can be combined with electron dense detecting agents for localization using electron microscopy or reacted with a second antibody carrying the marker that is directed against the first antibody (the antibody sandwich technique).

A number of early studies suggested that functionally different types of subcellular membranes were antigenically related or at least shared common

Table 6.5.   Intracellular transport kinetics of membrane proteins.

Membrane protein	Cell components				References
	Endoplasmic reticulum	Golgi apparatus	ER → GA	Plasma membrane	
	Minutes post chase				
Mixed membrane proteins	10	25	15	<60	Franke et al. (1971b)
HLA-A, HLA-B antigens	<10	30	15	<60	Krangel et al. (1979)
VSV protein[a]	<5	8–11	5	<22	Strous and Lodish (1980)
Nucleoside diphosphatase	7.5	15	7	<30	Eppler and Morré (1982)
H-2 antigens	7.5	15–30	15	<60	Croze and Morré (1981)

[a] Passage through the Golgi apparatus was determined indirectly by monitoring the sensitivity of the carbohydrate moiety to endoglycosidase H

antigens. Heberman and Stetson (1965) reported H-2 antigens on both surface and endoplasmic reticulum membranes but not on mitochondria of mouse liver homogenates. H-2 antigens have also been reported for nuclear membranes (Albert and Davis, 1973; see also Morré et al., 1979c). Wilson and Amos (1972) correlated six HL-A determinants with 5′-nucleotidase activity; the latter also being more widely distributed than just on the plasma membrane. Techniques of immunochemistry and immunocytochemistry have been employed widely in recent years to monitor the presence and flow kinetics of several membrane and secretory proteins from endoplasmic reticulum to Golgi apparatus to the cell surface (Table 6.5; Figs. 6.11 to 6.13; 8.2 and 8.4). Changing patterns in the amounts and distributions from one compartment to the next provide information important to the membrane flow-differentiation concept. This is especially relevant in terms of viral antigens or antigens associated with viral presence that are induced, readily amenable to localization and quantitation by immunological techniques, and provide decisive markers of flow-differentiation events (Appendix Table 4).

An important application of immunocytochemistry to the elucidation of Golgi apparatus function has been the use of antibodies directed against specific sugar domains of polysaccharides in plants (reviewed by Driouich and Staehelin, 1997). These studies have contributed substantially to the determination of the main sites of sugar additions with the plant Golgi apparatus contributing to the synthesis of cell wall matrix polysaccharides.

6.3.1. Evidence from Induced Systems

Valuable contributions to the understanding of membrane flow-differentiation have come from examples where alterations in membrane structure, composition and/or functional properties are associated with chemical transformation,

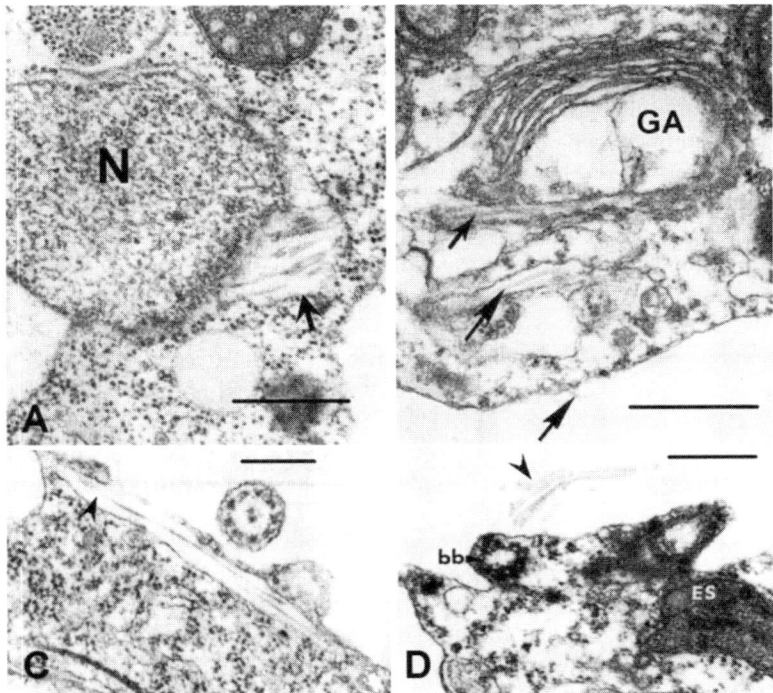

Fig. 6.11. Sections through reflagellating and colchicine-inhibited *Ochromonas minute* showing mastigonemes on, and being released to, the cell surface. (A) About 30 min after deflagellation, mastigonemes are seen in the Golgi apparatus, between the Golgi apparatus and the cell surface, and on the cell surface, ×26,500. (B) The apparent release of mastigonemes to the cell surface 60 min after deflaggellation, ×52,500. (C) Mastigonemes on the cell surface at the level of the basal body (bb) of the long flagellum 30 min after flagellation, ×48,000. (D) After 60 min, mastigonemes reappear near the flagellar base (*arrows*) ×32,000. ES=eyespot. Reprinted from Hill and Outka, 1974 and Morré and Mollenhauer, 2007 Int. Rev. Cytol., 262:191–218, Copyright 2007, with permission from Elsevier. Scale bar=0.5 μm.

viral infection, or other natural or externally induced developmental alterations. One of the most striking and poorly understood examples of a late differentiation event with the plasma membrane involves the myelination of peripheral nerves. Here Schwann cell plasma membranes are induced naturally to differentiate into a myelin sheath with attendant compositional, structural and morphological changes of considerable magnitude.

Of more immediate interest in the context of the present chapter are examples where the initial inductive event occurs in the nuclear envelope or endoplasmic reticulum such that the migration of the induced change toward the cell surface can be utilized as a route marker and to provide a time axis for flow. Such examples come from studies involving markers induced as a result of viral infection, studies with drug-induced enzymes, and from studies of surface coat markers.

There have been few real time visual markers of membrane flow-differentiation discovered. None are known for animal cells. Certain marine algae are coated by complex wall units known as scales which provide very distinctive markers. The scales are assembled within single Golgi apparatus cisternae and then secreted to the cell surface by release of an entire scale-containing vesicles. These vesicles are discharged at a rate of one in less than every 2 min and are large enough to be viewed with the light microscope. In one of these marine algae, *Pleurochrysis scherffeli*, Brown and coworkers (Brown, 1969) have monitored the kinetics of cisternal formation and discharge by direct observation of living cells.

Another type of visual marker is provided by the naturally-occurring mastigonemes of the flagellae of certain motile cells. Mastigonemes are microtubule-like hairs or appendages at the cell surface that are secured to portions of the flagellar plasma membrane (Fig. 6.11). Their function is unknown but they may aid locomotion of the organism in the direction of wave propagation along the flagellum. When flagellae are released mechanically, they are replaced, complete with mastigonemes, over a period of several hours. In the green alga *Ochromonas*, mastigonemes first appear within the perinuclear or endoplasmic space attached to membranes of the nuclear envelope or endoplasmic reticulum. Between 15 and 30 min after deflagellation, mastigonemes appear in secretory vesicles of the Golgi apparatus near the new flagellar base (Fig. 6.12). As the vesicles from the Golgi apparatus fuse with the plasma membrane at the new flagellar base, the nascent plasma membrane units are delivered with the mastigonemes oriented to yield two unbalanced files as in the original flagellum. Thus, in these and other organisms, mastigonemes originate within the endoplasmic space attached to the membrane of the nuclear envelope and/or endoplasmic reticulum, migrate via a directed flow mechanism involving membrane connections between contiguous membrane compartments, and are deposited to a specific region of the cell surface during flagellar regeneration.

Many examples of induced membrane constituents have been studied as a result of interest in virus-directed membrane assembly (Appendix Table 4). Virus envelopes are similar to cell membranes, structurally and chemically, and consist of lipids, proteins and carbohydrates. In some, glycoproteins project as spikes from the outer surface of the membrane with a carbohydrate-rich portion at the exterior and a hydrophobic portion embedded and anchored in the lipid bilayer of the membrane. On the inner surface of the virus envelope, there are sometimes a carbohydrate-free protein layer consisting of M (matrix) proteins.

Most of the enveloped viruses acquire their envelopes by a process of budding from cellular membranes. RNA viruses most frequently bud through the plasma membrane but other enveloped viruses may bud into cytoplasmic endomembrane spaces of the nuclear envelope, endoplasmic reticulum or Golgi apparatus. These latter examples provide many opportunities for the study of membrane flow differentiation in an induced system amenable to examination by pulse-chase techniques and direct visualization at the electron microscope level. Details are added by employing mutant strains, especially various temperature-sensitive mutants, where virus-directed products

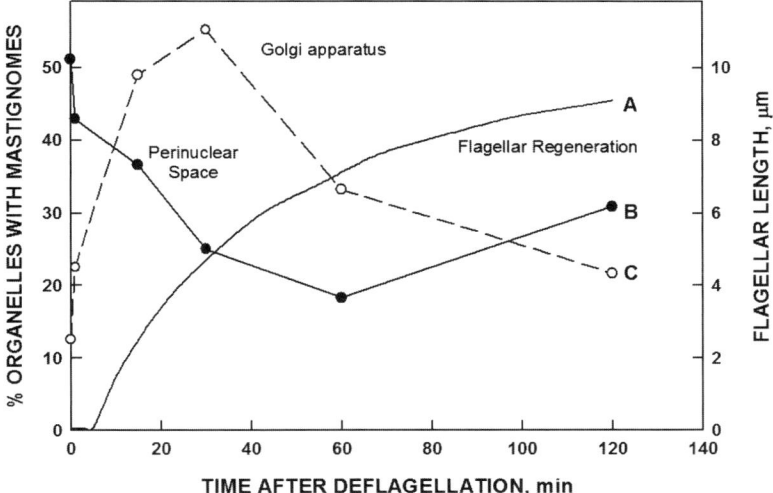

Fig. 6.12. Distribution of mastigonemes as determined in thin sections of *Ochromonas minute* fixed at intervals after deflagellation. Curve A: Reflagellation reference curve determined from light microscopy. Curve B: Percentage of midnuclear sections with greater than six mastigonemes per perinuclear space. Curve C: Percentage of Golgi apparatus with mastigonemes. Redrawn from Hill and Outka, 1974.

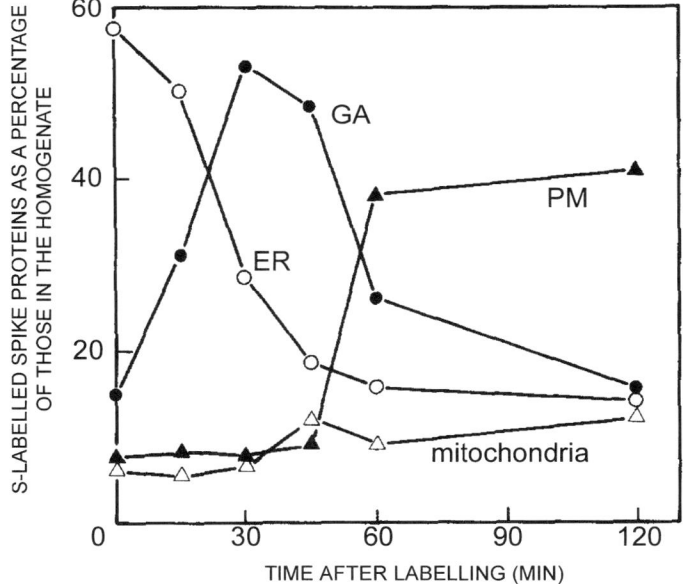

Fig. 6.13. Time course for incorporation of [^{35}S]methionine into Semliki Forest Virus-specified envelope proteins (E_1 and E_2) associated with endoplasmic reticulum (o), Golgi apparatus (•), plasma membrane (▲), and mitochondria (Δ) infected BHK cells. Membranes and virus were isolated from infected cells at various times after a 10 min pulse [^{35}S]methionine treated with spike protein-specific antibodies and subjected to gel electrophoresis. Individual proteins were analyzed for radioactivity. Redrawn from Green et al., 1981.

accumulate in certain portions of the pathway and transfer will continue only at permissive temperatures.

Evidence for a precursor–product relationship between viral proteins in endoplasmic reticulum and those located in the plasma membrane and subsequently incorporated into the envelopes of the budding virus has come from studies of cells infected with Semliki Forest virus (Green et al., 1981; Fig. 6.13), for example. When BHK cells were labeled with [^{35}S methionine], labeled viral spike proteins were located primarily in endoplasmic reticulum at the end of a 10 min pulse and were associated with plasma membranes after about 60 min. The viral proteins required about 15 min to exit the endoplasmic reticulum and 15 min to transverse the Golgi apparatus. In a similar study by Richardson and Vance (1976) with [^{3}H]leucine labeling, within 2 h more than half the label originally present in the endoplasmic reticulum had been transferred to the plasma membrane. By 11 h after the pulse, labeled virus proteins had disappeared from the plasma membranes and now the extracellular virus was maximally labeled (Fig. 6.13). Biochemical studies by Green et al. (1981) showed that most of the simple oligosaccharides of the viral spike proteins were modified to complex forms at the same time as the membrane-attached spike proteins passed through the Golgi apparatus.

Similar incorporation kinetics have been determined for other viruses. With Sendai virus, the delay of incorporation of glycopeptides between synthesis at endoplasmic reticulum and incorporation into the viral envelope is 15 to 30 min. With influenza virus, the antigens of the nucleocapsid can be demonstrated in the nucleus and cytoplasm in advance of their appearance at the cell surface by a variety of techniques including direct localization in the electron microscope by means of ferritin-labeled antibodies. In cell fractionation studies, the hemagglutinin antigen (HA) of influenza was found associated with the rough endoplasmic reticulum early after infection. Its concentration in smooth microsomes increased. Migration from rough to smooth membrane occurred within 10 min, however. In the smooth endomembranes, the large HA glycoprotein is cleaved into two smaller glycoproteins HA$_1$ and HA$_2$, These and other proteins of the influenza virion then accumulate at the surface membrane where they are found in association with purified plasma membrane preparations.

The intracellular pathway of biogenesis of the vesicular stomatitis virus transmembrane glycoproteins was investigated in situ by using indirect immunofluorescence of whole infected Chinese hamster ovary cells and immunoelectron microscopy of ultrathin frozen sections of infected cells by Bergmann et al. (1981). Transport of the glycoprotein was synchronized by using the temperature-sensitive virus mutant Orsay-45 and a temperature shift-down protocol. Sequential appearance of the glycoprotein in the rough endoplasmic reticulum, Golgi apparatus, and plasma membrane was demonstrated.

These and other studies suggest that proteins of the viral envelope migrate from rough through smooth membranes to the plasma membranes. Subsequently, the nucleocapsid aligns at these modified plasma membrane domains from which mature virions are released by budding for those viruses where virions are formed by budding from the plasma membrane. The passage through the smooth membrane fraction of the cell which includes the Golgi apparatus presumably reflects the requirement and time necessary for glycosylation of the glycoproteins and subsequent transport to sites of virus assembly. Extensive early indications of a potential role of the Golgi apparatus in the processing and segregation of viral glycoproteins are provided in Appendix Table 4.

In contrast to influenza and other viruses that bud from the plasma membrane of the host, nucleocapsids of herpes virus derive envelopes from the inner leaflet of the nuclear envelope. Nucleocapsid assembly occurs in the nucleoplasms. At the point of contact between the nuclear membrane and the nucleocapsid during encapsulation, the inner nuclear membrane often appears thicker and more intensely stained in electron micrographs. The viral particle is progressively enveloped by the nuclear envelope leaflet at this region and finally pinched off, leaving the nuclear membrane intact and the enveloped particle free in the perinuclear space between the inner and outer leaflets of the nuclear membrane.

Even though the nuclear membrane is the primary site of envelopment with herpes virus, viral antigens become associated not only with membranes of the nuclear envelope but with endoplasmic reticulum and plasma membranes as well, thus providing another example of migration of induced membrane constituents presumably from the nuclear envelope to the cell surface via internal endosomes.

A final example of an induced system is where drugs are used to stimulate the NADPH-linked mixed function oxidases of the microsomal drug detoxification system and the attendant membrane changes associated with this phenomenon (Fig. 4.9; Table 4.1). Administration of phenobarbital and other drugs to animals causes a marked elevation in drug detoxifying activities along with a marked proliferation of the smooth endoplasmic reticulum membranes that contain the enzymes. Both increased rates of synthesis and decreased rates of degradation are involved as well as synthesis and degradation of both proteins and phospholipids.

Ultrastructural and biochemical observations suggest that the smooth endoplasmic reticulum newly synthesized in response to drug administration arises at regions of transition between the rough endoplasmic reticulum and the smooth endoplasmic reticulum, possibly through continuous outgrowth from pre-existing rough membranes. An increase in the rough portion of the endoplasmic reticulum precedes chronologically the increase in the smooth portion, an observation confirmed by quantitative measurements (morphometry) from electron micrographs.

Rough and smooth endoplasmic reticulum are continuous with membranes of the Golgi apparatus and secretory vesicles in rat liver so that it is not surprising that the enzymes whose synthesis is stimulated by phenobarbital appear, after a time, in the Golgi apparatus. The lag time between the appearance of drug-induced enzyme in the smooth endoplasmic reticulum and its appearance in the Golgi apparatus has been determined to be 9 min.

6.4. Mechanisms of Membrane Differentiation

Since a principal function of the Golgi apparatus is to facilitate the differentiation of membranes from endoplasmic reticulum-like to plasma membrane-like, a complete understanding of mechanisms of membrane differentiation within the Golgi apparatus would be synonymous with the understanding of a major part of Golgi apparatus function. Involved are all aspects of the progressive conversion of the one type of membrane into the other. Whatever mechanisms are proposed must ultimately account for all of the changes in proteins, enzymatic activities, lipids and glycoconjugates that occur during the transformation.

Within the endomembrane system of rat liver and of mammary gland, the lipid/protein ratio of the membranes increases from less than 0.5 for endoplasmic reticulum to more than 0.7 for plasma membranes (Table 5.1). The proportion of phosphatidylcholine (lecithin) decreases, the proportion of spingomyelin increases.

The very large net increase in spingomyelin is accompanied by a parallel increase in cholesterol. Glycolipids and glycoproteins follow the pattern of increase from rough endoplasmic reticulum to plasma membrane, shown by sterols and spingomyelin. There is an average increase in chain length of fatty acids of phospholipids and lesser proportions of unsaturated fatty acids. Constituents, especially enzymes, characteristic only of nuclear envelope or endoplasmic reticulum must be deleted from the membrane while constituents characteristic of plasma membrane must be added. At the same time, many membrane constituents, those common to all of the endomembranes, must be conserved.

6.4.1. Biosynthetic Contributions to Membrane Differentiation

Synthesis of phospholipids, especially choline, ethanolamine, inositol and serine phospholipids, is a characteristic reaction of the endoplasmic reticulum. Nonetheless, the Golgi apparatus also has a modest capacity for synthesis of the same phospholipids whereas plasma membranes are unable to synthesize phosphatidylcholine, phosphatidylethanolamine and phosphatidylinositol. All endomembranes, including the plasma membrane, appear capable of synthesizing CDP-diglyceride, diglyceride, phosphatidylglycerol and phosphatidylserine, the latter by exchange. The site of spingomyelin synthesis remains to be established conclusively. However, findings involving the use of poorly penetrating inhibitors implicates the plasma membrane. It would also not be surprising to find that Golgi apparatus are capable, also, of synthesizing spingomyelin. Thus, to account for the characteristic lipid composition of the Golgi apparatus and plasma membrane it may be necessary only to transfer some phospholipids from endoplasmic reticulum via flow or some other mechanism and add to those, via direct synthesis, the additional phospholipids which appear in their composition. Except for spingomyelin, transfer from endoplasmic reticulum via carrier proteins of the cytosol is also a possibility.

To account for an increasing sterol content (Table 5.1), less information is available but biosynthesis emerges as a major contributor. The enzymes of sterol biogenesis are associated with smooth membranes (smooth endoplasmic reticulum, the Golgi apparatus and/or plasma membrane, i.e., those membranes also exhibiting highest sterol contents) but their precise intracellular locations remain to be determined.

Increases in glycolipids and changing patterns of glycolipids and glycoproteins that occur in Golgi apparatus and plasma membrane may be more complicated but as a first approximation reflect the intracellular distribution of the glycosyltransferases that catalyze the transfer of sugars from sugar nucleotide donors to appropriate acceptors. Although not localized there exclusively, these glycosyltransferases are concentrated in the Golgi apparatus (Chapter 5; Fig. 5.6) but may be present under certain conditions in endoplasmic reticulum and on the cell surface as well.

The mechanism by which enzyme activities are modulated during membrane differentiation may involve biosynthesis and insertion at specific sites of nascent polypeptide chains. However, protein changes are also subject to other types of modulations which are the principal subjects of the sections which follow.

6.4.2. Selectivity of Membrane Differentiation Mechanisms

An important feature of the concept of flow-differentiation of membranes is that of selectivity. Mechanisms must permit removals and additions of membrane constituents in such a manner that functionally appropriate constituents are transferred while functionally inappropriate ones are excluded. Endoplasmic reticulum membranes and plasma membranes are different – structurally, functionally, and in their chemical and enzymatic makeup. Golgi apparatus show many characteristics of both endoplasmic reticulum and plasma membrane but also have characteristics that are unique to Golgi apparatus. With some membrane constituents, those common to all endomembranes, direct transfer from one compartment to the next seems to occur. However, constituents unique only to endoplasmic reticulum/nuclear envelope or Golgi apparatus must either be deleted or fail to transfer in the conversion to plasma membrane whereas still other constituents characteristic only of plasma membranes must be added.

6.4.3. Golgi Apparatus Polyribosomes – A Means to Achieve Selective Addition of Proteins?

A mechanism whereby membrane proteins may be added to Golgi apparatus during membrane flow-differentiation has been suggested from the presence of clusters of *free* polyribosomes associated with Golgi apparatus of rat liver and other tissues (Chapter 5; Fig. 5.5). These Golgi apparatus-associated polyribosomes are not membrane-bound as are those attached to rough endoplasmic reticulum. Rather, they are a class of "free" polyribosomes located within the zone of exclusion (Chapter 2) or specialized cytoplasmic zone in which Golgi apparatus reside. Specific incorporation of radioactive amino acids in a cell-free system for protein synthesis by Golgi apparatus-associated polyribosomes is as high or higher than that of free ribosomes from the cytosol in rat liver. The peptides synthesized correspond to electrophoretic bands absent from endoplasmic reticulum but present in plasma membrane or correspond to peptides unique to the Golgi apparatus.

A possibility currently under exploration is that the Golgi apparatus-associated polyribosomes are programmed to translate messenger RNA for the glycosyltransferases that appear to function uniquely or predominantly in Golgi apparatus of some cell types. The latter occurs with lactose synthetase of mammary gland. This enzyme is a galactosyl transferase involved in the biosynthesis of lactose, the dominant disaccharide constituent of milk.

The existence of Golgi apparatus-associated polyribosomes may eventually help to explain how incorporation of enzymes/peptide chains can occur at Golgi apparatus independently of endoplasmic reticulum and other cell components to facilitate membrane differentiation. To what extent other classes of cytoplasmic polyribosomes may contribute to membrane biogenesis is not known. Mollenhauer and Morré (1978a) have reported polyribosomes associated with forming acrosome membranes during the acrosomal stage of spermiogenesis which are similar in appearance and apparent function to the Golgi apparatus-associated polyribosomes.

6.4.4. Examples of Selective Enzyme Deletion

The removal of enzymatic activities from membranes during flow-differentiation could occur by (a) failure to transfer; (b) inactivation; or (c) removal of the peptide chain or portions of it from the membrane. The latter apparently accounts for the loss of at least one glycosyltransferase from membranes leaving the Golgi apparatus to become incorporated into the plasma membrane in mammary gland. This enzyme is most concentrated in Golgi apparatus and secretory vesicles of the gland but is reduced or absent at the apical plasma membrane where the secretory vesicles are added to discharge the milk secretions (see Chapter 8). What seems to happen is a proteolytic cleave of the transferase with conversion from a form of high molecular weight firmly associated with the membrane to a soluble form of lower molecular weight. The proteolytic activity responsible for the cleavage of the glycosyltransferase from the membrane was identified and partially characterized by Paulson and Colley (1989).

Experimental evidence for membrane alteration by selective deletion of glycosyltransferase activities has come from studies where derivatives of the apical plasma membrane, the membranes surrounding globules of milk fat, have been isolated carefully under conditions unfavorable to proteolysis (low temperature, rapid isolation). Under these conditions, some of the membrane-associated high molecular weigh form of the lactose synthetase is retained in association with the membrane. When such membranes are then incubated at room temperature, this residual lactose synthetase activity is released to the supernatant. And upon subsequent centrifugation–resuspension cycles appears in the supernatant as the lower molecular weight form of the enzyme.

With sialyltransferase, a somewhat different mechanism has been suggested by Verbert et al. (1977) where the enzyme becomes masked or cryptic following discharge from the Golgi apparatus to the plasma membrane.

Other potential examples of selective enzyme deletion are provided by various so-called "microsomal" electron transport chain components enriched in endoplasmic reticulum but present in varying amounts in the Golgi apparatus and even the plasma membrane (Chapter 8). These include NADH-cytochrome c (ferricyanide) oxido-reductase, NADPH-cytochrome c oxido-reductases, cytochrome b_5 and cytochrome P-450 (P-420). As with the lactose synthetase of mammary gland carefully isolated plasma membranes of rat liver can contain up to 30% of both the cytochrome b_5 and of the sum of cytochrome P-450 + P-420 content of endoplasmic reticulum. Studies of Jarasch et al. (1979) suggest that these microsomal cytochromes are subject to degradation in a manner similar to that suggested for lactose synthetase. For example, if plasma membranes containing P-450 are incubated at room temperature, the activity is lost suggesting some sort of inactivation process and that the absence of P-450, for example, in certain plasma membrane preparations may be the result of its degradation during membrane isolation.

With the microsomal cytochrome b_5 and its reductases, more of these activities are retained by both Golgi apparatus and plasma membranes. The cytochrome b_5 extracted from either endoplasmic reticulum or plasma membranes appears to be identical. The reductases, while having similar kinetic constants, appear to differ in their sensitivity to certain inhibitors depending upon cellular location. An apparent NADH-ferricyanide oxidoreductase of plasma membrane, of secretory vesicles and of mature elements of the Golgi apparatus resists the rigors of cytochemical localization while the corresponding activity of endoplasmic reticulum and forming Golgi apparatus elements does not. The latter can be demonstrated only in unfixed and broken cells where entry of reactants is unimpeded while the former can be demonstrated under a variety of conditions including following mild glutaraldehyde fixation. This cytochemical difference not only provides a marked visual demonstration of membrane differentiation in rat liver (Fig. 6.1) but points to somewhat more subtle mechanisms operative in enzyme modifications during the flow-differentiation of membranes.

Any activity concentrated in either the endoplasmic reticulum or Golgi apparatus and depleted or absent from the plasma membrane is a potential candidate for removal from the membrane by selective enzyme deletion. One such example, little studied, is the enzyme thiamine pyrophosphatase. Used widely as a cytochemical marker for the Golgi apparatus, the enzyme may be more widely distributed and, depending on the developmental stage, may be expressed in either the endoplasmic reticulum, the plasma membrane or both in addition to the Golgi apparatus. In a time-related cytochemical study of first cleavage embryos of the frog, *Xenopus*, Sanders and Singal (1975) reported reaction product from the hydrolysis of thiamine pyrophosphate initially within the vesicles and cisternae of the Golgi apparatus but subsequently on the furrow surface. Thus, while the mechanisms are unknown, thiamine pyrophosphatase may eventually provide an example of an enzyme activity that is either conserved and delivered to the plasma membrane, or deleted from the membrane (or masked or inactivated) depending on the tissue and the stage of development.

6.4.5. Summary of Flow-Differentiation Mechanisms

Figure 6.14 provides a summary of some of the various possibilities according to the examples already presented. These include:

1. *Selective transfer*: Some membrane constituents, e.g., cytochrome b_5 and the corresponding reductase, appear to be in part conserved and transferred from endoplasmic reticulum to Golgi apparatus with kinetics corresponding to those of other membrane proteins. Other proteins of the membrane that appear to be transferred as well include H-2 alloantigens of the mouse liver, nucleoside phosphatase, and nucleotide phosphorylase. In virus-infected cells, selective transfer of viral envelope proteins also takes place (see evidence for membrane flow from induced systems; Appendix Table 4).

2. *Selective removal*: Loss through proteolytic actions of the galactosyltransferase of lactose formation from Golgi apparatus of mammary gland provides one example.

3. *Transfer followed by masking*: The cryptic or masked sialyltransferase of Verbert et al. (1977) may provide an early example of a change corresponding to this category.

4. *Transfer followed by inactivation or degradation*: Cytochrome P-450 would appear to be in part transferred but then inactivated although the mechanism of inactivation, probably enzymatic, has not been elucidated.

5. *Selective addition*: A protein translated on free polyribosomes (e.g., NADH-cytochrome c reductase), could potentially be inserted into membranes at any stage of the flow-differentiation pathway. Those translated on membrane-associated polyribosomes must, by necessity, be inserted only at the rough endoplasmic reticulum. As Golgi apparatus appear to have associated with them in the cytoplasm a special class of free polyribosomes, the translation products of these Golgi apparatus-associated polyribosomes are excellent candidates to be added selectively at the Golgi apparatus. The products have not been identified but conceivably might include certain of the glycosyltransferases that appear restricted in their enzymatic activity to membranes of the Golgi apparatus. In fact,

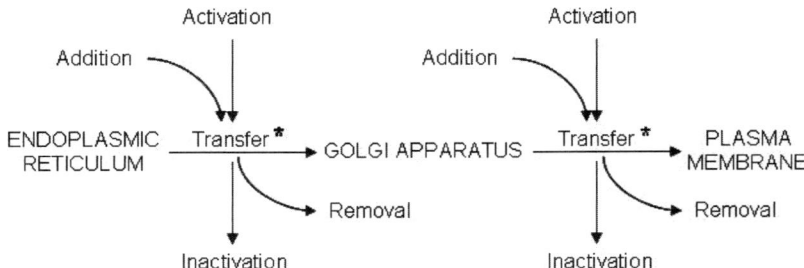

Fig. 6.14. Summary of changes in enzyme content and activity during membrane differentiation. Reproduced from Morré and Mollenhauer, 1974 with permission from McGraw-Hill.

such enzymes could be selectively inserted at any level of the stacked Golgi apparatus cisternae and thereby, be most important in directing the events of membrane differentiation.

6. *Selective activation*: An example of selective activation of a Golgi apparatus enzyme is again provided by lactose synthetase. It requires a specific protein, a so-called specifier protein known as α-lactalbumin, for activity. In the absence of α-lactalbumin, the product is mainly *N*-acetyl-aminolactose with UDP-galactose as donor and *N*-acetylglucosamine as acceptor for the galactose with the resultant formation of the milk sugar, lactose.

Activation of the ion-stimulated ATPase of the plasma membrane occurs by a different mechanism but with the same overall result. The enzyme consists of two subunits, one glycosylated, the other not. The glycosylated subunit is translated on membrane-associated polyribosomes and is processed in the Golgi apparatus. The non-glycosylated subunit is translated on free polyribosomes and arrives at the plasma membrane by another, unknown route. However, once inserted into the plasma membrane, the two subunits combine and the functional holoenzyme results.

Thus, flow-differentiation of membranes emerges as a complex process involving a number of different mechanisms whereby membrane composition is progressively altered but in terms of specific constituents and of enzymatic activities. It may be most appropriate to regard it as a process in the same sense as we regard respiration and small molecule transport as processes.

6.4.5.1. Membrane Differentiation In vitro

In this chapter, evidence has been summarized consistent with the notion that flow-differentiation of membranes within the Golgi apparatus is largely a function of Golgi apparatus enzyme systems. Consequently, these aspects of the process should be carried out by isolated Golgi apparatus in an appropriate cell-free environment. In fact, most of the partial reactions of membrane differentiation (e.g., glycoprotein and glycolipid glycosylation, protein and glycoprotein processing, phospholipid biosynthesis and modification, enzyme inactivation, protein synthesis and insertion into membranes from Golgi apparatus polyribosomes, etc.) have been carried out and demonstrated to occur in cell free systems using isolated preparations of Golgi apparatus from rat liver.

A cytochemical demonstration of in vitro membrane differentiation has also been possible using Golgi apparatus isolated from a plant source (hypocotyls of soybean, *Glycine max*). Plasma membranes of plant cells exhibit a characteristic staining reaction by combining with phosphotungstic acid at low pH (PACP stain). This reaction is seen in vivo on only a few of the most mature Golgi apparatus elements. Normally only the membranes of mature or migrating secretory vesicles are reactive while the stacked cisternae are completely negative (cf. Fig. 6.5). The membrane constituents responsible for the reactivity extracted from the membrane by organic solvents are presumed to represent glycolipids. Frantz et al. (1973) showed that when either in situ or when freshly isolated, membranes of cisternal stacks of soybean did not stain by the procedure. However, after incubation of the pellets for either 2 hours at 25°C or 8 hours at 0°C the membranes acquired PACP reactivity. The staining capacity first appeared at the mature face or pole and subsequently progressed toward the forming pole or face. This demonstration of membrane differentiation in vitro shows not only further evidence of its occurrence but confirms the suspected gradient of activities responsible across the stacked cisternae from the forming to the maturing pole.

6.5. Functional Significance of Flow-Differentiation of Membranes

The occurrence of membrane differentiation with the Golgi apparatus has implications and potential significance not only to the fulfillment of the major Golgi apparatus functions in membrane biogenesis and secretion but to the origin of specific surface characteristics of cells and tissues as well and as a model of cell, tissue and organ differentiation. While most conspicuous in the Golgi apparatus where it progresses across the stacks of cisternae, flow-differentiation may very well be a general phenomenon of all transitional membrane elements.

1. *Membrane biogenesis*: A major feature of membrane flow-differentiation is that it allows for specialization of membrane types within the endomembrane system through progressive modification of an existing membrane type according to defined developmental sequences. Additionally, it provides a mechanism whereby membrane constituents may be either selectively transferred along a series of cell components to account for the origin of the plasma membrane of the cell surface as well as various specialized internal membranes including Golgi apparatus elements, primary lysosomes, and so-called smooth elements of the endoplasmic reticulum.

2. *Secretion*: To be discussed in detail in a later chapter, the endomembrane system provides a major intracellular route for export of products of synthesis within which a variety of macromolecules are sequestered, concentrated, modified and externalized by one cell type for subsequent utilization by other cells. This transfer process requires that originally dissimilar membrane types acquire the ability to fuse at specific sites (coalescence of transition vesicles, fusion of secretory vesicles with the plasma membrane), a process undoubtedly facilitated by or perhaps even dependent upon membrane differentiation.

3. *Origins of specific surface characteristics of cells and tissues*: An important corollary of the membrane flow-differentiation hypothesis is that to the extent that the endomembrane system is involved in the origins of cell surface membranes, those surface characteristics that determine self and non-self influence morphogenesis, impart immunological specificity, are affected by oncogenic transformation (Chapter 12), etc. are determined, in effect by activities of the endomembrane system. This is illustrated visually by the work of Varga et al. (1976a, b) where receptors for melanocyte

stimulating hormone in mouse melanoma cells are exclusively associated with discrete areas of the cell surface that overlay the Golgi apparatus directly. Thus, while one normally focuses on Golgi apparatus roles in membrane biogenesis and secretion, it may be that the most important functions of membrane flow-differentiation are those that relate to the formation of new membranes for cell replication, growth and changes in surface characteristics associated with the differentiation of cells.

6.5.1. Cell, Tissue and Organ Differentiation

While ultimately under genetic control, membrane differentiation as well as cellular differentiation is a programmed series of events where one event may influence or trigger subsequent events in the sequence. While we know very little of the complete sequence whereby normal or abnormal differentiation events are controlled and regulated, it seems likely that flow-differentiation of membranes provides one mechanism whereby these changes may be expressed as one of the many opportunities for their regulation and control.

6.6. Dynamic Aspects of the Flow-Differentiation of Membranes – Membrane Flow

The membrane flow hypothesis as stated originally (Franke et al., 1971b; Morré et al., 1979b) simply requires that "the biogenesis of certain membranes be accomplished by the physical transfer of membrane material from one cell component to another in the course of their formation or normal functioning." Synthesis of membrane proteins is restricted to ribosomes and to ribosome-rich regions of the cell. With rat liver, it is estimated that about two-thirds are synthesized on polysomes bound to membranes of the endoplasmic reticulum while the remainder may be synthesized by free polyribosomes (Elder and Morré, 1976a). Protein and lipid glycosylation are predominantly activities of endoplasmic reticulum and Golgi apparatus as are terminal enzymes for biogenesis of membrane phospholipids. How then do plasma membranes acquire their membrane constituents? Through what mechanism are membrane constituents synthesized in one cell component (e.g., endoplasmic reticulum) transferred or transported to sites of utilization as constituents of another cell component (e.g., plasma membrane)? Evidence favors a mechanism where membrane constituents synthesized in endoplasmic reticulum (or nuclear envelope) are transported physically to the cell surface via vesicles or by lateral movement within the membranes of specialized transition elements in various combinations. A vesicle is a common feature of the last transfer step to the plasma membrane and during transit modifications in membrane composition take place to provide for differentiation from endoplasmic reticulum-like to plasma membrane-like. These dynamic aspects of passage of membrane through the Golgi apparatus region of the cell in large measure may determine not only Golgi apparatus function but its origin and maintenance as a discrete cell component as well. Two independent groups have applied fluorescence tagging and advanced video microscopy techniques to confirm the membrane maturation model from visual observations (Losev et al., 2006; Matsuura-Tokita et al., 2006).

6.6.1. General Morphological Basis for Membrane Flow

The structural basis for a flow mechanism of membrane biogenesis is provided from electron micrographs from secretory and other cell types that show a common pattern of membrane relationships and interconnections as well as correlative studies of membrane differentiation that compare and contrast different membrane types in situ and in isolated fractions. Based on these investigations, endomembrane components may be arranged in a compositional and ontogenetic sequence as follows: Nuclear envelope–rough endoplasmic reticulum–smooth endoplasmic reticulum–endoplasmic reticulum-derived transition elements–Golgi apparatus and related transition elements–Golgi apparatus-derived secretory vesicles–plasma membrane. In this sequence, the potential generating elements, endoplasmic reticulum and nuclear envelope, are most distinct from the plasma membrane and have the greatest potential for biogenesis of membrane constituents.

Since membrane flow is a dynamic phenomenon, structural elements must be considered not only in the context of the major membrane elements that comprise the system but, in addition, the various types of tubules and vesicles which may serve to interconnect the major part of the system. Many of the associations that provide continuity within the system may be transient, e.g., intervening tubules or vesicles, that are continuous perhaps only fleetingly and either become assimilated into a larger structure or separated soon after contact.

Intermittent vesicular bridges are one type of dynamic transition element that permits transfer of material (both membrane and secretory product) in the absence of permanent structural continuities. Their functions are somewhat analogous to that of a series of navigational locks that permit passage from one level to the next in a waterway without free intermixing of the two compartments to be interconnected. This is especially important at the cell surface where permanent connections between internal and external compartments might lead to an unregulated loss of materials from the cell to the environment and vice versa.

In a few examples, flow of membranes or of membrane constituents can be visualized by purely morphological criteria. Among the most accurate determinations of flow rate through a Golgi apparatus are the real time visual observations of the alga *Pleurochrysis scherffelii* (Brown, 1969). Here Golgi apparatus participate in the fabrication of complex carbohydrate scales; synthesis of each scale is sequential across the stacked cisternae beginning at the forming face. Secretion of each scale is accompanied by the loss of an entire Golgi apparatus cisternae at a rate of about one cisterna per minute. Since the cisternae migrate to and fuse with the plasma membrane, replacement cisternae must form at the same rate as cisternae are utilized in scale secretion to account for the continued functioning of the Golgi apparatus.

The time required for the build-up of a new Golgi apparatus cisternae has been estimated to be approximately 3 to 4 min from studies with embryogenic suspension cultures of carrot. Here, vesicle release is transiently blocked by the sodium selective ionophore monensin (Pressman 1986, Chapter 10) so that cisternae accumulate for a time at the maturing face (Figs. 6.15 and 6.16). In the presence of 10 mM monensin, an average of one additional cisternae per stack was formed within the first 2–4 min of monensin treatment. In some experiments, a second cisternae was formed within about 6 min (Fig. 6.17). Thereafter, large vacuoles began to appear in the cytoplasm adjacent to the Golgi apparatus. By 1 h of monensin treatment, the regions of the cells containing dictyosomes were populated by large numbers of vacuoles (up to 20 or more per electron microscope section). These large vacuoles were interpreted as swollen dictyosome cisternae that separated from the stack but had not migrated from the Golgi apparatus zone in the monensin-treated cell.

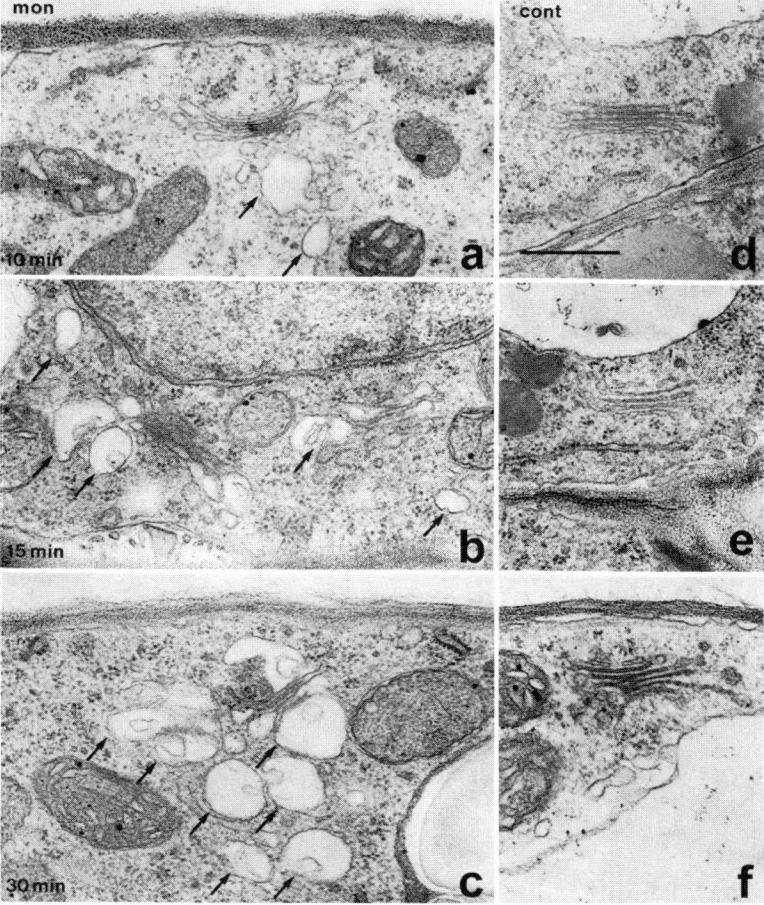

Fig. 6.15. Golgi apparatus morphology of cultured carrot cells treated with monensin. Cisternal swelling becomes more pronounced and the number of separated and swollen cisternae (*arrows*) increases from one or two at 10 min to more than 6 (up to 10 per dictyosome) by 30 min. Reproduced from Morré et al., 1983a with permission from Wissenschaftilche Verlagsgesellschaft mbH. Scale bar = 0.5 μm.

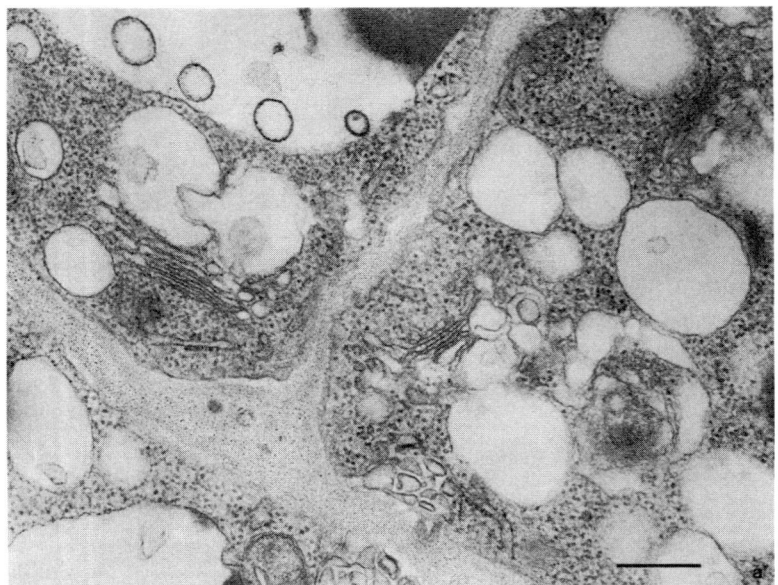

Fig. 6.16. As in Fig. 6.15 except after 1 h of monensin treatment. Each stack is surrounded by numerous swollen cisternae generated but unable to migrate away from the stack. Reproduced from Morré et al., 1983a with permission from Wissenschaftliche Verlagsgesellschaft mbH. Scale bar = 0.5 μm.

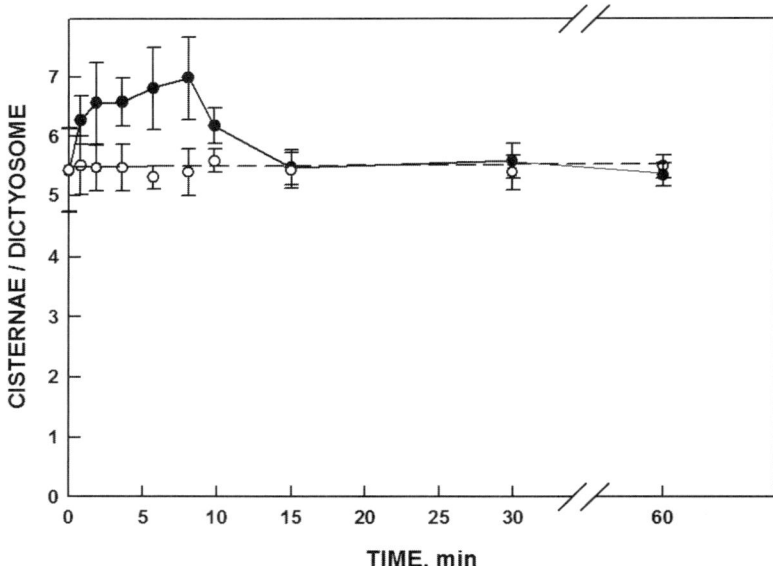

Fig. 6.17. Numbers of cisternae per dictyosome as a function of time of treatment ± 10 μm monensin. Results are averages from three experiments ± standard deviations among experiments. Each experiment consisted of counts of the number of cisternae per dictyosome of all dictyosome profiles observed in thin sections of ten exterior cells selected at random to yield a per cell average based on 30 cells. Points marked by an asterisk are significantly different from controls ($p < 0.01$) as determined by Student's t test. Reproduced from Morré et al., 1983a with the permission from Wissenschaftliche Verlagsgesellschaft mbH.

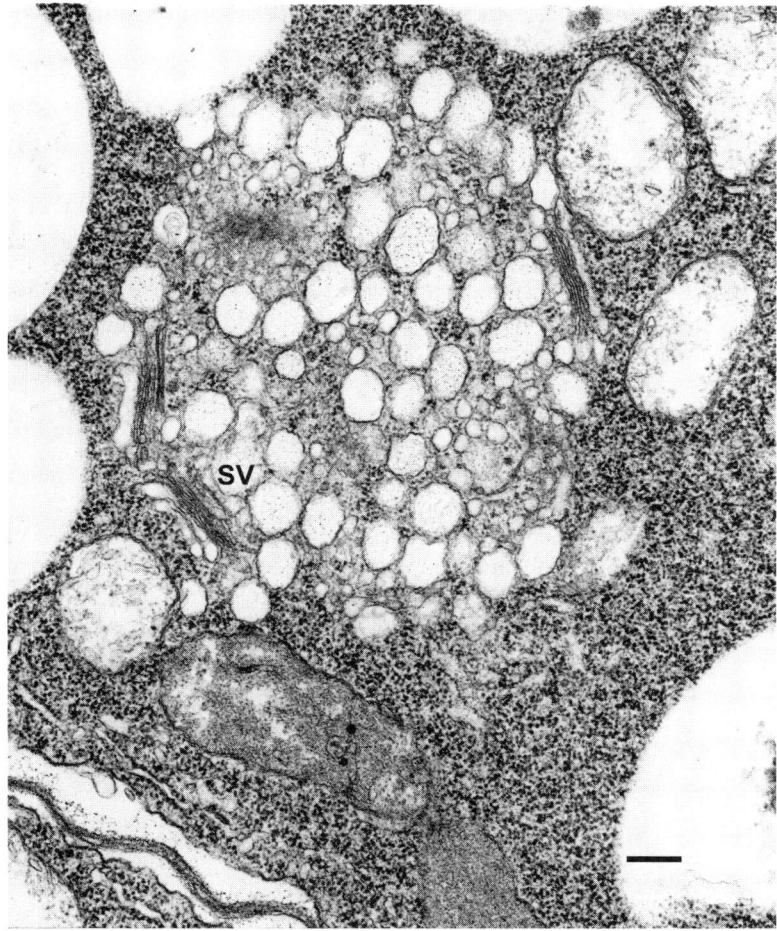

Fig. 6.18. A portion of an epidermal cell from a maize seedling treated with 100μg/mL cytochalasin B in 1% DMSO for 2h. The secretory vesicles (SV) are seen to aggregate around the stacks and appear unable to migrate to the apical cell surface ×14,000. Reproduced from Mollenhauer and Morré, 1976b, with permission from the authors. Scale bar = 0.5μm.

Evidence for membrane flux through the Golgi apparatus also comes from studies where secretory vesicle migration away from the stack is blocked by cytochalasin B (Fig. 6.18). The vesicles accumulate with time in increasing numbers in support of a flow mechanism.

In certain algae, mastigonemes, microtubule-like appendages attached to algal flagella, provide visual markers for determining the timescale of flow of a membrane constituent. Here the passage through the endoplasmic reticulum and Golgi apparatus takes about 30 min and passage through the Golgi apparatus about 20 min. As summarized in the sections which follow, estimates of product and membrane flow through the Golgi apparatus generally fall within the range of 10–40 min (e.g., Table 6.6).

6.6.2. Kinetics of Membrane Flow

Amino acids are rapidly incorporated into membrane proteins of rough endoplasmic reticulum and nuclear envelope with subsequent appearance in other membrane fractions. Invariably in such kinetic studies, radioactive proteins appear first in rough endoplasmic reticulum, followed by smooth endoplasmic reticulum, Golgi apparatus, and plasma membrane in chronological order of the appearance of maxima in membrane labeling. The lag times vary from a few minutes for smooth endoplasmic reticulum to more than 20 min for plasma membrane. Such studies have now been carried out with purified and identified membrane proteins, mixed membrane proteins which have been extracted with high salt plus detergent to remove extrinsic, secretory, and loosely bound proteins, and individual polypeptide "bands" of mixed membrane proteins. Despite the observation that secretory proteins frequently attain a specific activity greater than that of membrane proteins, membrane proteins do become heavily labeled so that their movement during membrane flow may be monitored.

For membrane proteins of Golgi apparatus, both rapid and delayed phases of incorporation are seen (Fig. 6.19). Input from two sources, flow from endoplasmic reticulum and direct insertion of proteins, possibly including some newly synthesized by Golgi apparatus-associated or cytoplasmic polyribosomes, has been proposed to explain the unusual pattern of labeling of Golgi apparatus membranes. Participation of more than one class of polyribosome, one attached to rough endoplasmic reticulum and another operating later in the flow pathway, help explain the frequent observation that, at longer incorporation times, membrane proteins of smooth microsomes, Golgi apparatus, and plasma membrane attain higher specific radioactivities than those of rough endoplasmic reticulum on a total protein basis. When labeled membrane proteins are separated by gel electrophoresis, the electrophoretically separated protein bands do not exhibit kinetics resembling typical secretory proteins nor are they homogeneous in their labeling and decay characteristics. From these data, a mechanism involving the synchronous bulk flow of membranes from endoplasmic reticulum to the plasma membrane seems unlikely. Labeling of Golgi apparatus cisternae fractionated by preparative free-flow electrophoresis (Chapter 3) was determined for Golgi apparatus from livers of rats receiving a pulse of 1 mCi [^{35}S] methionine for 7.5 min followed by a chase with unlabeled methionine in 10,000-fold excess. Cis cisternae were maximally labeled 5 min post-pulse, the two median cisternae were maximally labeled at 10 and 15 min post-pulse while trans cisternae were labeled at 20 min post-pulse. Trans face secretory vesicles were labeled more slowly with maximum radioactivity found at 20–30 min post-pulse (Morré et al., 1984a).

Results with mixed membrane proteins in liver have been confirmed from analyses of the flow kinetics of single intrinsic membrane proteins (Fig. 6.19A–C; Table 6.5). H-2 alloantigens are one example of a glycoprotein, firmly associated with membranes, has been studied decisively in purified membrane fractions by

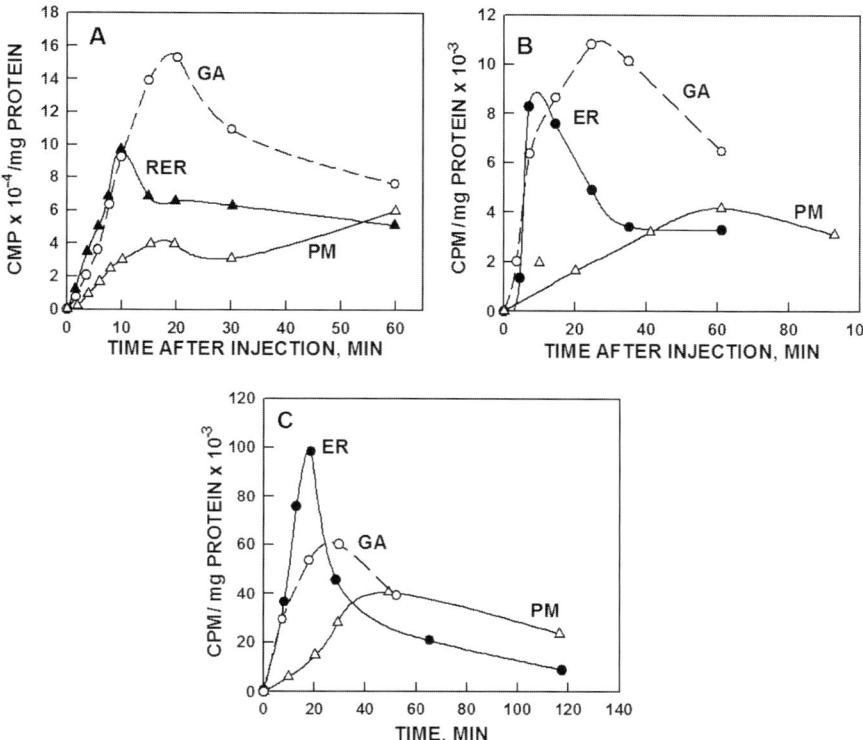

Fig. 6.19. Flow kinetics of migration of membrane protein from endoplasmic reticulum to the plasma membrane via the Golgi apparatus. (A) Kinetics of appearance of radioactivity from L-[guanido-^{14}C]arginine into a fraction of mixed membrane proteins from rat liver. Rats were injected with labeled amino acid and after the times indicated, rough endoplasmic reticulum (RER), smooth endoplasmic reticulum (SER), Golgi apparatus (GA), and plasma membrane (PM) fractions were isolated and analyzed. Membranes were extracted with buffers containing 1.5 M KCl and 0.1% deoxycholate and the radioactivity of the insoluble material was determined directly as a measure of intrinsic mixed membrane proteins (Franke et al., 1971b). (B) Flow kinetics of membranes of elongating tubes of germinating tobacco pollen. Pollen tubes germinated for 3–4 h were pulsed for 7.5 min with 100 μCi[^{3}H]leucine followed by removal from the pulse medium by centrifugation and transfer to fresh medium containing excess unlabeled leucine. Homogenates were prepared and fractionated at the times indicated. Fraction identity was confirmed by electron microscopy (Kappler et al., 1986; Noguchi and Morré, 1991a, b). (C) Transit of [^{3}H]-labeled proteins in slices of livers from 10-day-old rats. The slices were pulse-labeled with 30 μCi[^{3}H]leucine for 7.5 min followed by removal of the leucine. The rapidly washed slices were then placed in fresh medium containing a 1,000-fold excess of unlabeled leucine. At the time indicated, portions of the slices were removed, the different cell fractions were prepared from the same homogenates and specific radioactivities of each of the fractions were determined (Morré et al., 1987b). Reproduced from Morré and Mollenhauer (2007).

quantitative immunoprecipitation techniques. With [^{35}S]methionine as precursor, the order of labeling is endoplasmic reticulum, Golgi apparatus, and plasma membrane. Rapid turn-over components of H-2 antigens during passage through both endoplasmic reticulum and Golgi apparatus suggest sequential transfer with ultimate accumulation in the plasma membrane (Fig. 6.20).

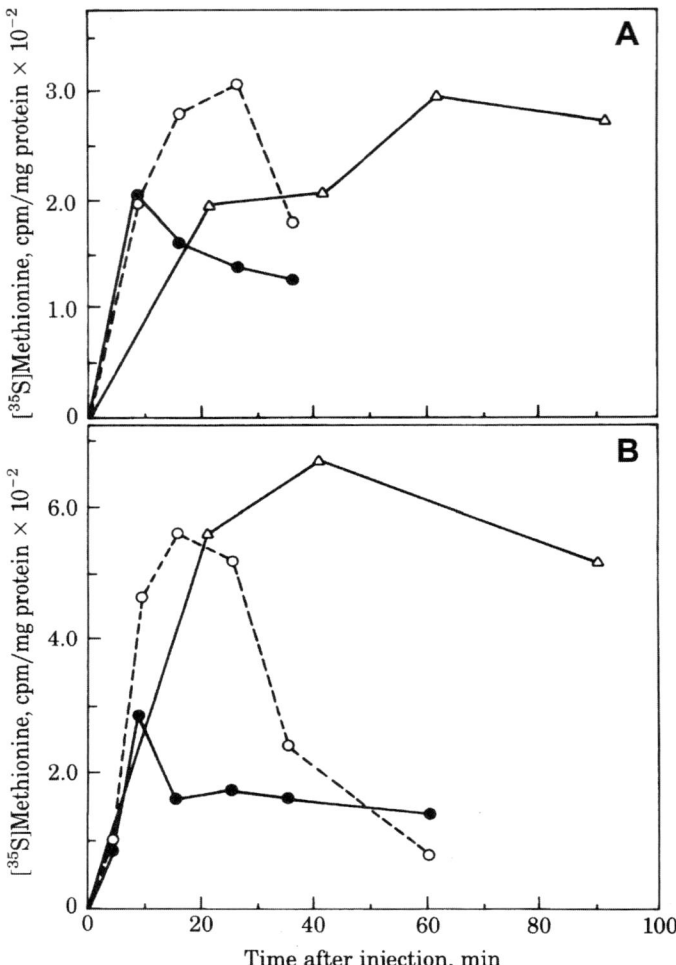

Fig. 6.20.   Kinetics of incorporation of radioactivity from [^{35}S]methionine into H-2 antigen immunoprecipitates from 1% NP-40-solubilized ER (•) Golgi apparatus (o), and plasma membrane (Δ) fractions from mouse liver comparing two different antisera in (A) and (B). Standard pulse-chase methodology was used to label H-2 antigens. Membranes of each fraction were solubilized and precleared, and H-2 antigen was immunoprecipitated. The immunoprecipitates were washed extensively with phosphate-buffered saline, and solubilized in NCS tissue solubilizer (Amersham), and radioactivity was determined directly. The specific activity plotted as cpm/mg of protein refers to cpm in the immunoprecipitate per mg of protein of pooled, starting cell fractions from eight individual mice. Reproduced from Croze and Morré, 1981 with permission from the National Academy of Sciences.

A nucleoside diphosphatase from rat liver endomembranes also exhibited similar flow kinetics with the same labeling and turn over order as observed for H-2 antigens (Fig. 6.21).

Autoradiographic studies with labeled galactose in mucin secreting cells, e.g., goblet cells of the rat colon, provided one of the first estimates of the flow rate of membrane precursors from the Golgi apparatus to the plasma membrane

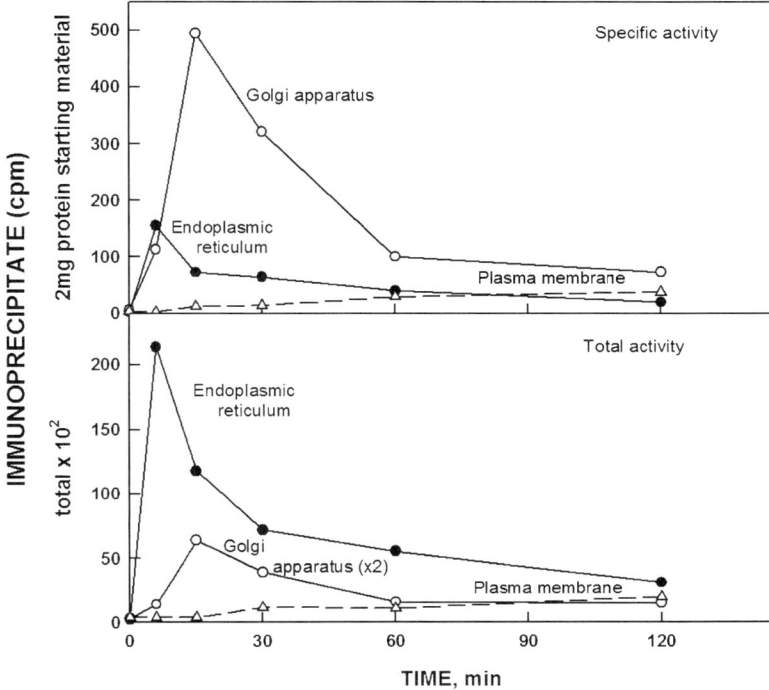

Fig. 6.21. Kinetics of incorporation and rapid turnover of radioactivity from [^{14}C-guanidol]-L-arginine into immunoprecipitated and gel-purified nucleoside diphosphate phosphatase (band 2) of rat liver. In the *upper* figure (A) specific activity refers to comp in the immunoprecipitate per 2 mg protein of pooled starting fraction. In the *lower* figure (B), data are adjusted for total amount of starting fraction from 10 g liver. Reproduced from Eppler and Morré, 1982 with permission from Elsevier.

(Neutra and Leblond, 1966a, b). They estimated that every two to four min, a distal Golgi apparatus cisterna was converted into mucin granules for transport to the cell surface (Chapter 8). In more recent studies, other sugars including *N*-acetylmannosamine, has been utilized as specific precursors to monitor sites of sialic acid incorporation and migration to the cell surface. Results indicate that sialic acid residues are added in the Golgi apparatus with no significant direct incorporation at the cell surface. Transfer to the plasma membrane, again, seems to be via Golgi apparatus vesicles. A similar conclusion has been reached from cell fractionation studies (Fig. 6.19; Chapter 3).

6.6.3. Bulk Flow of Membrane Lipids

Pulse-labeling and turn-over experiments with [^{3}H]glycerol in rat liver (Glauman et al., 1975) and *Acanthamoeba* (Chlapowski and Band, 1971) have suggested the possibility of bulk flow (as opposed to delivery via cytosolic phospholipid exchange or transfer proteins) of some membrane lipids from endoplasmic reticulum to Golgi apparatus to plasma membrane. These studies were not definitive and do not rule out other forms of transfer and/or de novo synthesis. More recent evaluations of bulk lipid flow within the endomembrane system are lacking. Incorporation of [^{35}S] into sulfolipid of Golgi apparatus and plasma membrane fractions of kidney more clearly demonstrated synthesis in the Golgi apparatus and subsequent transfer to plasma membrane (Fleischer et al., 1974; Zambrano et al., 1975).

Thus, for rat liver, the total time required for formation and maturation of a single cisternae of the Golgi apparatus can be estimated from labeling and turnover of Golgi apparatus membrane proteins to be 15 to 20 min (Table 6.5). Since there are about 4 cisternae per dictyosomal stack, this means that a cisterna is converted into vesicles on the average of one every 4 to 5 min with the concomitant buildup of a new cisterna at the opposite face. This is approximately the rate of cisternal buildup observed in carrot cells following a monensin block (Fig. 6.17). This is slightly slower than the rate of cisternal release observed by light microscopy for the scale-forming alga *Pleurochrysis scherffelii* of one in slightly less than every 2 min and somewhat faster than that observed for another scale-forming alga *Hymenomonas carterae* of one cisternae every 6 min. However, all estimates point to a rate of cisternal turnover of one every 1 to 6 min, with 4 min being near the mean. This indicated that the average life of a Golgi apparatus stack during active membrane flow is about 20 min.

If the Golgi apparatus is in such a rapid state of flux, what is the purpose of the cisternal stack? We do not know the answer to this question but one possibility is that it serves as a convenient form of holding pattern to permit the completion of membrane maturation from ER-like to plasma membrane-like. New cisternae are formed at one face as they are gradually displaced toward the opposite or maturing face over the 20 min life time of the Golgi apparatus cisternae, the necessary membrane modification that precede secretory vesicle formation can take place.

6.6.4. Evidence from Induced Systems

A few additional examples exist where a membrane constituent has been induced to form and the appearance of this new constituent used to mark the progress of membrane flow. Drugs that induce the appearance of the NADPH-like mixed function oxidases of drug detoxification provide a thoroughly investigated example. Administration of phenobarbital and other drugs to animals causes a marked proliferation of smooth endoplasmic reticulum with elevated drug-detoxifying activities. The lag time between administration of the drug and the first appearance of the induced enzyme in endoplasmic reticulum was 10 to 20 min and 7 min later appeared in the Golgi apparatus (Fig. 4.9; Table 4.1).

Coat proteins of viruses that are induced following infection become incorporated into the plasma membrane of host both for enveloped viruses that acquire their envelopes by budding through the plasma membrane as well as those that do so by budding through the nuclear envelope or other internal membranes (Appendix Table 4).

In all examples thus far studied, the viral proteins and glycoproteins reaching the cell surface appear to do so via a flow mechanism (Fig. 6.13). The order of appearance of these induced membrane proteins is nuclear envelope (endoplasmic reticulum), smooth internal membrane (= Golgi apparatus?), and plasma membrane. The kinetics of passage through these various compartments is similar to that observed for intrinsic membrane proteins normally made by the cells.

6.6.5. Energetics of Membrane Flow-Differentiation and Problems of Regulation

There have been almost no studies on the nature of the driving forces or transducing mechanisms whereby chemical potential is converted into a vectorial lateral displacement of membrane constituents or vesicles during the flow-differentiation of membranes. In contrast to membrane differentiation, flow dynamics have not been studied in cell-free systems and the possibility of doing so are not readily apparent. It has been possible to demonstrate the transfer of the G protein of vesicular stomatitis virus from a compartment where it is sensitive to hexosaminidase B to a compartment where it becomes hexosaminidase B resistant (Chapter 10) but it is not certain that this is not a processing response in contrast to some mechanism of vesicular transfer. Clearly, the question of regulation of membrane flow and differentiation must await more information concerning mechanisms.

6.7. Summary

The flow-differentiation or membrane maturation model of Golgi apparatus function (Fig. 6.22) embodies basic concepts of cisternal formation at one face of the Golgi apparatus from membranes originating at the endoplasmic reticulum and utilization of cisternae in vesicle formation at the opposite face for delivery to the plasma membrane as existing cisternae are displaced from one position within the stack to another. Dynamic aspects indicated initially by autoradiography and cell fractionation studies have been corroborated by newer approaches of fluorescent labeling and with living cells.

Much remains to be learned about the mechanisms of flow-differentiation of membranes and their regulation, however. Only a few examples at each of the various levels show the types of mechanisms (retention, addition, deletion, etc.) which may operate to affect the synthesis and differentiation of membranes and to effect the ultimate conversion of endoplasmic reticulum membranes to plasma membrane. In concert with membrane differentiation, membrane flow (= flow-differentiation of membranes) emerges as a fundamental and general cellular process critical to growth, development and homeostasis soundly based on both morphological and biochemical observation of animal, fungal and plant cells. Bulk migration of membranes or of extended membrane domains helps account for the delivery of new membranes during secretion and growth, in the renewal of cell surfaces during development or in response to environmental stimuli, or to replace existing membranes lost to degradation and turnover. The endoplasmic reticulum is the principle site of biogenesis of membrane materials, especially proteins and phospholipids. All available evidence suggests that these materials

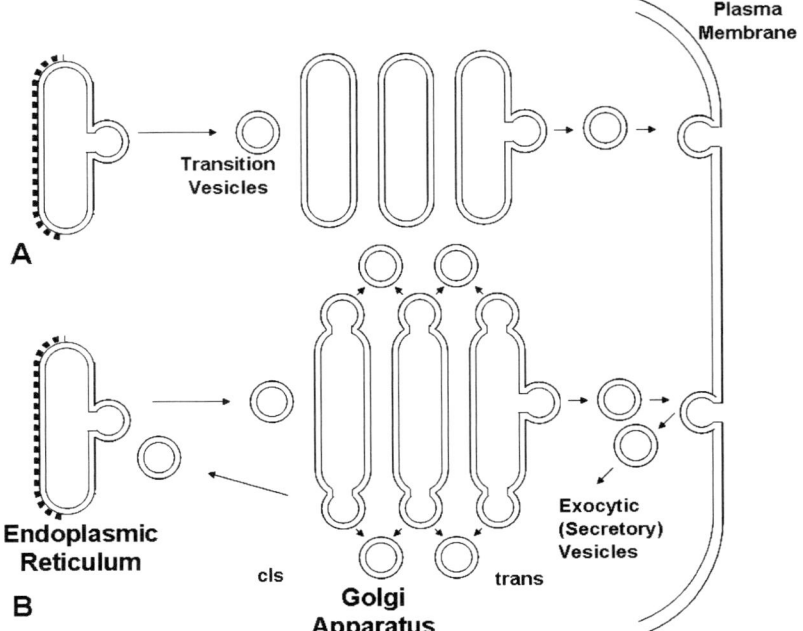

Fig. 6.22.   Comparison of the flow differentiation (A) and the vesicle shuttle (B) models of intra-Golgi apparatus transport. (A) According to the flow differentiation model, cisternae are formed at the *cis* face of the Golgi apparatus by coalescence of transition vesicles that are derived from the endoplasmic reticulum. The membranes are then modified at successive levels within the Golgi apparatus stack. Eventually, the membrane is discharged through formation of secretory vesicles and the release of a cisternal remnant. Consequently, the entire Golgi apparatus turns over, with the buildup of new cisternae at the *cis* face and the release of mature plasma membrane-like units at the *trans* face. (B) According to the vesicle shuttle model, the Golgi apparatus does not turn over in its entirety. Rather, migration through the system is accomplished by shuttle vesicles that move material from the endoplasmic reticulum to the Golgi apparatus and back and from one cisterna to the next and back, within the Golgi stack. The vesicle membrane that was discharged to the plasma membrane would be returned by compensatory endocytosis. (Modified from Morré and Keenan, 1997). Reproduced from Morré and Mollenhauer (2007) with permission of Elsevier.

are transferred from endoplasmic reticulum to other cytoplasmic membranes via transition elements or small transition vesicles derived from such transition elements. The transfer process appears to be selective so that some constituents are transferred while others are not. The operation of vesicular transfer mechanisms does not exclude other mechanisms, e.g., carrier proteins for phospholipids as one example, from operating in a parallel fashion to effect similar transfer, for example, to mitochondria. Additionally, some constituents, e.g., glycosyltransferases and products of glycosyl transfer, appear to be added at the Golgi apparatus or secretory vesicles while other constituents may be altered or removed. Membrane flow contributions to plasma membrane formation are especially obvious in epithelial cells of lactating mammary glands, or elongating neurons or pollen tubes, where contributions of secretory vesicles to the cell surface during growth are not only obvious but considerable.

7

Biochemistry

7.1. Introduction

This chapter reproduces in modified form a contribution on Golgi apparatus biochemistry prepared for the centennial year of Golgi apparatus discovery by Thomas W. Keenan (Keenan, 1998). The chapter contents overlap topics of other chapters in the book but many are given in greater detail and from a different prospective such that the sense of the original text and references has been retained.

Biochemical studies of the Golgi apparatus were possible only when isolated fractions in useful yield and fraction purity were developed in the late 1960s and early 1970s first from plants (Morré and Mollenhauer, 1964) and rat liver (Morré et al., 1969, 1970a) and later from other tissues and eventually from cultured cells. Detailed methodology is provided in Chapter 3. Early studies focused on marker enzymes as a means of assessing fraction purity and enrichment and glycosyltransferases, the latter already inferred from an abundant literature reporting cytochemical and electron microscope–autoradiographic studies that revealed the Golgi apparatus to be a site of protein glycosylation (Peterson and LeBlond, 1964).

Schneider and Kuff (1954; see also Kuff and Dalton, 1959) were the first to report isolated Golgi apparatus but insufficient material was obtained to permit biochemical characterization. The first Golgi apparatus fractions in useful yield and fraction purity were from plants (Morré and Mollenhauer, 1964; Morré et al., 1965). The fractions were found to be enriched in CDP-choline cytidyl transferase activity (Morré et al., 1965) and these findings paved the way for the use of this enzyme as a plant Golgi apparatus marker and for a central role of the Golgi apparatus in phospholipid metabolism (Morré, 1970b). Morré and colleagues (Morré et al., 1969, 1970a; Morré, 1971) next reported a method for isolation of Golgi apparatus from rat liver that yielded fractions of intact Golgi apparatus stacks of greater than 80% purity, as assessed both by morphometry and marker enzyme assay. In this method, low-shear homogenization was combined with rapid isolation of the Golgi apparatus from the homogenate. The low-shear homogenization minimized fragmentation of the Golgi apparatus, enabling the stacked Golgi apparatus cisternae to be enriched by differential centrifugation at moderate centrifugal force. Final purification was then achieved by a brief centrifugation over a simple sucrose density gradient. Rapid isolation/purification minimized unstacking and alterations of cisternal morphology and reduced loss of peripheral tubules (Morré, 1971). Golgi apparatus fractions isolated by this method resembled closely in their ultrastructural characteristics, the Golgi apparatus in situ (Morré et al., 1970a). The general method has been used for isolation of Golgi apparatus from a variety of animal and plant tissues (Fleischer et al., 1969 (rate zonal), Leelavathi et al., 1970; Schachter et al., 1970; Fleisher et al., 1974; Hino et al., 1978a) (isolation of unstacked cisternae).

Alternatively, methods were developed that relied on isolation first of a microsomal fraction, and then separation of Golgi apparatus elements from the microsomal fraction. Smooth microsomes separated from the total microsomal fraction of pancreatic acinar cells were enriched in elements of the Golgi apparatus (Meldolesi et al., 1971). Ehreinreich et al. (1973) incorporated a step that involved dosing rats with ethanol prior to animal sacrifice. The ethanol increased the number of lipoprotein particles in the Golgi apparatus elements thereby decreasing their density. Three fractions of Golgi apparatus-derived elements, corresponding to lipoprotein particle-loaded vesicles, lipoprotein particle-loaded cisternae, and cisternae free of lipoprotein particles were isolated from a microsome fraction of livers from the ethanol-intoxicated rats. Other density-based separations have been applied to obtain light and heavy Golgi apparatus of fractions (Deutscher et al., 1983) and further separation of light fractions into tubule and lipoprotein particle-containing secretory vesicle subfractions (Merritt and Morré, 1973; Ovtracht et al., 1973). Subfractionation of Golgi apparatus along the polarity axis (i.e., for separation of cis-, medial-, and trans-cisternae) has been accomplished by preparative free-flow electrophoresis (Morré et al., 1983b, 1984a). Golgi apparatus fractions isolated by a rapid method of Morré et al. (1972) were unstacked by incubation with a mixture of α- and β-amylases to remove previously added stabilizing dextrans (Chapter 3) and mild mechanical shear. Unstacked preparations were then separated into a series of fractions of increasingly electrophoretic mobility, with cis-most cisternae having the lowest mobility and trans cisternae having the greatest mobility. The basis for the electrophoretic separation has been traced to cis to trans gradient in a diffusion potential, electronegative on the outside, rather than to negatively charged surface molecules located at the internal surface of the cisternae and having little or no influence on electrophoretic mobility (Morré et al., 1994a).

7.2. Enzymology of the Golgi Apparatus

Cytochemistry provided a sound basis for predicting several enzymatic activities that would be found in Golgi apparatus fractions. Thiamine pyrophosphatase was detected based on cytochemical detection of electron-dense deposits of lead phosphate formed from thiamine pyrophosphate hydrolysis in the presence of lead nitrate (e.g., Allen and Slater, 1961; Novikoff and Goldfisher, 1961; Fig. 7.1) especially in elements at the trans-most part of the cisternal stacks (Goldfischer, 1982). Approximately 10% of the total thiamine pyrophosphatase activity was localized in Golgi apparatus fractions (Cheetham et al., 1971). Both by direct assay, and cytochemistry a cis to trans gradient of thiamine pyrophosphatase activity was found in rat liver Golgi apparatus fractions isolated by preparative free-flow electrophoresis.

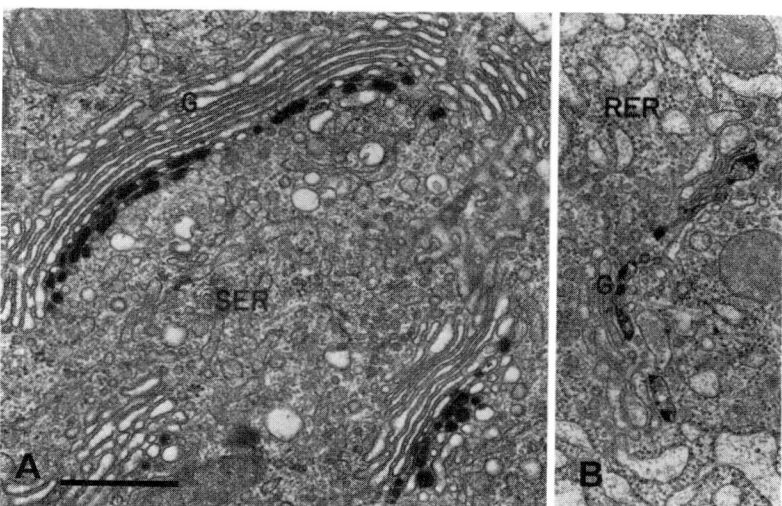

Fig. 7.1. Golgi regions of a rat epididymal epithelial cell (A) and a pancreatic acinar cell (B) prepared for the cytochemical demonstration of TPPase. The reaction product is localized specifically within the inner, concave cisternae of the Golgi apparatus. Reaction product is absent in regions of smooth- (SER) and rough (RER)-surfaced endoplasmic reticulum in these cell types. Reproduced from Cheetham, Morré, Pannek and Friend, 1971, Journal of Cell Biology, 49:899–905. Copyright 1971 The Rockefeller University Press. Scale bar = 0.5 μm.

From ultrastructural characteristics, it was predicted that Golgi apparatus were involved in membrane differentiation, specifically in transforming membranes from endoplasmic reticulum-like to plasma membrane-like (Grove et al., 1968; Morré et al., 1971d, 1974a; Morré and Mollenhauer, 1974a, b; Chapter 6). This prediction would require that isolated Golgi apparatus contain enzymes found in both endoplasmic reticulum and in plasma membranes. Numerous enzymatic activities characteristic of endoplasmic reticulum or plasma membranes are present in isolated Golgi apparatus in specific activities too high to be ascribed to contaminating membranes (Cheetham et al., 1970). Included are ion-stimulated nucleoside triphosphatases, nucleoside di- and monophosphatases, acid phosphatase, glucose-6-phosphatase, arylsulfatases, NADH dehydrogenases, and several enzymes involved in lipid biosynthetic pathways (reviews Morré et al., 1971a, 1974a, b). In subfractionation studies with rat liver enzymes enriched in the endoplasmic reticulum are associated primarily with the more dense and less electrophoretically mobile Golgi apparatus elements whereas those enzymes typical of plasma membrane are more concentrated in less dense, more electrophoretically mobile Golgi apparatus elements (e.g., Bergeron et al., 1973; Bergeron, 1979; Hino et al., 1978b; Morré et al., 1984a). Alkaline phosphatase also is present in Golgi apparatus (Tokumitsu and Fishman, 1983).

The two exo-α-D-mannosidases known as mannosidases I and II, enzymes that remove mannose residues from the $Glc_{1-3}Man_{4-9}GlcNAc_2$-N-linked oligosaccharide chains of glycoproteins also are Golgi apparatus located (Tabas and

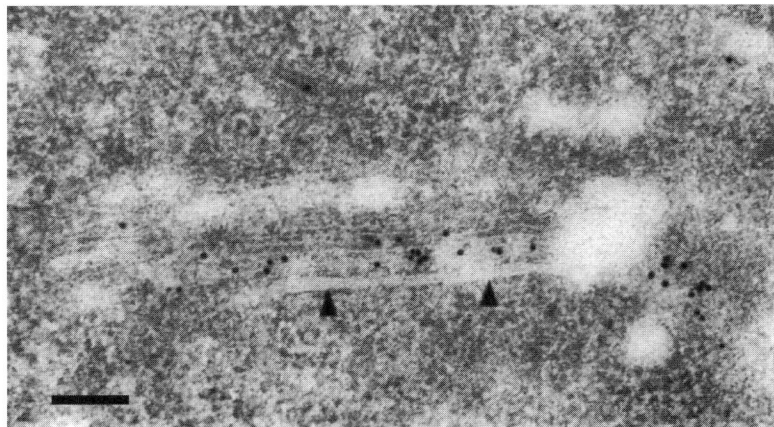

Fig. 7.2. Immunoelectron microscopic demonstration of mannosidase II in a rat liver hepatocyte. Immunolabeling for mannosidase II, as indicated by the highly electron-dense gold particles, is present throughout the cisternal stack and also observed in the trans-tubular network (or trans-Golgi network) of the GA. Ultrathin frozen tissue section, protein A-gold technique. Reproduced from Roth, 1997 with permission from Springer-Wien. Scale bar = 0.2 μm.

Kornfeld, 1979; Tulsiani et al., 1982; Fig. 7.2). G-glycoprotein processing endo-α-mannosidases that release Glc_{1-3}-Man oligosacchrides are found in Golgi apparatus (Spiro, 1994) as has tyrosylprotein sulfotransferase (Huttner, 1987). The latter enzyme concentrated at the trans Golgi apparatus face interacts with a large number of proteins, many but not all of which are secretory proteins which have been found to undergo modification by sulfation of one or more tyrosine residues. Consistent with this function, Golgi apparatus from rat liver have been found to translocate adenosine 3′-phosphate 5′-phosphosulfate (PAPS, the donor of sulfate for sulfotransferase reactions) (Schwarz et al., 1984). Furin, a serine endoprotease expressed in numerous cells and tissues that cleaves precursor proteins on the carboxyl side of Arg–X–Lys and Arg–Arg sequences is found with Golgi apparatus (Thomas, 1994). Furin is involved in processing of several secretory and plasma membrane proteins including proalbumin in hepatic Golgi apparatus (Misumi et al., 1991). Two other proteases closely related to furin designated PC2 and PC3 have been found within secretory granules of neuroendocrine tissues (Smeekens, 1994).

A NADH dehydrogenase activity of Golgi apparatus, assayed with cytochrome c or ferricyanide as the electron acceptor was 25 to 30% of the specific activity of endoplasmic reticulum (Jarasch et al., 1979; Huang et al., 1979). This activity was insensitive to inhibition by rotenone and other inhibitors of mitochondrial respiration (Jarasch et al., 1979). An NADH dehydrogenase was confirmed as a constituent of the Golgi apparatus by cytochemistry (Morré et al., 1974a, b, 1978b; Fig. 7.3) and by subfractionation of Golgi apparatus (Morré et al., 1974; Hino et al., 1978b; see also Borgese and Meldolesi, 1980). Cytochrome b_5, cytochrome P-450 and cytochrome P-420 of the microsomal electron transport system have been found in Golgi apparatus at levels of about 30% of those

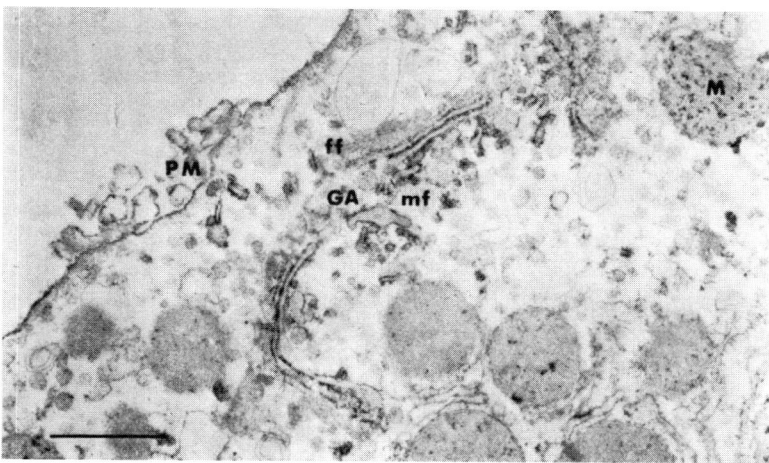

Fig. 7.3. Cytochemical detection of NADH ferricyanide reductase of hepatocyte Golgi apparatus based on formation of deposits of copper ferrocyanide. A differentiation between reaction products was localized to the plasma membrane (PM), transface cisternae of the Golgi apparatus and mitochondria (M). Reproduced from Morré, Vigil, Frantz, Goldenberg and Crane, 1978a with permission from Wissenschaftliche Verlagsgesellschaft mbH. Scale bar = 0.5 μm.

found in endoplasmic reticulum (Jarasch et al., 1979). The NADH dehydrogenase of Golgi apparatus, and of clathrin-coated vesicles, from rat liver can utilize monodehydroascorbate (ascorbate free radical) as an electron acceptor (Sun et al., 1984).

Adenosine triphosphatases (ATPases), especially monovalent cation-stimulated ATPases, long have been known to be plasma membrane constituents involved in ion transport. ATPases associated with Golgi apparatus have certain specific functions in this cell component. There is evidence that an ATPase of Golgi apparatus functions as a proton pump, resulting in acidification of forming or formed secretory vesicles (Zhang and Schneider, 1983; Barr et al., 1984). Proton translocation across the trans Golgi apparatus cisternae is driven not only by ATP but by hydrolysis of inorganic pyrophosphate as well (Brightman et al., 1992). Golgi apparatus-derived vesicles from intestine accumulate calcium, and this uptake is vitamin D-dependent (Freedman et al., 1977; Walters and Weiser, 1984). Golgi apparatus preparations from liver (Hodson, 1978) and mammary gland (Baumrucker and Keenan, 1975; West, 1981; Walters, 1984; Virk et al., 1985) were found to have a calcium-stimulated ATPase that drives calcium accumulation into the membrane-bounded compartment. Milks of many species are high in calcium (for example, cow's milk is 30 mM in calcium) and much of this calcium in bound to caseins, a group of phosphorylated proteins found in milk (Jenness, 1988). Thus, a specific function of the calcium pump ATPase of Golgi apparatus of milk-secreting cells apparently is to accumulate calcium for secretion. Histochemistry has confirmed that calcium is concentrated within cisternae and secretory vesicles of the Golgi apparatus from lactating mammary gland can accumulate citrate, an abundant anion of milk, again a concentration gradient (Zulak and Keenan, 1983). But the mechanism by which citrate is accumulated has not been explored.

ATP is transported into lumenal compartments of the Golgi apparatus, and there is evidence that this translocation of ATP is coupled to the exit of AMP from the lumen (Capasso et al., 1989). ATP in the lumenal compartments of Golgi apparatus serves as a donor of phosphate for phosphorylation of proteins (Capasso et al., 1989). While phosphorylation of secretory proteins in Golgi apparatus has been studied little in comparison with the numerous studies of phosphorylation and dephosphorylation of cascade enzymes and molecules in signaling pathways, several phosphorylated secretory proteins are known. These include vitellogenin, some proteoglycans, and the caseins of milk (for references see Capasso et al., 1989). Caseins have been shown to be phosphorylated by protein kinase(s) resident in Golgi apparatus and with active sites facing the lumenal spaces (e.g., Bingham et al., 1972; MacKinlay et al., 1977; Szymanski and Farrell, 1982).

Golgi apparatus has been shown to stain intensely with carbohydrate-selective staining reagents (e.g., LeBlond, 1950; van Heyningen, 1965; Rambourg et al., 1969). Electron microscope–autoradiographic studies verified that sugars were added to proteins as they transversed the Golgi apparatus (LeBlond and Bennett, 1977) suggestive of the presence of glycosyltransferases. Indeed, some glycosyltransferases were sufficiently concentrated in Golgi apparatus preparations to serve as absolute markers for Golgi apparatus purification. Among these were galactosyl- (Morré et al., 1969; Fleischer et al., 1969) and N-acetylglucosaminyl- (Morré et al., 1969; Wagner and Cynkin, 1969) and sialyl (Schachter et al., 1970) transferases from liver. The galactosyltransferase for synthesis of lactose was enriched in Golgi apparatus from lactating mammary gland (Keenan et al., 1970).

Subsequently a number of sugar nucleotide: glycoprotein glycosyltransferases were found to be concentrated in Golgi apparatus (Fig. 7.4). All three glycosyltransferases that add the trisaccharide unit terminating the oligosaccharide chain of many secreted glycoproteins were concentrated in Golgi apparatus. These transferases were UDP-GlcNAc transferase, UDP-Gal transferase, and CMP-NeuNAc transferase (Schachter et al., 1970; Munro et al., 1975). In addition, GDP-fucose:glycoprotein glycosyltransferase also was found to be enriched in Golgi apparatus fractions (Munro et al., 1975). Evidence was obtained that these glycosyltransferases had different distributions with respect to the polarity axis of the cisternae comprising Golgi apparatus stacks. N-acetylglycosaminyl transferase, the first enzyme to act in adding the terminal trisaccharide unit to many glycoproteins, was concentrated in cisternae derived from cis or medial regions of the stack. Galactosyl- and neuraminyltransferases, which act after N-acetylglucosamine is added, were concentrated in cisternae from trans regions (e.g., Dunphy et al., 1985; Bretz et al., 1980; Elhammer and Kornfeld, 1984; Hartel-Schenk et al., 1991; Morré et al., 1983a). Subfractionation studies also led to the demonstration that the UDP-N-acetylglucosaminyltransferase involved in biosynthesis of the phosphomannosyl recognition marker in lysosomal enzymes (Pohlmann et al., 1982; Deutscher et al., 1983) and the mannose-6-phosphate receptor for lysosomal enzymes (Brown and Farquhar, 1984; Fig. 7.5) was concentrated in cis Golgi apparatus cisternae.

In plant cells, the assembly of cell wall matrix polysaccharides is a principal function of the Golgi apparatus (reviewed by Driouich and Staehelin, 1997; Staehelin and Moore 1995). N- or O-linked oligosaccharide sidechain-containing

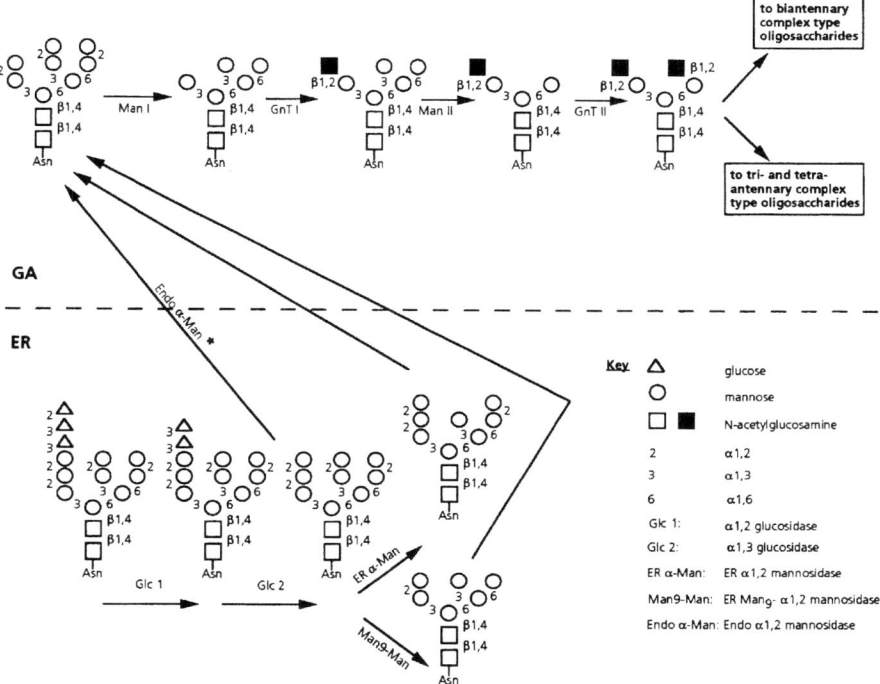

Fig. 7.4. Schematic presentation of processing reactions in the endoplasmic reticulum (ER) and Golgi apparatus (GA) for asparagine-linked oligosaccharides. Endo-α-mannosidase converts $Glc_1Man_{9-4}GlcNAc_2$-oligosaccharides to $Man_{8-3}GlcNAc_2$ oligosaccharides, the former being a processing intermediate of glucosidase II, is not indicated in the scheme. Reproduced from Roth, 1997 with permission from Springer-Wien.

glycoproteins also are assembled within the Golgi apparatus of plant cells (reviewed in Driouich et al., 1994).

7.3. Glycosphingolipid Synthesis

Before methods for isolation of Golgi apparatus were developed, it was assumed that gangliosides (glycosphingolipids that contain one or more molecules of a sialic acid in their glycan chain) were present only in plasma membranes. However, it was found that gangliosides also are characteristic constituents of Golgi apparatus membranes (Keenan et al., 1972b; Matyas and Morré, 1987). Several glycosyltransferases involved in assembling the oligosaccharide chain of gangliosides and non-ganglioside (neutral) glycosphingolipids were found to be concentrated in liver Golgi apparatus (Keenan et al., 1974; Wilkinson et al., 1976; Richardson et al., 1977; Fleischer, 1977; Figs. 7.6 and 7.7). Several of these glycosphingolipid glycosyltransferases of Golgi apparatus subsequently were extensively purified and characterized (e.g., Kaplan and Hechtman, 1983; Melkerson-Watson and Sweeley,

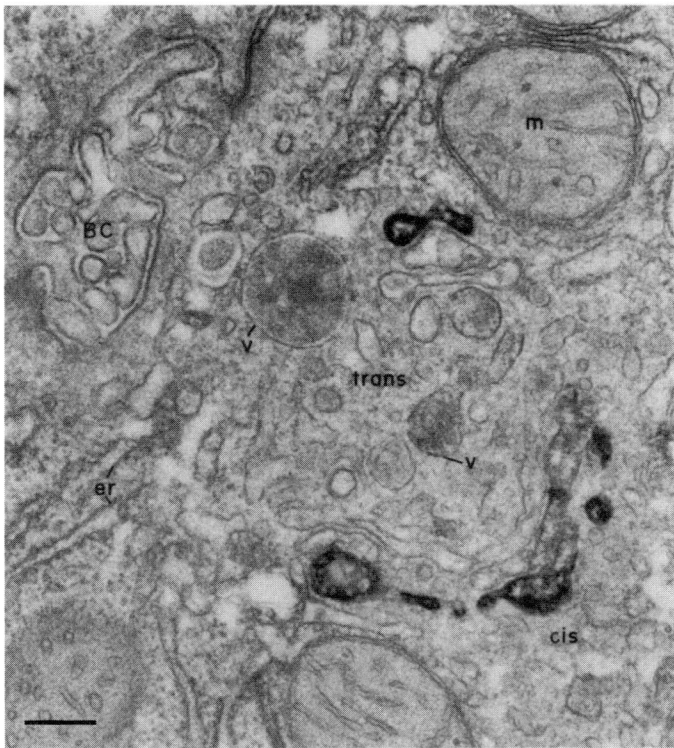

Fig. 7.5. Immunodetection of mannose-6-phosphate receptors in rat hepatocytes. The immunore-active receptors are concentrated in the cis-most Golgi apparatus cisterna. VLDL-filled secretory vacuoles (v), which do not contain reaction product, are present on the trans aspect. In hepatocytes one or several cis Golgi apparatus cisternae may contain immunoreactive receptors. BC, Bile canaliculus; M, mitochondria. Reproduced from Brown and Farquhar, 1984 with permission from MIT Press. Scale bar = 0.2 μm.

1991). There is evidence that single UDP-GalNAc and CMP-NeuNAc transferases of Golgi apparatus can catalyze addition of the respective sugar to different lipid acceptors (Pohlentz et al., 1988). Further, as with glycosyltransferases active with glycoproteins, there appears to be subcompartmentalization of glycosphingolipid sialyltransferases within different cisternae of Golgi apparatus (Trinchera and Ghidoni, 1989). Synthesis of both neutral glycosphingolipids and gangliosides requires the sequential glycosylation of the precursor ceramide N-acylated sphin-gosine. Whether all the glycosyltransferases involved in the biosynthetic pathway for the most complex glycosphingolipids are concentrated in Golgi apparatus remains to be established. There is evidence, however, that the glycosyltransferase catalyzing the first glycosylation reaction with ceramide also is a resident enzyme of the Golgi apparatus (Durieux et al., 1990; Futerman and Pagano, 1991). But, this glycosyltransferase may reside in other cell compartments as well (Warnock et al., 1994).

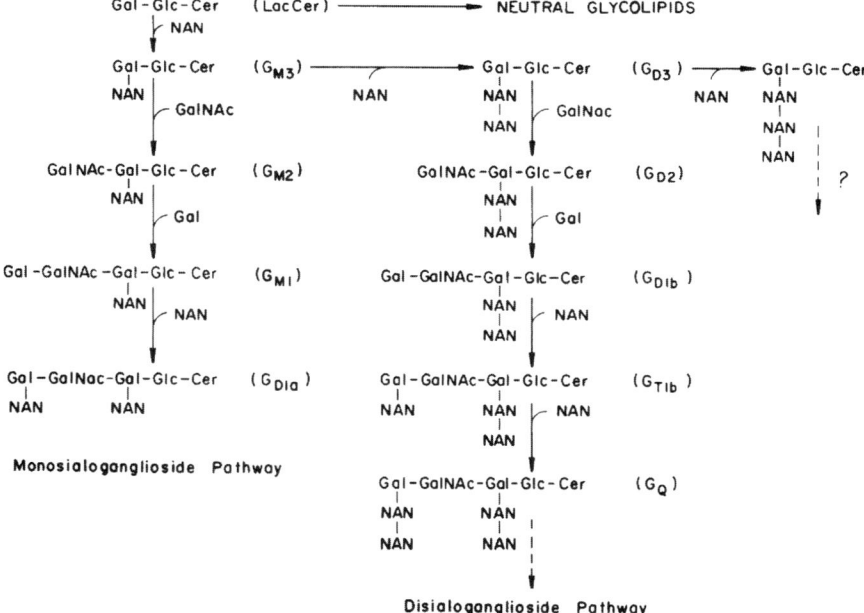

Gal   GalNAc

Gal

Glc

Cer

NAN

Fig. 7.6. Structure of the monosialoganglioside G_{MI}. The most complex of the monosialoganglio-sides, G_{MI}, consists of a ceramide moiety (Cer) linked through a glycosidic bond to an oligosaccharide. The oligosaccharide consists of glucose (Glc), galactose (Gal), N-acetylgalactosamine (GalNAc), and sialic acid (NAN = N-acetylneuraminic acid). Reproduced from Morré, Kloppel, Merritt and Keenan, 1978a with permission from John Wiley and Sons, Inc.

```
Gal-Glc-Cer     (LacCer) ──────────────────►  NEUTRAL GLYCOLIPIDS
      │ NAN
      ▼
Gal-Glc-Cer     (G_M3) ──────────────►  Gal-Glc-Cer   (G_D3) ──┬──►  Gal-Glc-Cer
  │                                          │                      │
 NAN          NAN                           NAN       NAN          NAN
      │ GalNAc                               │ GalNac                │
      ▼                                      ▼                      NAN
GalNAc-Gal-Glc-Cer   (G_M2)          GalNAc-Gal-Glc-Cer   (G_D2)     │
  │                                          │                      NAN
 NAN                                        NAN                      │
      │ Gal                                  │ Gal                   ▼   ?
      ▼                                      ▼
Gal-GalNAc-Gal-Glc-Cer   (G_MI)      Gal-GalNAc-Gal-Glc-Cer   (G_Dlb)
  │                                          │
 NAN         NAN                            NAN       NAN
      │                                      │
      ▼                                      ▼
Gal-GalNac-Gal-Glc-Cer   (G_Dla)     Gal-GalNAc-Gal-Glc-Cer   (G_Tlb)
  │            │                             │         │
 NAN          NAN                           NAN       NAN    NAN
                                                      │
Monosialoganglioside  Pathway                        NAN
                                                      ▼
                                     Gal-GalNAc-Gal-Glc-Cer   (G_Q)
                                       │            │
                                      NAN          NAN
                                       │            │
                                      NAN          NAN
                                                   ▼
                                     Disialoganglioside  Pathway
```

Fig. 7.7. The ganglioside biosynthetic pathways of rat liver. Cer, ceramide (N-acylspingosine); Gal, galactose; GalNAc, N-acetylgalactosamine; NAN, N-acetylneuraminic acid (sialic acid). Based on studies of Keenan et al. (1974) and Merritt et al. (1978a, b). Reproduced from Morré et al., 1978a with permission from John Wiley and Sons, Inc.

7.4. Nucleotide Sugar Transporters

Glycosylation of proteins and lipids within the lumen or on lumenal membrane surfaces requires not only the glycosyltransferases, but also the sugar nucleotide donors. Further, the nucleotide product of the glycosyltransferase reaction must be transported out of the membrane-bounded compartment. That the nucleotide product must leave the Golgi apparatus compartment was inferred from the fact that, at least in some cases, this product of the transferase reaction is inhibitory to glycosyltransferases. In studying lactose synthesis by Golgi apparatus from mammary gland, Kuhn and White (1976) obtained evidence for specific transport of UDP-Gal across the Golgi apparatus membrane; the UDP formed upon transfer of galactose was degraded to UMP and transported back out of the lumenal compartment (Kuhn and White, 1977). Brandan and Fleischer (1982) found that the nucleoside diphosphatase of Golgi apparatus has its active site exposed on the lumenal face of the membrane, that it functioned to hydrolyze UDP to UMP and Pi, and that the UMP formed moved rapidly out of the lumen of the Golgi apparatus. Much is now known about transport of nucleotide sugars into Golgi apparatus, and transport of nucleotides and nucleosides out of the Golgi apparatus compartment (reviews, Perez and Hirschberg, 1986; Hirschberg and Snider, 1987; Hirschberg, 1997a, b; Hirschberg et al., 1998; Fig. 7.8). Golgi apparatus membranes have specific transporters for UDP-galactose. UDP-N-acetylglucosamine, UDP-N-acetylgalactosamine, GDP-mannose, GDP-fucose, and

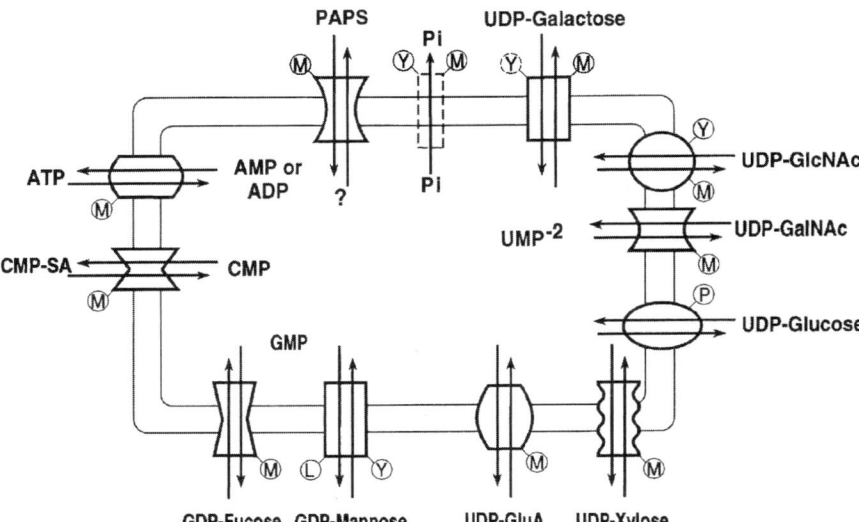

Fig. 7.8. Nucleotide sugar, nucleotide sulfate, and ATP transporters of Golgi apparatus membranes of mammals, yeast, protozoa, and plants. They are antiporters with the corresponding nucleoside monophosphate except for PAPS, and ATP, where it is either AMP, ADP, or both. M, mammals; Y, yeast; P, plants; L, *Leishmania*. Reproduced from Hirshberg, 1997b with permission from Springer-Wien.

CMP-sialic acid. Once the glycosyltransferase reaction has occurred, the nucleotide diphosphate products are hydrolyzed to phosphate and nucleotide monophosphates within the lumenal compartment. The transporters for at least some of the nucleotide sugars are part of an antiproton system, with entry of the sugar nucleotide coupled to discharge of the nucleotide monophosphate.

7.5. Golgi Apparatus Markers for Medial Cisternae

The so-called medial or intercalary cisternae from the mid region of the Golgi apparatus stack have had few markers which distinguish them from other Golgi apparatus compartments. One marker for medial cisternae of largely unknown function in the Golgi apparatus is the enzyme that releases inorganic phosphate from α-NADP (α-nicotinamide adenine dinucleotide phosphate) described cytochemically by Smith (1980) and subsequently from biochemical studies by Navas et al. (1986; Fig. 7.9).

7.6. Lipid Composition

Golgi apparatus membranes are about 40% lipids of the total lipids, phospholipids comprised about 55% and other nonpolar lipids about 35% (Yunghans et al., 1970). The distribution pattern of phospholipids was found to be intermediate between that of endoplasmic reticulum and that of plasma membranes (Keenan and Morré, 1970). Sphingomyelin accounts for about 4% of the total

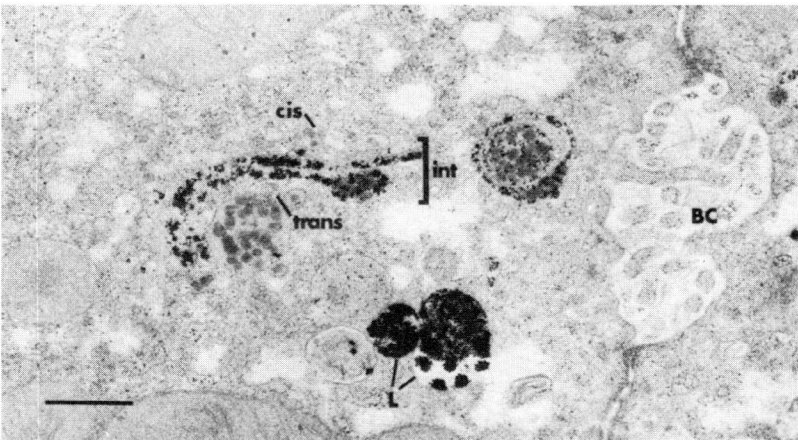

Fig. 7.9. Cytochemical localization of NADP phosphatase in Golgi apparatus of rat liver parenchyma. Reaction product was localized in two intermediate (int) or intercalary cisternae in the midregion of the stack. The cis-most (cis) and trans-most (trans) as well as all other cellular structures, except for lysosomes (L), were negative. BC, Bile canaliculus. Navas, Minnified, Sun and Morré, 1986, BBA, 881:1–9, Copyright 1986 with permission from Elsevier. Scale bar = 0.5 μm.

phospholipid and phosphatidyl choline accounts for about 60%. In contrast in the plasma membrane sphingomyelin accounts for about 20% of the total phospholipids and phosphatidylcholine is about 40%. Reflective of the intermediate phospholipid composition of the Golgi apparatus, sphingomyelin accounts for about 12% and phosphatidyl choline accounts for about 45% of the total phospholipid. The other major phospholipids, phosphatidyl serine (3 to 4%), phosphatidyl inositol (8 to 9%), and phosphatidylethanolamine (17 to 19%) are nearly the same for endoplasmic reticulum, Golgi apparatus, and plasma membrane. Higgins and Hutson (1984) found large differences in the accessibility of the head groups of phosphatidylcholine and phosphatidylethanolamine to membrane-impermeant reagents in cis- and trans-enriched Golgi apparatus fractions. This suggests that lipids may become reoriented in the plane of the membrane during membrane maturation across the stack.

In addition to relatively high levels of cholesterol, liver Golgi apparatus have large amounts of triacylglycerols (e.g., Keenan and Morré, 1970; Howell and Palade, 1982). Appreciable amounts of the triacylglycerols in liver Golgi apparatus fractions appear to be associated with lipoprotein (VLDL) particles of the lumenal compartments. While some of the cholesterol of liver Golgi apparatus is in the contents and not the membranes, it appears that Golgi apparatus membranes are enriched in cholesterol compared to the endoplasmic reticulum.

Bretscher and Munro (1993) have postulated that cholesterol enrichment in the Golgi apparatus serves as a potential mechanism to retain enzymes in the membranes of the Golgi apparatus. It had long been recognized that there was a correlative relationship between the sphingomyelin and cholesterol contents (Patton, 1970) and the total phospholipid and cholesterol contents (Razin, 1974) of cellular membranes. More recently it was recognized that cholesterol and sphingolipids (sphingomyelin, neutral glycosphingolipids, gangliosides) cluster to form microdomains (van Meer, 1989) or rafts (Simons and Ikonen, 1997) that move within the fluid bilayer of the membrane. Rafts appear to play roles in binding of certain proteins to membranes, and in membrane trafficking (Simons and Ikonen, 1997).

Ubiquinones, primarily ubiquinone-9 (Nyquist et al., 1970; Zambrano et al., 1975), vitamin A, primarily as retinol esters (Nyquist et al., 1971a), and vitamin K (Nyquist et al., 1971b) also occur in lipid extracts from Golgi apparatus. All of these lipid molecules were found to be concentrated in Golgi apparatus, relative to amounts found in other cell fractions, a role of Golgi apparatus peripheral tubules in ubiquinone synthesis and sequestration has been postulated (Mollenhauer and Morré, 1998).

7.7. Phospholipid Biosynthesis

Isolated Golgi apparatus synthesize a number of the phospholipids found in Golgi apparatus membranes. The terminal enzymes involved in biosynthesis of phosphatidylcholine, phosphatidylinositol, phosphatidylserine and phosphatidyl glycerols were present in Golgi apparatus preparations from liver. By studying the incorporation of lipid precursors, Higgens and Fieldsend (1987) obtained evidence that synthesis of phosphatidyl choline occurs in Golgi apparatus in vivo as well. Golgi apparatus from liver also has been shown to have phospholipase A activity concentrated in cis- elements, and acyltransferase activity located primarily in the cis- and medial regions (Moreau and Morré, 1991; Lawrence et al., 1994). These activities give to Golgi apparatus elements the ability to remodel the fatty acids of at least certain phospholipids.

The major pathway for sphingomyelin synthesis is by direct transfer of the phosphorylcholine group from phosphatidylcholine to ceramide involving the enzyme sphingomyelin synthase. The bulk of this enzymatic activity has also been found in Golgi apparatus (van Helvoort et al., 1997). The ceramide acceptor for sphingomyelin (and glycosphingolipid) synthesis presumably is transferred from endoplasmic reticulum, the site where ceramides are synthesized (van Meer, 1989). In cell-free systems, ceramides can be transferred from endoplasmic reticulum to Golgi apparatus, but in contrast to protein and phospholipid transfer, this ceramide transfer apparently is not via transition vesicles (Moreau et al., 1993). The possibility the Golgi apparatus may synthesize ceramides has not been explored rigorously. Ceramide synthesis and its stimulation by exogenously supplied D-sphingosine was demonstrated in homogenates and purified membrane fractions from rat liver, but Golgi apparatus was inactive (Walter et al., 1983). Significant D-sphingosine or ceramide-dependent formation of glycosylceramide, or glycosylceramide-dependent formation of lactosylceramide also could not be demonstrated with isolated Golgi apparatus.

7.8. Protein Composition of the Golgi Apparatus

Ongoing quantitative proteomic analyses have begun to reveal the detailed distribution and subcellular location of proteins within components of the secretory pathway (Gilchrist et al., 2006). With 1400 proteins analyzed, findings are consistent with the membrane maturation model of Chapter 6 along with previously validated concepts of Golgi apparatus resident proteins and of proteins transiently involved in various trafficking steps.

Thus, many, but not all, proteins that reach the plasma membrane transit through the Golgi apparatus. At least some of the membrane proteins of lysosomes transit through the Golgi apparatus as well (Brown and Farquhar, 1984). But, localization of specific enzymes, especially the glycosyltransferases, to defined regions of the Golgi apparatus made it evident that some membrane-associated proteins well may be retained for a time within the Golgi apparatus. Distinguishing between resident and transitory membrane proteins often is problematic. Insulin (Posner et al., 1978; Schilling et al., 1979), prolactin (Josefsberg et al., 1979), and acetylcholine (Fambrough and Devreotes, 1978) receptors all are present and functional in Golgi apparatus membranes. But, whether these receptors are in transit to plasma membrane or are resident to and play some biological role in Golgi apparatus is still uncertain. With some of the Golgi apparatus glycosyltransferases it was believed that the membrane-spanning domain, or specific amino acid sequences within the membrane-spanning domain, may serve as retention signals (Nilsson and Warren, 1994; Paulson and Colley, 1989). However, amino acid sequences of the large number of glycosyltransferases that have been cloned and sequenced (Stanley, 1994) provide no evidence for a common retention signal. Two other hypotheses already discussed to explain retention of the proteins with the Golgi apparatus were that Golgi apparatus proteins formed oligomers that were not mobile within the membrane, or that decreased membrane thickness due to increased cholesterol content led to immobilization of resident Golgi apparatus enzymes (Nilsson and Warren, 1994).

Several glycosylated intrinsic membrane proteins have been identified in Golgi apparatus (Capasso et al., 1988; Yuan et al., 1987). By immunostaining, some appear to be localized specifically to Golgi apparatus, and are segregated into different regions of the Golgi apparatus stack (Yuan et al., 1987). Peripheral proteins associated with Golgi apparatus bind to microtubules (Allan and Kreis, 1989; Bloom and Brashear, 1989; Torii et al., 1997). A protein designated GM130, based on its mass (130 kDa) and association with Golgi apparatus matrix proteins remaining after extraction with Triton X-100 and low salt, and localized to cis Golgi apparatus elements, binds to p115 vesicle docking protein (Nakamura et al., 1997).

Use of cell-free and permeabilized systems for reconstitution of steps in intracellular transit has led to identification of a large number of proteins that interact, at least temporally, with Golgi apparatus membranes (Chapter 10). Many of these studies have been with respect to vesicle-mediated anterograde transport from endoplasmic reticulum to cis regions of Golgi apparatus, between Golgi apparatus cisternae, and from trans regions of Golgi apparatus to the plasma membrane. There is abundant evidence for vesicular-mediated anterograde and retrograde transport between endoplasmic reticulum and Golgi apparatus, and transport from Golgi apparatus to plasma membrane. But, whether intra-Golgi apparatus anterograde transport involves free vesicles or coated ends of tubules has been questioned (Weidman, 1995; Morré and Keenan, 1997). Nonetheless, many of the identified proteins clearly play roles in transit, whether or not all transit steps occur via vesicles. Proteins identified as functioning in transit or retrieval and that interact, at least temporally, with Golgi apparatus membranes include the COP I and COP II proteins, which form coats at vesicles or membrane regions undergoing vesiculation. One of the COP I proteins is ADP-ribosylation factor (ARF), a ubiquitous small GTP binding protein and member of the Ras superfamily of GTP-binding proteins. Several members of the Rab family of small GTP-binding proteins, including rabs 1a, 1b, 2, 6, and 10 associate with Golgi apparatus elements as well (Zerial et al., 1994; Fig. 7.10). NSF (N-ethyl-maleimide-sensitive fusion protein) and SNAPs (soluble NSF attachment proteins) also associate peripherally with membranes of Golgi apparatus (Rothman, 1994). Also involved in transit, and associated with Golgi apparatus membranes, are a group of integral membrane proteins known collectively as SNAREs (SNAP receptor proteins), which serve as vesicle and target membrane-specific receptors (Rothman and Wieland, 1996). The list of transit-functional proteins associated with Golgi apparatus continues to increase. For example, Hay et al. (1997)

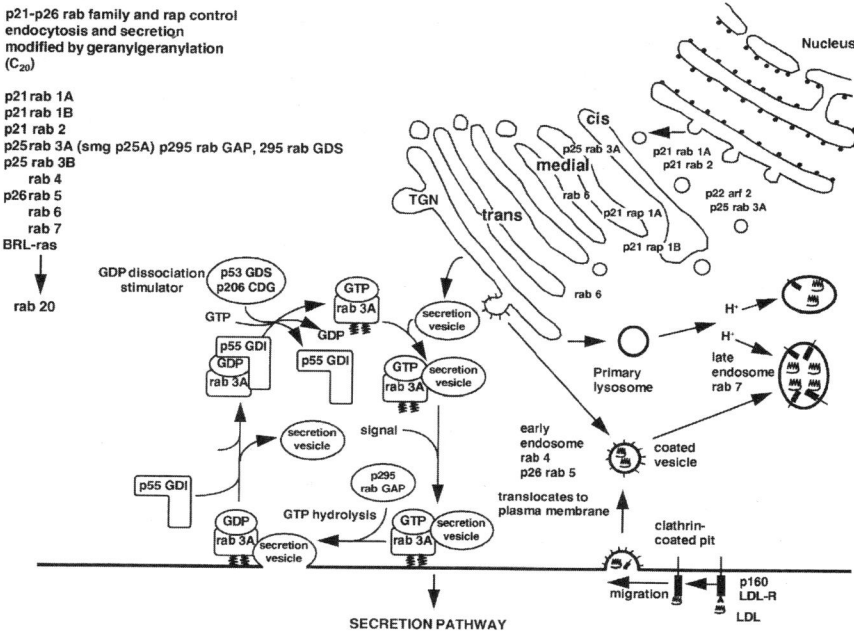

Fig. 7.10. Intracellular localization of some ubiquitously expressed Rab proteins. Rabla, Rablb, and Rab2 have been localized to and function in transport between the endoplasmic reticulum and GA. Rab6, Rab8, and Rab12 are associated with the GA. Rab6 is distributed from medial-Golgi to the trans-Golgi network, Rab8 is localized to the trans-Golgi network, post-Golgi vesicles, and the plasma membrane and regulates traffic between these compartments. In the endocytic pathway, Rab5a is associated with the plasma membrane, clathrin-coated vesicles, and early endosomes and regulates endocytic transport, whereas Rab4a, which is present on early endosomes, plays a role in recycling from early endosomes to the cell surface. Rab5b, Rab5c, and Rab4b appear to share the functional properties of Rab5a and Rab4a, respectively. Rab7 and Rab9 are both associated with late endosomes, but only Rab9 controls transport to the trans-Golgi network. Rab18, Rab20, and Rab22 have been localized to early endosomes, Rab24 is associated with both the Er/cis-Golgi region and late endosomes and is postulated to function in autophagy related processes. *Abbreviations*: ER, endoplasmic reticulum; CGN, cis-Golgi network; TGN, trans-Golgi network; EE, early endosomes; LE, late endosomes. *Arrows* specify membrane traffic between compartments.

identified two previously unrecognized integral membrane proteins of cis Golgi apparatus by virtue of their involvement in a vesicle–receptor system.

While there is a seemingly exponentially expanding list of proteins known to be associated, peripherally or integrally, with Golgi apparatus membranes (Fig. 7.11), few are permanent residents. Clearly one can envision at least two different temporal classes of Golgi apparatus membrane-associated proteins. One class would be those membrane proteins in transit to other locations like lysosomes and the plasma membrane. Another class would be those cytosolic proteins that associate with, and dissociate from, Golgi apparatus. Glycosyltransferases are usually included in any discussion of permanent protein residents of Golgi apparatus. But

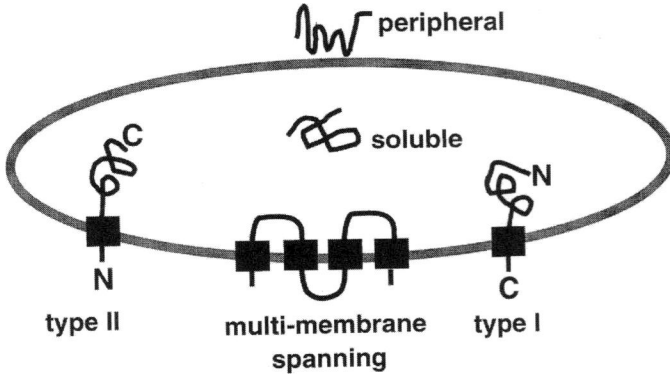

Fig. 7.11. Resident Golgi apparatus proteins include type I and type II membrane proteins, multimembrane-spanning proteins, peripheral membrane proteins, and soluble luminal proteins. Reproduced from Gleeson, 1998 with permission from Springer Science + Business Media.

whether glycosyltransferases truly are permanently resident in Golgi apparatus is not at all clear. Soluble forms of glycosyltransferases have been found in a number of secreted fluids, and there is some evidence that these soluble enzymes arise through proteolytic cleavage of the membrane-bound enzymes (Paulson and Colley, 1989). Thus, even glycosyltransferases whose physiological function is in the Golgi apparatus may not be permanent residents of Golgi apparatus membranes.

Since the Golgi apparatus has different functions than does endoplasmic reticulum and the plasma membrane, two of the cell components that the Golgi apparatus functionally interconnects, Golgi apparatus would be expected to contain proteins that are specific to or functional only in the Golgi apparatus. How these proteins arrive at and become functional in Golgi apparatus but are inactivated or excluded from transit to the plasma membrane has been little studied. In addition, a number of post translational modifications of proteins occur within the Golgi apparatus including glycosylation, sulfation, phosphorylation and proteolysis. On cell fractionation some cytosolic polysomes remained associated with the Golgi apparatus. Based on the products of run-off translations, proteins synthesized by these Golgi apparatus-associated polysomes were different, based on SDS-polyacrylamide gel mobility, than proteins synthesized on the bulk of the cytosolic polysomes or by polysomes bound to endoplasmic reticulum (Elder and Morré, 1976a). The possibility that nascent polypeptides destined for Golgi apparatus, or perhaps for the plasma membrane, may target polysomes to Golgi apparatus rather than plasma membrane, however, remains to be established.

7.9. Summary

The study of Golgi apparatus biochemistry was possible only with the availability of isolated Golgi apparatus fractions in useful yield and fraction purity. Enzymatic activities demonstrated from cytochemistry were confirmed and sugar nucleotide: glycoprotein glycosyltransferases were found to be concentrated in the Golgi apparatus as predicted from autoradiography and from electron microscopy with carbohydrate-selective staining reagents. A Golgi apparatus localization of enzymes for synthesis of glycolipids and for matrix polysaccharides followed in rapid succession. The requirement of sugar nucleotide donors at the luminal side of the Golgi apparatus was met by sugar nucleotide transporters. Lipid and protein compositions were determined along with roles in phospho- and sulfolipid biosynthesis. The findings of the biochemical era of Golgi apparatus investigation that began in the mid 1960s and extended into the late 1990s provided important preliminary evidence both for the compositional uniqueness of the Golgi apparatus membranes but also for an intermediate character reflecting both their origins from the endoplasmic reticulum and their final destinations at the plasma membrane.

8

Function in Secretion

The functioning of the Golgi apparatus as part of an integrated and interassociated endomembrane system can be appreciated most readily from the important role of the Golgi apparatus in the elaboration of protein secretions for export to the cell's exterior. The type of secretion that involves the Golgi apparatus is most appropriately referred to as exocrine secretion where materials elaborated within the cell are packaged into discrete transport packets or vesicles and discharged. Discharge is accomplished by fusion of the membranes of the vesicle with the surface membrane or plasma membrane of the cell.

To illustrate the importance of secretory mechanisms involving the Golgi apparatus, it is sufficient just to list some of the many products processed via the classical endoplasmic reticulum–Golgi apparatus–secretory vesicle–plasma membrane export route. The list includes not only many important digestive enzymes, blood constituents and hormones but most of the major food and fiber commodities of agriculture and commerce (except for starch). In this latter category are included fat and protein components of milk, cell wall materials, latex (the source of natural rubber), natural oils, collagen, and mucopolysaccharides of connectives tissues in meat, egg yolks, and storage lipids and proteins of seeds.

It is quite possible that secretory processes are one of the major factors that have allowed the success of multicellular organisms. Differentiated cells by definition become specialized to the point of being dependent upon other cells for essential constituents they no longer produce and benefit the organism as a whole by efficient production of constituents not produced by other cells. In order for the products of specialized cells to reach other cells requiring them, they must first be secreted from the producing cells.

8.1. Role of the Golgi Apparatus in Secretion

The role of the Golgi apparatus in secretion is manifold and varied. Among the specific tasks carried out are the following:

1. Receiving products of synthesis from other parts of the cell prior to their packaging for export (chiefly from the endoplasmic reticulum).
2. A concentration of secretory products. "Concentration" is usually understood to mean that the secretory material becomes visibly more dense and closely packed.

D. James Morré and Hilton H. Mollenhauer, *The Golgi Apparatus*.
© Springer 2009

3. Chemical modification (processing) of the materials to be secreted. Addition and removal of sugars from glycoproteins would be one example.
4. Synthesis of secretory materials. Particularly for pure polysaccharides secreted by many plant cells, the Golgi apparatus is the primary site of synthesis.
5. Sorting of secreted materials some of which may be routed to lysosomes or other types of vacuoles while other materials may be directed for export to the cell surface.
6. The membrane of the secretory vesicle is frequently plasma membrane-like in composition and cytochemical characteristics, and is capable of fusing with the plasma membrane to effect the discharge of secretory materials. In secretory cells, a major consequence of membrane flow-differentiation appears to be the elaboration by the Golgi apparatus of these specialized membranes that surround secretory vesicles.

8.2. A General Model for Golgi Apparatus Functioning in Secretion

The major transport vehicle for Golgi apparatus secretion is the secretory vesicle, secretory granule or condensing vacuole. These three terms are basically synonyms for the same structure but with strict temporal distinctions. Secretory vesicles are usually released continuously from the Golgi apparatus and appear to migrate rapidly and more or less directly to the plasma membrane as diagrammed in Fig. 8.1. The kinetic relationship between sites of synthesis (endoplasmic reticulum), site of packaging (Golgi apparatus) and delivery to the circulation are shown in Fig. 8.2 for albumin of rat liver.

Condensing vacuoles remain in the vicinity of the Golgi apparatus for some time while additional secretory materials are added. When filled, they are referred to as secretion granules. Mature secretion granules may remain in the cytoplasm for a considerable time often accumulating in large numbers. The content of the granules is usually brought about or at least accelerated in response to a stimulus such as a hormone.

An important property of secretion is that it is *vectorial*, i.e., there is net movement of the secreted product from sites of synthesis to the cell surface and usually to a particular region of the cell surface, i.e., that portion bounding a gland lumen or sinusoidal space. A second important property of secretion is that it is *compartmentalized*. Secreted materials are synthesized in such a manner that they accumulate at the luminal surfaces of internal membranes. They then remain within this so-called endoplasmic space until discharged. Free mixing with or even migration through the cytoplasm has been ruled out as a general possibility for secretory proteins, glycoproteins, polysaccharides and mucopolysaccharides.

Both vectorial and compartmentalized transfer of materials is achieved by an array of membrane-bounded vesicles as well as tubules of various sorts. Chief among these are the secretory vesicles or granules cited above that may operate over long distances but also included are the small transition vesicles that operate over short distances as well as transport tubules that may provide direct continuity among conjoining elements.

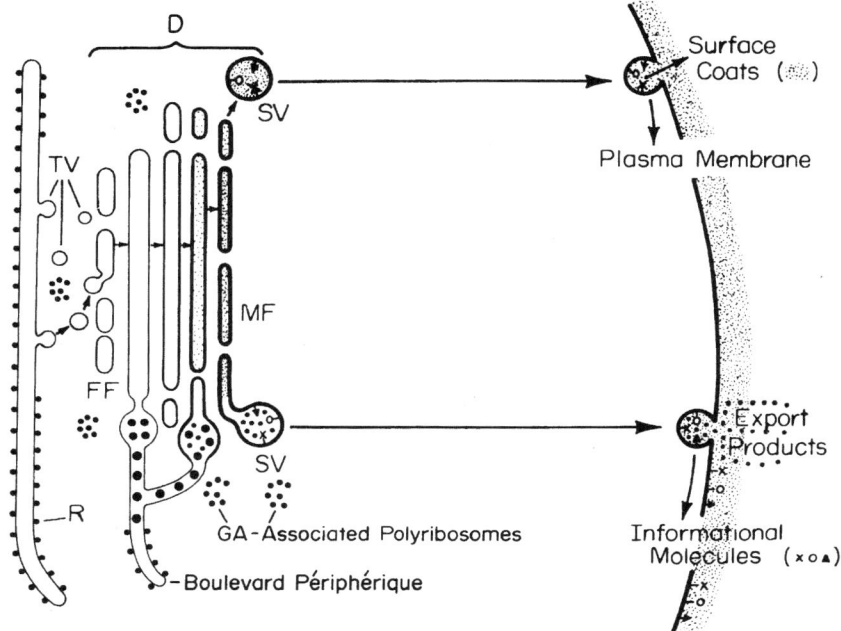

Fig. 8.1. Diagrammatic representation of endomembrane functioning in exocytosis and secretion. Reproduced from Morré and Ovtracht, 1981 with permission from Elsevier.

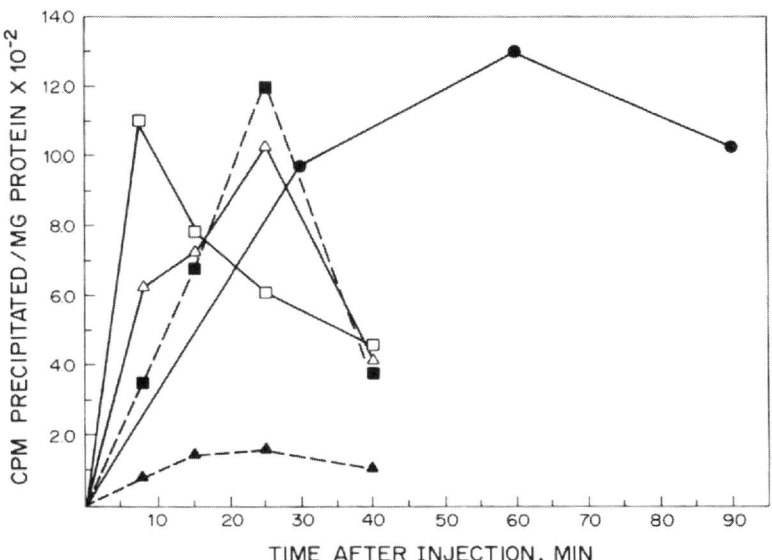

Fig. 8.2. Kinetics of transport of albumin after a 7.5-min pulse-chase of [³H]leucine by rat liver as determined by cell fractionation and specific immunoprecipitation of albumin in endoplasmic reticulum, Golgi apparatus, secretory vesicles, and cisternae. Albumin was extracted from serum using cold alcoholic trichloroacetic acid. Endoplasmic reticulum (□—□), Golgi apparatus (■- - -■) and serum (•—•), and Golgi apparatus sub-fractions secretory vesicles (△—△) and cisternae, (▲- - -▲). Reprinted from Frantz, Croze, Morré and Schreber, 1981, BBA, 678:1–9, Copyright 1981 with permission from Elsevier.

Vesicles formed from one compartment must migrate to and fuse with a second compartment. Vesicle formation and migration presumably requires an energy source and transduction mechanism as well as guidance. Vesicle migration is usually directed by guide elements or activity gradients. Considerably less is known about the energetics of vesicle migration. Some possibilities are discussed later in this chapter.

In the sections that follow secretory pathways are described and compared from several of the more commonly studied examples. Information derived from these examples is then summarized in the form of some general structural models.

8.3. Specific Examples of Golgi Apparatus Secretion

8.3.1. Enzyme and Proenzyme Secretion by Acinar Cells of Pancreas and Parotid Gland

Studies of acinar cells of the pancreas and parotid gland have exerted a major influence on the progress of understanding of the role of the Golgi apparatus in secretion. Beginning with the classical studies of Caro and Palade (1964) and extending to detailed metabolic studies with tissue slices, secretory processes in these tissues have been studied in greater detail than from any other source. Among the products of discharge are digestive enzymes and pro-enzymes important to the digestive functions carried out by these two exocrine glands.

These proteins are synthesized by polyribosomes associated with the abundant rough endoplasmic reticulum of these cells but are subsequently concentrated in irregularly-shaped secretory vesicles at or near the Golgi apparatus that are termed condensing vacuoles. Once formed, the condensing vacuoles often remain in the vicinity of the Golgi apparatus where they assume a spherical shape as they accumulate the diluted secretory products coming from the rough endoplasmic reticulum and develop into mature granules (Fig. 8.3). The filling and condensing process may require several hours. The precise route followed by the secretory proteins in reaching the granules is not known nor is it known whether or not the condensing vacuoles are even attached to the Golgi apparatus cisternae during much of the filling period. The kinetics for parotid gland are shown in Fig. 8.4.

For pancreas, it is clear that the newly synthesized proteins appear first in the lumens of the endoplasmic reticulum and do not move through the cytoplasm as free proteins. When [^{3}H]leucine is administered to animals or to pancreatic slices, autoradiographic grains are not observed over the stacked cisternae of the Golgi apparatus but rather are restricted primarily to the Golgi apparatus periphery. At least in unstimulated glands, the central saccules of the Golgi apparatus appear to be bypassed. The favored hypothesis is that this bypass is accomplished by small transition vesicles that bud off the endoplasmic reticulum, migrate to the forming granule and fuse with it to deliver both secretory proteins and membranes. Empty vesicles are then suggested to recycle back to the endoplasmic reticulum so that the condensing vacuoles may accumulate secretory product without receiving an excess of membrane. An alternative route is provided by direct delivery of secre-

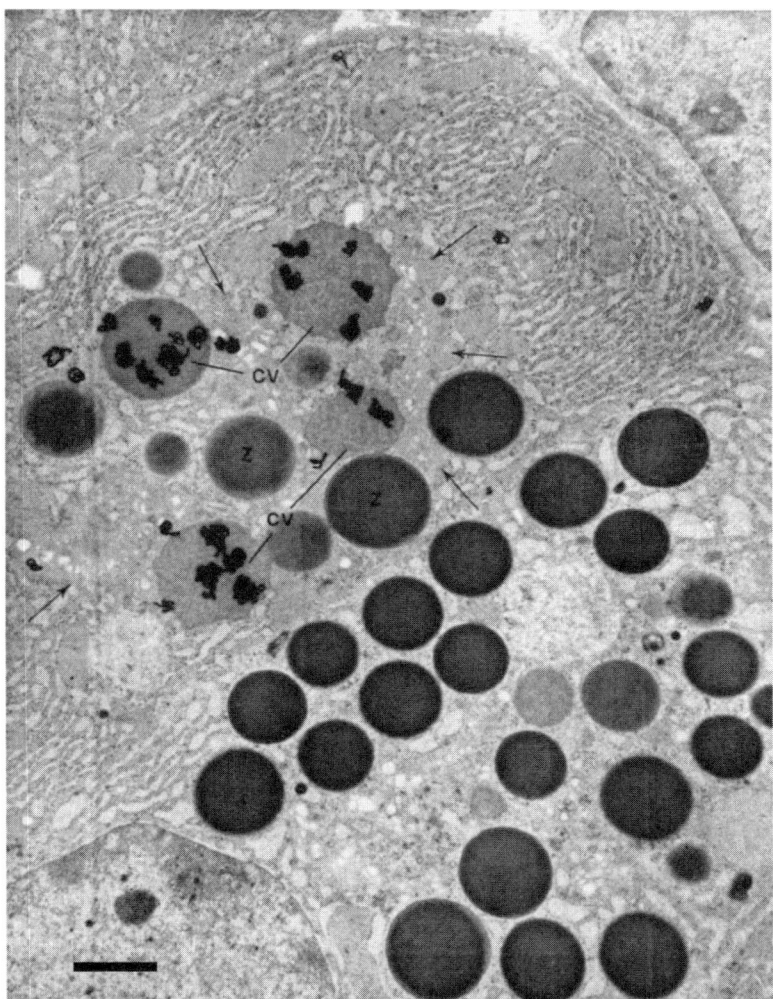

Fig. 8.3. Portion of an acinar cell of the guinea pig pancreas pulse labeled with leucine ^{14}C to show concentration of radioautographic grains over the condensing vacuoles (CV) of the Golgi apparatus. Arrows indicate the peripheries of the Golgi apparatus which are unlabeled. Also unlabeled are the mature zynogen granules (lower right). Reproduced from Jamieson and Palade, 1967b, Journal of Cell Biology, 34:597–615. Copyright 1967, The Rockefeller University Press. Scale bar =1μm.

tory product to vesicles via tubular connections with endoplasmic reticulum (as diagrammed in Fig. 8.1) thereby obviating the need for concomitant membrane transfer and/or recycling.

The formation of the granule membrane offers an interesting biosynthetic problem. Biochemical analyses tend to rule out direct formation of granule membranes by bulk transfer of membrane directly from endoplasmic reticulum. Both acinar cells and chromaffin cells of the adrenal medulla that contain catecholamines accumulate large numbers of mature granules in the absence of

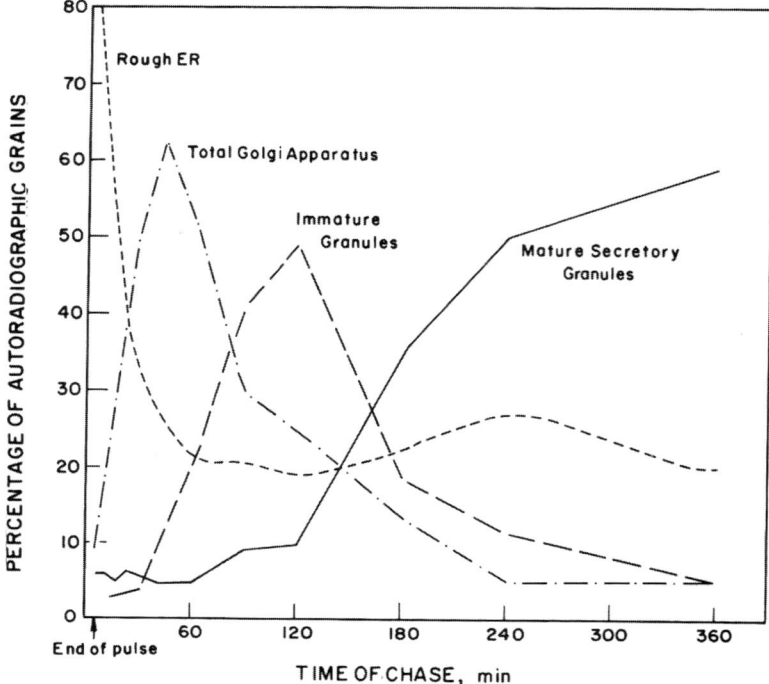

Fig. 8.4. Kinetics of transport of labeled secretory proteins by acinar cells of the rabbit parotid glad. Results are based on counts of autoradiographic grains at successive intervals after a labeling period. More than 75% of the incorporated label moves as a wave as it passes from rough endoplasmic reticulum to the Golgi apparatus to condensing vacuoles and, finally, mature zymogen granules. Reproduced from Castle, Jamieson and Palade, 1972, Journal of Cell Biology, 53:290–311. Copyright 1972, The Rockefeller University Press.

secretagogues, compounds that promote granule discharge. From these cells, mature granules can be isolated readily in quantities sufficient to permit fractionation into granule membranes and granule contents. Gel patterns of membrane proteins reveal only a few major proteins and suggest that the major polypeptide chains of granule membranes and of microsomes are not identical. However, some evidence indicates that some zymogen granule membrane precursors are glycosylated and must, therefore, pass through the Golgi apparatus.

Once granule discharge has been stimulated and the granule membranes fuse with the plasma membrane to discharge their content, much of the granule membrane does not make a permanent contribution to the plasma membrane. Most often, especially in stimulated secretion, the surface contributed by the granule membrane would result in a large excess of surface membrane at the gland lumen unless some compensatory mechanism of membrane retrieval or breakdown were operative. Cope and Williams (1973) applied morphometric methods to an understanding of this problem after isoproterenol (a secretagogue)-induced discharge of granules in the parotid gland of the rabbit. They found that 1,340 μm^2 of granule membrane fused with the plasma membrane during a 95% depletion of granules.

Of this, $1{,}150\,\mu m^2$ was eliminated in about 2 h. During the period of elimination, the increase in intracellular smooth membranes was small with no evidence that the zymogen granule membrane was stored as smooth membrane fragments either in the region of Golgi apparatus or elsewhere in the cytoplasm.

8.3.2. Secretion of Lipoprotein Particles by Liver Parenchymal Cells and Adsorptive Cells of the Small Intestine

Detailed studies of fat absorption by the intestine and subsequent entry into the circulation via the liver have stressed the importance of endoplasmic reticulum and Golgi apparatus in the intracellular synthesis, assembly, and transport of lipoproteins and lipoprotein particles. In liver, the secretory vesicles of the Golgi apparatus contain particles of very low-density lipoproteins (VLDL) or low-density lipoproteins (LDL) (Mahley et al., 1971; Figs. 8.5 and 8.6). These particles are the principal vehicles for triglyceride and cholesterol transport from liver to extrahepatic tissues.

The polypeptides of the lipoproteins are synthesized on polyribosomes and associated messenger RNA at the rough endoplasmic reticulum. Electron microscope studies show that in isolated livers, the strongly osmiophilic particles, which approximate the size range of plasma VLDLs, appear within minutes after perfusion with free fatty acids. They appear first within the smooth endoplasmic reticulum, and Golgi apparatus, but within 15 min are found in the capillary space of Disse. From there, the particles proceed directly to the circulation.

In intestinal fat-absorbing cells, the process of triglyceride resynthesis and release of chylomicra follows the same pattern. Triglyceride resynthesis occurs in the smooth endoplasmic reticulum. Under conditions of feeding a diet rich in fat, these membranes proliferate apparently via conversion from rough endoplasmic reticulum. Transfer of lipid from smooth endoplasmic reticulum to Golgi apparatus vesicles then occurs. The final step in fat transport is the release of chylomicra to the extracellular space by fusion of secretory vesicles with the lateral plasma membrane.

During passage of lipoprotein particles through the Golgi apparatus of liver, some processing may occur. This is evidenced from a reduction in particle diameter between those found in the smooth endoplasmic reticulum and those found in the mature secretory vesicles of the Golgi apparatus and discharged into the capillary space. The latter are about 30 nm smaller in diameter, corresponding to a 50% decrease in volume and more closely approximate the size range of LDL rather than the VLDLs first seen in the smooth endoplasmic reticulum. Little is known about the transformation except that it appears to be accompanied by an overall loss of triglyceride through the action of a lipase (Hess et al., 1979).

To account for the appearance of liver-derived VLDL-sized particles in the circulation a Golgi apparatus bypass has been postulated (Morré, 1981; Twaddle et al., 1981). The bypass would involve direct discharge of larger particles, without processing, by special ER elements located at or near the base of the cell near the capillary space, and without passage through the Golgi apparatus. The bypass mechanism has been studied in isolated, perfused livers (Twaddle et al., 1981) but has proven difficult to confirm biochemically since the special elements of

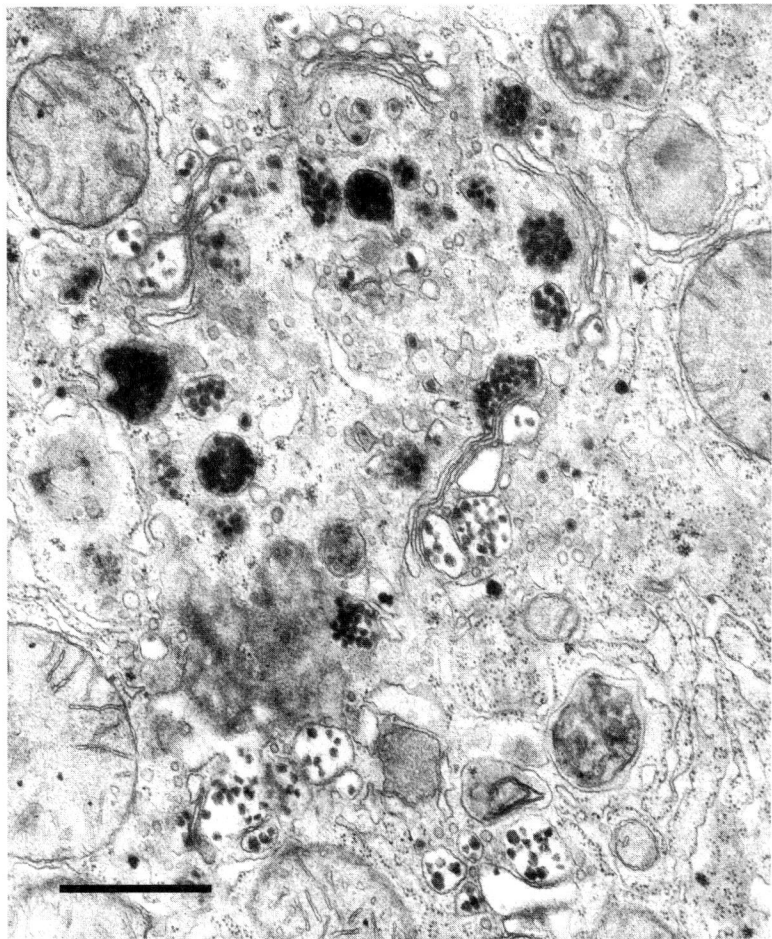

Fig. 8.5. Electron micrograph of Golgi apparatus of rat liver illustrating lipoprotein carrying vesicles and tubules and absence of lipoprotein particles from the flattened portions of the cisternal stacks. From Morré 1987. Bar = 1 μm.

endoplasmic reticulum from the base of the cell cannot be distinguished from that associated with the Golgi apparatus once the livers have been homogenized.

8.3.3. Mucin Secretion

Secretions of mucins rich in carbohydrates are generally moved to the cell surface via vesicles derived from Golgi apparatus (Figs. 8.7 and 8.8; Rambourg et al., 1987). The study of this process has been amenable to study by a variety of approaches although isolation of Golgi apparatus from such cells has proven difficult in many instances. Among the most widely used approaches is that of autoradiography after administration of appropriate radioactive precursors. In a classic series of studies, Neutra and LebBlond (1966a, b) used autoradiography in

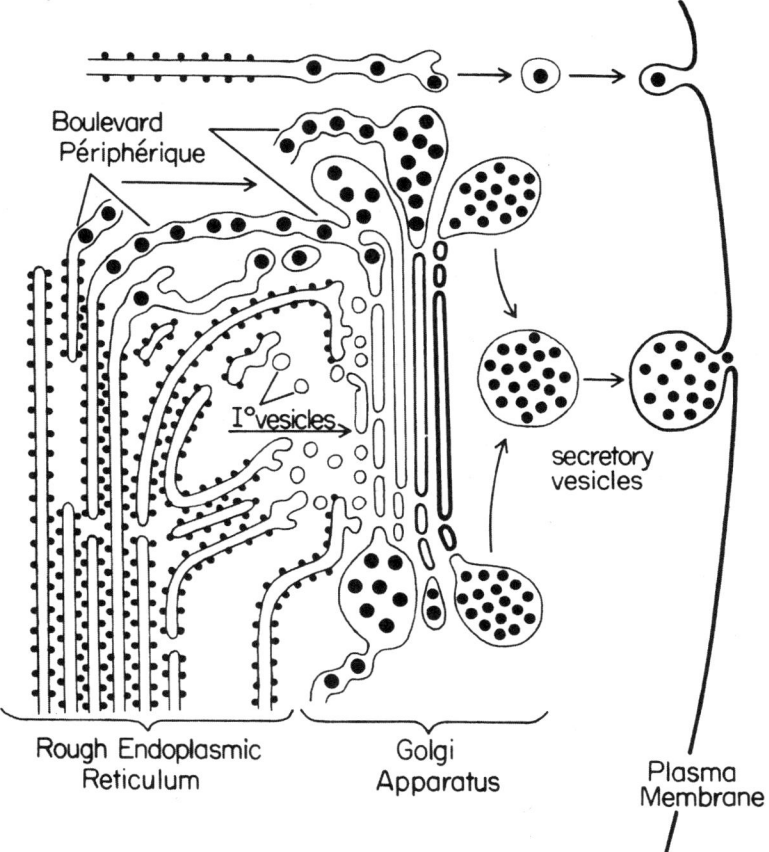

Fig. 8.6. Diagram summarizing routes of synthesis and secretion of serum lipoproteins in rodent liver. Reproduced from Ovtracht et al., 1973 and Morré et al., 1974a with permission from the authors.

combination with electron microscopy to monitor mucin secretion in goblet cells of the colon. They concluded that every 2 to 4 min a distal or mature cisterna of the Golgi apparatus was converted to mucin granules for transport to the surface of the cell. More important, their studies showed the early incorporation of sugars (e.g., galactose) almost exclusively over the Golgi apparatus and the subsequent movement or migration of the label to the cell surface. The actual localization of the mucins in the secretory vesicles has come from cytochemistry where a variety of mucin specific staining procedures have been applied to the electron microscope to establish unequivocally the presence of mucins in the granule content.

The kinetics of mucin formation has been studied almost exclusively from autoradiographic investigations. For example, Michaels and LeBlond (1976) followed the appearance and disappearance of label from various cellular regions of the ascending colon of adult mice following an intravenous injection of [³H]fucose. The label appeared first in the Golgi apparatus where its amount decreased rapidly with time indicating export away from this cell component where

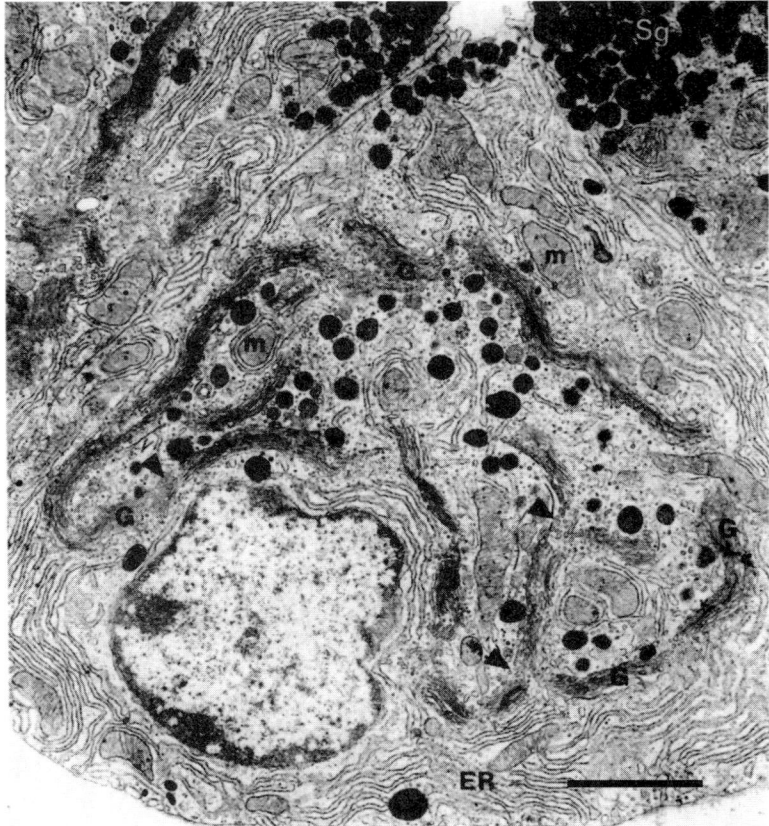

Fig. 8.7. Electron micrograph of a secretory mucous cell of the Brunner's gland illustrating the extensive Golgi apparatus (G) characteristics of mucous-secreting cells. Mucous-containing secretory granules (sg) accumulate ready for discharge upon appropriate stimulation of the gland. m= mitochondria. Reproduced from Rambourg Clermont, Hermo and Segretain, 1987 with permission from John Wiley and Sons, Inc. Scale bar = 5 μm.

it was formed. By 20 to 30 min, a peak of labeling occurred in smooth "carrier" vesicles and by 4 h, the label was concentrated maximally at the cell surface. The synthesis and secretion of cell surface mucins and mucopolysaccharides is represented diagrammatically in Fig. 8.8 for columnar cells of the mouse colon. The carrier vesicles are assumed to arise from the Golgi apparatus and migrate to the cell surface. The fibrillar contents form the mucin layer of the microvillar surface while the vesicle membranes add new plasma membrane to the cell surface. A role for the Golgi apparatus in the synthesis of matrix polysaccharides (proteoglycans and glycosaminoglycans) also has been indicated (Roth, 1977).

A mucin-type secretion is characteristic as well of many types of plant cells whose principal function seems to be the production of polysaccharide rich slimes or mucins. Unequivocal evidence that polysaccharide-containing vesicles originate from the Golgi apparatus has been provided for cap cells of the root

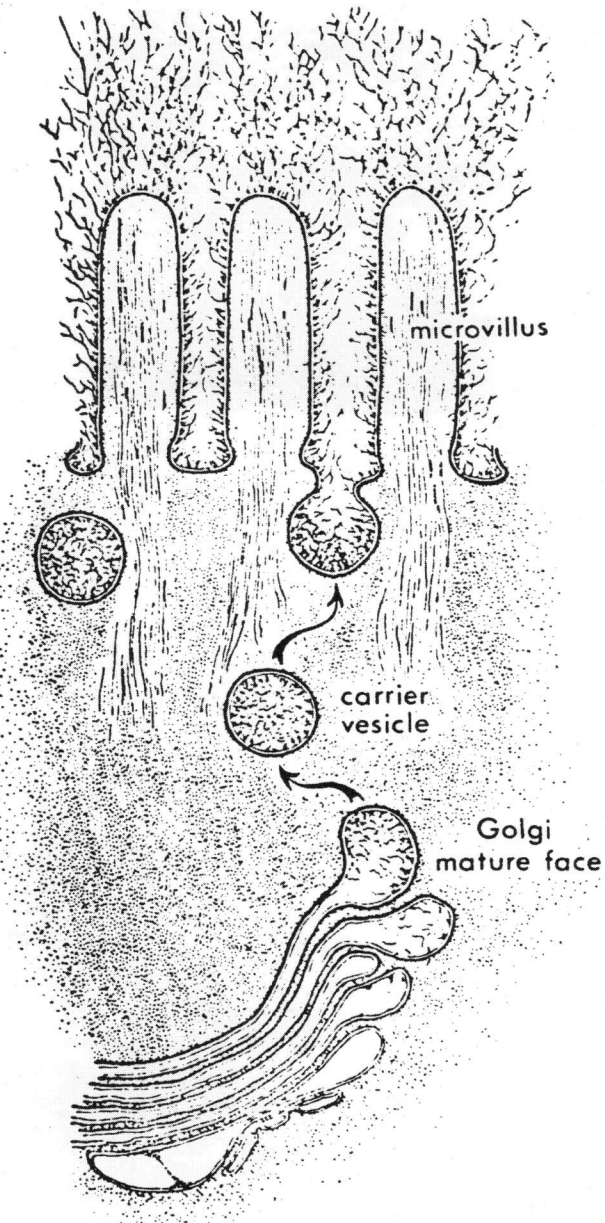

Fig. 8.8. Diagrammatic representation of synthesis and secretion of the cell surface of columnar cells of the mouse colon. Autoradiographic evidence shows that carrier vesicles arise from cisternae of the Golgi apparatus and migrate to the cell surface. Their membranes fuse with the plasma membrane and the fibrils within the vesicle become the cell coat. Reproduced from Michaels and Leblond, 1976 with permission from the Societe Francaise de Microscopie Electronique.

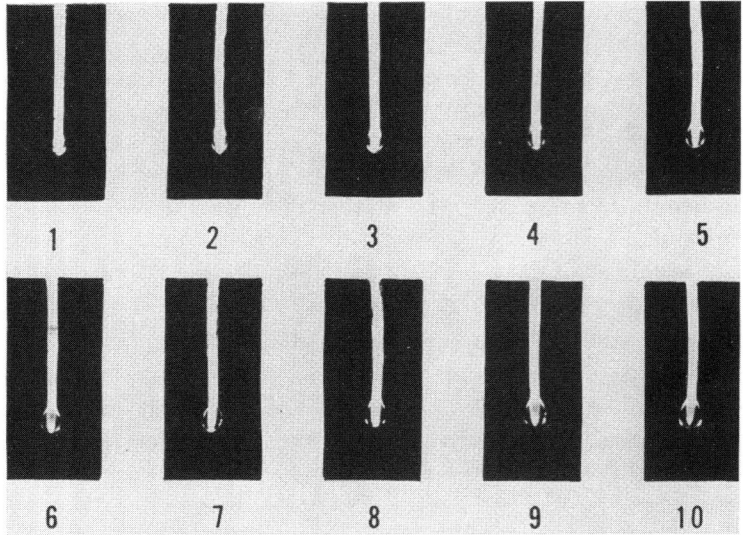

Fig. 8.9. Rating index for visual estimation of amount of secretory product calibrated in mg hydrated polysaccharide per root tip. Reproduced from Morré, Jones and Mollenhauer, 1967 with permission from Springer-Wien. Actual size.

tip of maize where cytochalasin B was used to block vesicle migration to the cell surface but not vesicle formation (Fig. 6.18; Chapters 6 and 11). Under the conditions of the cytochalasin B block, large numbers of vesicles accumulated in the immediate proximity of the Golgi apparatus. If allowed to proceed normally, the carbohydrate-rich contents of the secretory vesicles are released to the cell's exterior, pass through the cell walls and accumulate as a hydrated slime droplet adhering to the root tip (Figs. 8.9 and 8.10). The size of the adhering droplet can be estimated and used as a rapid method for quantitating the amounts of mucin secreted with time. Approximately 4 h is required to produce a droplet at which time the droplet is released from the root tip into the medium and the formation of a new droplet ensues (Fig. 8.11). The process of product delivery to the plasma membrane is diagrammed in Fig. 8.12 for outer cap cells of the maize root.

In other slime-producing plant cells, Schnepf and Busch (1976) have calculated from morphometric data that approximately 3% of the plasma membrane is replaced by secretory vesicles each minute in mucous glands of *Mimulus tilingii*. The trapping slimes of insect trapping plants and mucilages of seed coats (e.g., *Plantago*) and seedpods (e.g., okra) also involve the Golgi apparatus and transport via secretory vesicles.

8.3.4. Cell Walls and Cell Wall Units

The cell surface of plants is characterized by a biphasic cell wall. A microfibrillar or discontinuous phase is assembled predominantly from one type of polysaccharide (β-1,4-glucan, β-1,3-glucan, β-1,3-xylan, β-1,4-mannan, or chitin) via a process involving plasma membrane-bound enzymes and surface-based

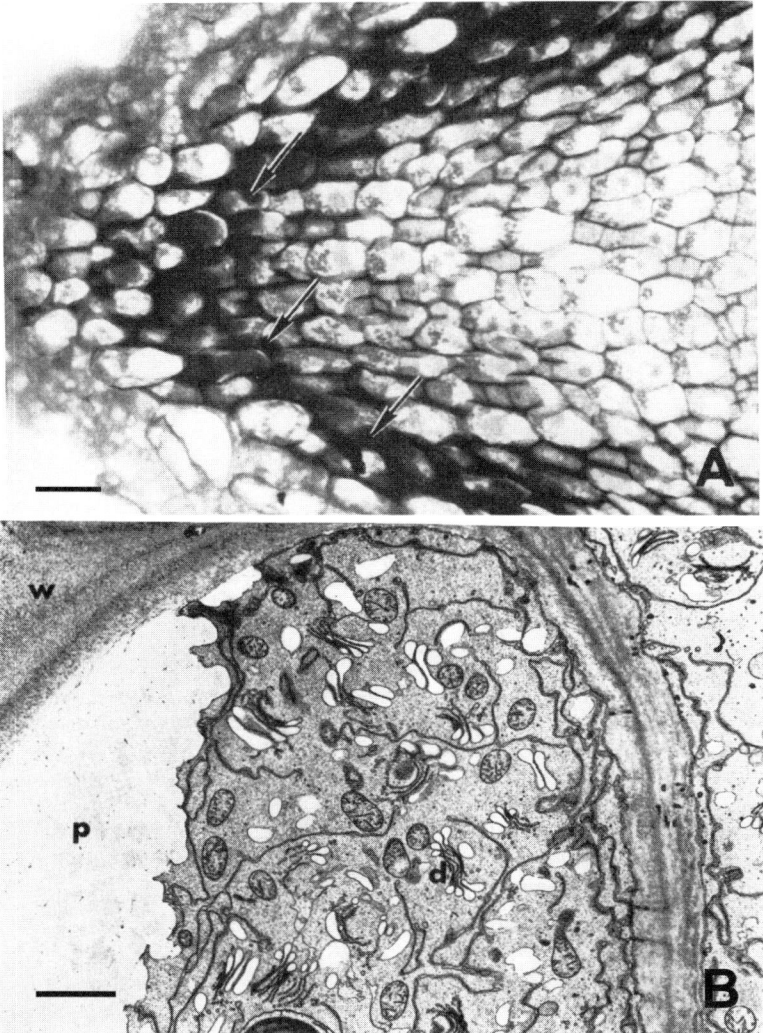

Fig. 8.10.   Polysaccharide secreting cells of the maize root tip. (A) Light micrographs of root caps after periodic acid-Schiff staining. Secretory product accumulations are indicated by arrows. (B) Electron micrographs showing changes in secretory product accumulations (p) adjacent to an exterior cell wall (w) in outer root cap cells. Reproduced from Morré, Jones and Mollenhauer, 1967 with permission from Springer-Wien. Scale bar = 5 μm.

assembly and orientation mechanisms. The matrix or continuous phase is a mixture of pectins and hemicelluloses derived from predominantly mixed polymers of uronic acids, pentoses, and hexoses that differ from extraneous wall components of similar composition such as slimes and mucilages which pass through the wall. Matrix materials are secreted initially, perhaps via endomembrane vesicles derived either from Golgi apparatus, endoplasmic reticulum or both, but proof of this is lacking except for dividing and tip-growing cells (see below). Continued synthesis

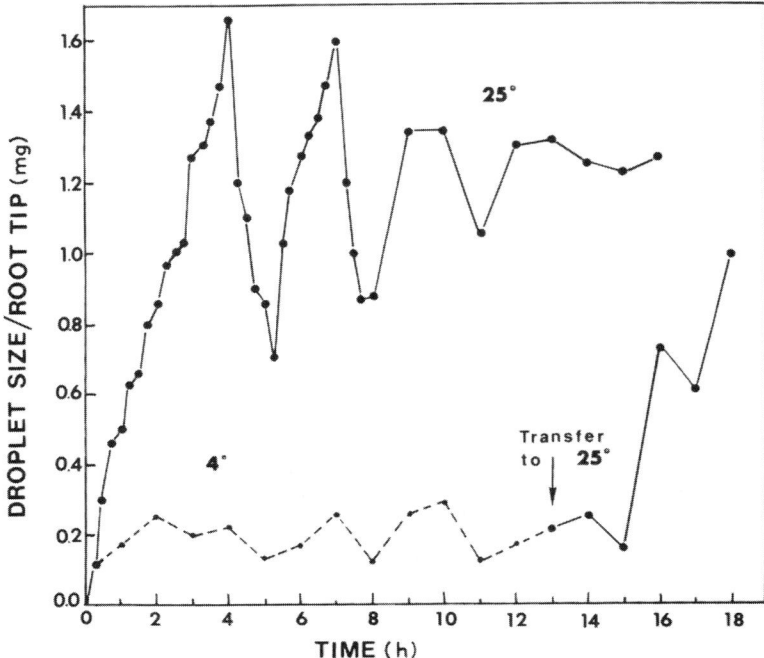

Fig. 8.11. Kinetics of polysaccharide secretion by root cap cells of the maize root tip at 4 and 25° C through four synchronous secretion cycles. Roots were synchronized by removal of existing polysaccharide droplets and transfer to new media. Reprinted from Mollenhauer, Morré and Van Der Woude. Copyright 1975 with permission from Elsevier. Scale bar = 2 μm.

of matrix polysaccharide at the cell surface following incorporation of the vesicle membrane into the plasma membrane remains as a definite possibility to explain deposition of matrix materials in the absence of further vesicular additions to the growing wall. Additional evidence for secretion of cell wall materials by Golgi apparatus vesicles comes from certain algae where walls are fabricated, as discrete units, within cisternae and vesicles derived from the Golgi apparatus.

1. *Cell plate formation.* During the formation of the cell plate or first wall of a dividing plant cell, vesicles of endomembrane origin, including vesicles from the Golgi apparatus, fuse to form the separation between the two daughter cells. The membranes of the vesicles form the new plasma membrane while the contents of the vesicles coalesce to contribute to the new cell wall, the so-called middle lamellae (Fig. 9.2).

2. *Tip-growing cells.* Walls of all plant cells that elongate by tip growth appear to be derived in large measure from contents of secretory vesicles. New plasma membrane is obtained predominantly, if not exclusively, from membranes of secretory vesicles and vesicle contents contribute to the growing cell wall. Examples include pollen tubes, rhizoids, fungal hyphae and plant hairs (Fig. 8.13). In pollen tubes of Easter lily, for example, it has been calculated that, for each single cell, Golgi apparatus produce and export in

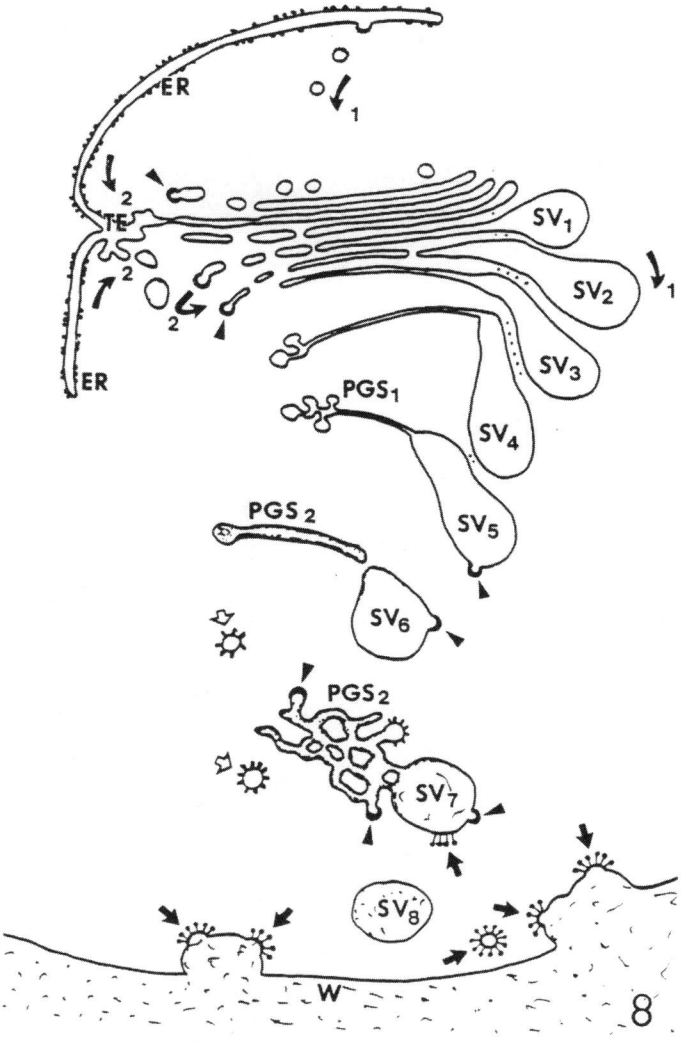

Fig. 8.12.   Diagrammatic interpretation of membrane/product movement through the Golgi apparatus of outer root cap cells of maize. At each level in the stack one or two large secretory vesicles
are attached to the central parts of the cisternae by tubules. The secretory vesicles (SV_1-SV_8) increase
in size from cis (forming) to trans (maturing) poles of the dictyosome. At maturity, the most trans
cisterna separates from the stack to form the first post-Golgi apparatus structure (PGS_1). Small vesicle
buds form on the peripheral edges of the sloughed cisterna concomitant with a decrease in cisternal
size. These cisternae seem to disappear either by vesiculation and loss of membrane or by transition
into the second post-Golgi apparatus structures PGS_2 or both. The secretory vesicles, (SV_4, SV_5)
eventually also separate from the stack and, at about this time, a second post-Golgi structure (PGS_2)
can be observed. PGS_2 also disappears with the concomitant formation of tubules and vesicular buds.
Eventually, the secretory vesicle (SV_8), now devoid of appendages, fuses with the plasma membrane
and discharges its product out of the cell. Reproduced from Mollenhauer and Morré, 1991 with
permission from the John Wiley and Sons, Inc.

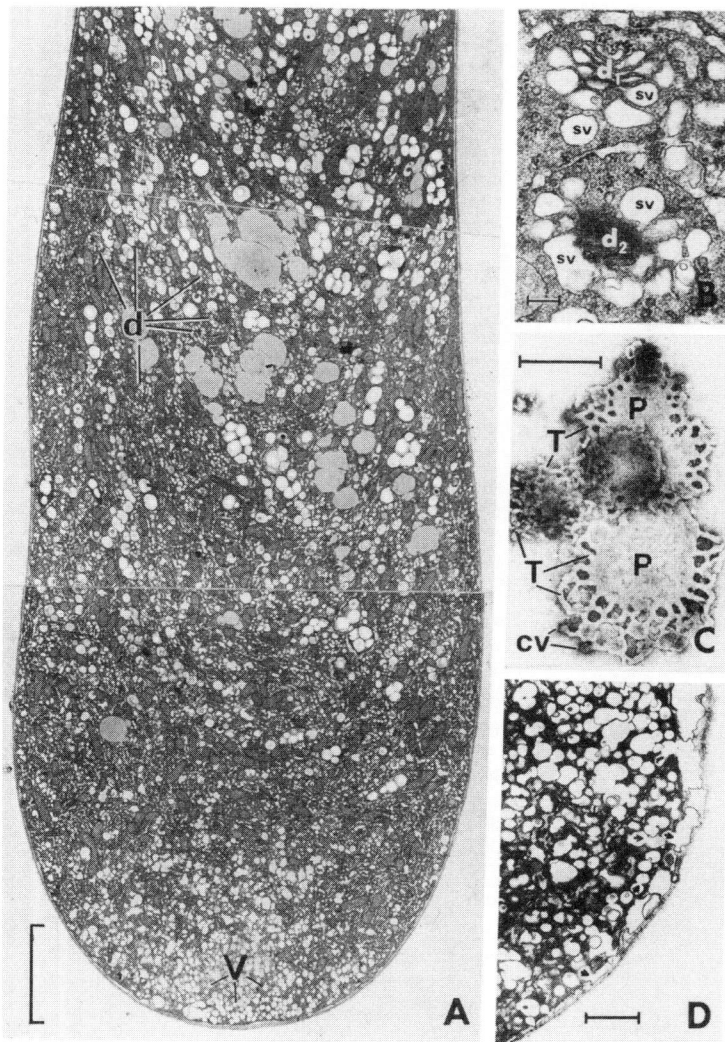

Fig. 8.13. Tip-growing regions of the pollen tube of Easter lily (*Lilium longiflorum*). (A) Approximately median longitudinal section showing secretory vesicles (V) concentrated at the apex of the tube and the numerous dictyosomes (d) of the distally located Golgi apparatus, from which the vesicles arise. Montage of three electron micrographs. Glutaraldehyde-acrolein-osmium tetroxide fixation. Bar, 5 μm. (B) Dictyosomes (d) in cross section (d_1) or sectioned tangentially (d_2) to show the secretory vesicles (sv) attached to the central platelike portion of the cisternae via the system of peripheral tubules. Bar, 0.2 μm. (C) A portion of a dictyosome from germinating pollen, isolated and negatively stained with potassium phosphotungstate, to show the central platelike region (P) and the system of peripheral tubules (T). The small cisterna from near the forming face (top) is almost entirely tubular, while the cisternae nearer the maturing face have more extensive platelike regions. Coated vesicles (cv) attached to the cisternal tubules are a consistent feature of all dictyosome cisternae. Bar, 0.5 μm. (D) Enlargement of the pollen tube apex to illustrate images of vesicle fusion (small arrows) commonly observed in this region. Bar, 1 μm. In this and other tip-growing cells, additions of Golgi apparatus-derived vesicles provide an important mechanism for surface growth. Adapted from Morré and Van Der Woude, 1974 with permission from Elsevier.

excess of 1,000 secretory vesicles per minute needed to generate $300\,\mu m^3$ of new plasma membrane per minute during pollen tube elongation (Table 11.1). Perhaps as much as 50% of the total cell wall material is also derived in this system from secretory vesicles (Morré and Van Der Woude, 1974), largely in the form of the so-called matrix polysaccharides whereas the glucans of the microfibrillar phase appear to synthesize independently of secretory vesicles as in the situation where secretory vesicle fusion with the growing tip is blocked by inhibitory concentration of calcium ions.

3. *Cell wall units of algae.* Many algae walls are fabricated, at least in part, within cisternae and vesicles derived from the Golgi apparatus. An especially instructive example is provided by the coccoliths of the Chrysophycean algae where a stepwise assembly of a complex wall unit occurs in Golgi apparatus (Figs. 8.14 and 8.15). The coccoliths or surface scales that are secreted at the cell surface to form the wall are elaborate structures with organization depending on growth phase of the organism. In the *Pleurochrysis* phase of growth, for example, the scales consist of a radial system of non-cellulosic microfibrils and a spirally arranged cellulosic microfibrillar system. The spirally-arranged, cellulosic microfibrils are covalently linked to protein and both the radial and spiral systems of microfibrils are surrounded by an amorphous matrix deposited upon and within the fibrillar networks. In the *Cricosphera* phase, two different scale types are produced. One has a peripheral network of calcium carbonate crystals deposited on the rim. These calcified scales are known as coccoliths. In all three types of scales, the radial microfibrils are assembled first, unfold and are followed by the deposition of the spiral bands of cellulosic microfibrils. Finally, the covering of amorphous material is added, followed by the calcification of the rim if the scale is one of these destined to become coccoliths. The entire process takes place within the confines of a single cisterna of the Golgi apparatus. The classic work of Brown (1969) indicates that the complex surface scales are synthesized and discharged along with Golgi apparatus cisternae at a rate of one in less than every 2 min. In this group of organisms, single intact cisternae are discharged from the dictyosome and function as secretory vesicles and fuse with the plasma membrane to discharge the scales at the cell surface. The kinetics of appearance of scales at the cell surface has been monitored. Additionally, the scale-containing vesicles are sufficiently large so that their formation within and discharge from the cytoplasm has been observed and recorded by cinematography. Since both the origin of the vesicle from the Golgi apparatus and its final fusion with the plasma membrane has been viewed in living cells, the total contribution of the Golgi apparatus in wall formation in these cells cannot be disputed.

Unfortunately, little is known about the biochemical events in the biosynthesis of these complex surface units in the scale forming algae. The intricacy of scale organization and the precision with which each scale is formed and deposited in an orderly fashion at the cell surface represents one of the finest examples of the capabilities of the Golgi apparatus to not only sequester and secrete complex substances but to direct their organization as well.

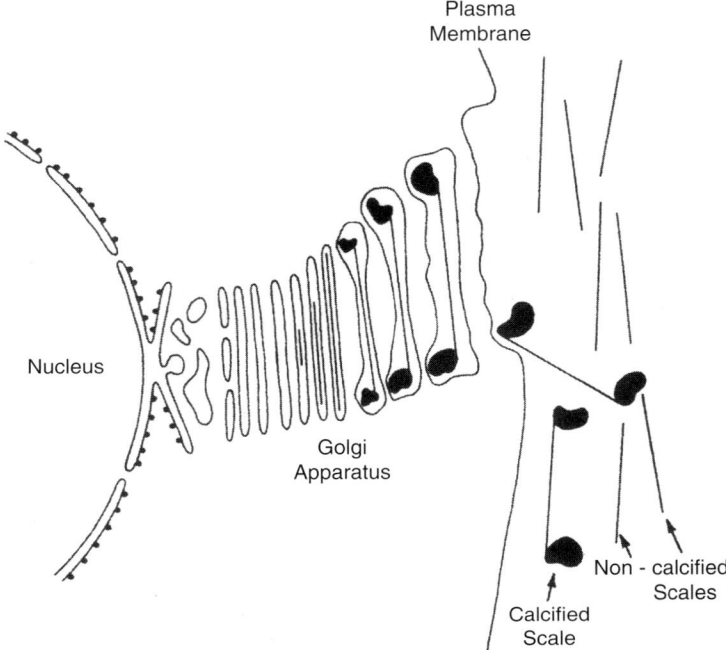

Fig. 8.14. Diagram illustrating the process of scale formation in haptophycean algae. The scales are formed within cisternae of the Golgi apparatus. The cisternae then separate from the stack and the cisternal membranes fuse with the plasma membrane during discharge of individual scales. Calcification occurs at the scale margin (solid black projections) prior to discharge of the scale to the cell surface. Reproduced from Morré and Mollenhauer, 1976 with permission from Springer Science + Business Media.

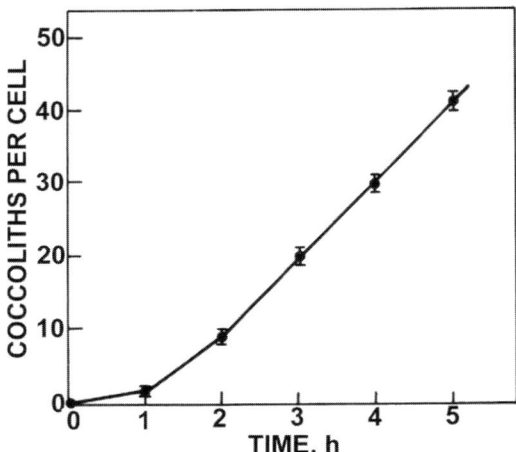

Fig. 8.15. Kinetics of secretion of calcified scales (coccoliths) in *Hymenomonas carterae*. Calcium-deleted cells were transferred to a medium favoring coccolith formation. After an initial lag of 1-2 h, coccoliths were produced at an average rate of 10/11 cell per h. Adapted from Williams, 1974 with permission from the author.

4. *Primary wall of elongating cells of higher plants.* A major role of the plant Golgi apparatus is in the synthesis and delivery to the cell wall of matrix polysaccharides (pectins and hemicelluloses, i.e., xyloglucans and glucurono-xylans) (Driouich and Staehelin, 1997; Staehelin and Moore, 1995; Gibeau and Carpita, 1994) and in the glycosylation of cell wall hydroxyproline-rich glycoproteins and the arabinogalactan proteins (Driouich et al., 1994; Johnson and Chrispeels, 1987). The carbohydrate moieties of these proteins are composed primarily of O-linked arabinose and galactose moieties attached to hydroxyproline and serine residues. Unequivocal evidence is lacking for participation of contributions from Golgi apparatus or other endomembrane vesicles to the formation of the microfibrillar or cellulosic components of primary walls or even of extensive contributions to the cell wall matrix during the grand phase of cell expansion in higher plants except for tip-growing cells already mentioned.

5. *Secondary walls of higher plants.* Once higher plant cells have slowed in their growth and formation of secondary wall begins, some types of secondary thickenings, particularly in vessel cells (xylene and phloem), have been attributed to Golgi apparatus vesicles. However, the bulk of the cellulosic wall appears to be formed without a clear involvement of either the Golgi apparatus or any other vesicle-mediated type of secretion.

8.3.5. Hormones

Hormones, by definition, are substances produced in a cell of one particular type and that exert an action on another, perhaps distant, type of cell. In order to carry out this peculiar physiological role, hormones must be secreted.

While evidence for Golgi apparatus secretion of low molecular weight hormones especially the steroid hormones is lacking, other polypeptide hormones are secreted by an endomembrane route involving the Golgi apparatus. Evidence has come largely from electron microscope cytochemistry where antibodies to hormones complex with electron dense markers (e.g., ferritin) or with enzymes that generate electron dense precipitates under appropriate conditions of incubation (horseradish peroxidase).

Examples of protein hormones secreted via pathways involving the Golgi apparatus include insulin, adrenocorticotrophic, thyroglobulin, and prolactin. Not only are the hormones themselves processed and secreted in a sequence of events involving the Golgi apparatus-derived secretory apparatus but often the receptors for the hormones appear to follow a similar route of delivery to the cell surface.

8.3.6. Fat-Soluble Vitamins and Essential Oils

Golgi apparatus of rodent liver are enriched in vitamin A (retinol) (Nyquist et al., 1971a) and retinol binding proteins as well as vitamin K (plastoquinone) (Nyquist et al., 1971b) and coenzyme Q (ubiquinone) (Nyquist et al., 1970). While a specific function for these vitamins in the Golgi apparatus remains to be established, a transport role is also possible. In plants, secretory cells of the oil ducts of many species are characterized by prominent arrays of smooth tubular endoplasmic reticulum

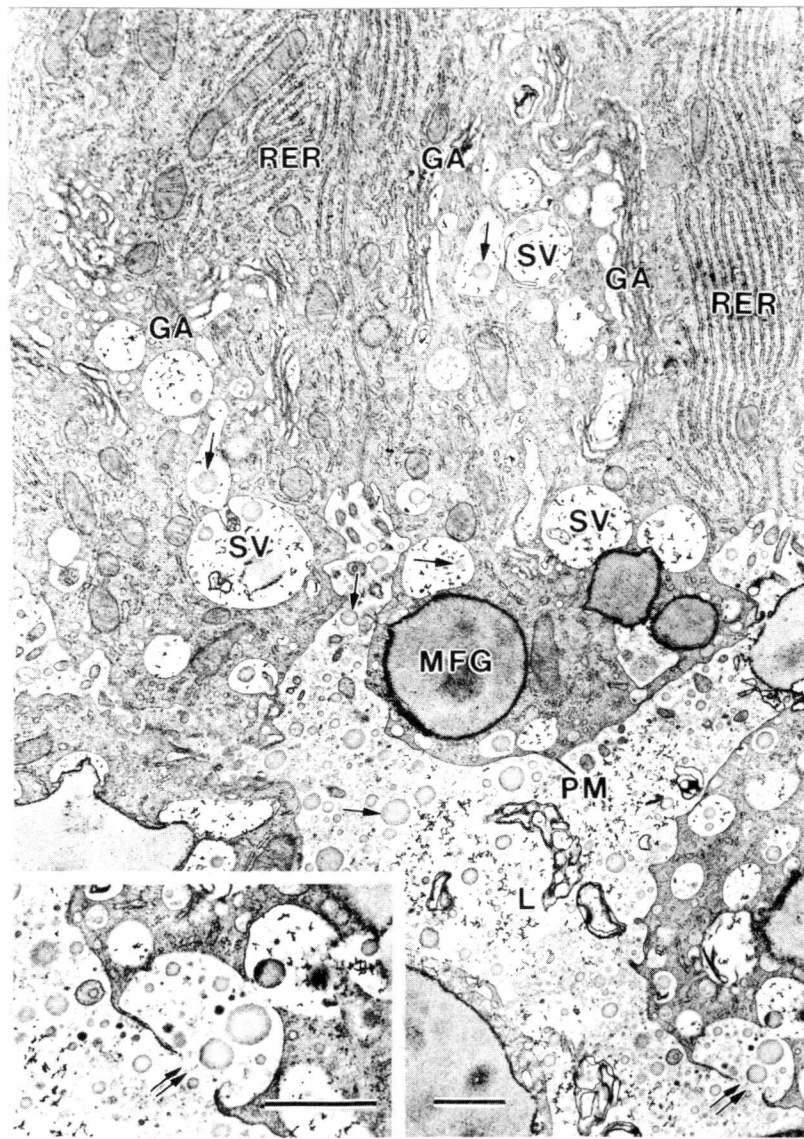

Fig. 8.16.   Electron micrograph of a portion of the epithelium of a lactating rat mammary gland. Large secretory vesicles (sv) containing structures morphologically recognizable as casein micelles (arrows) are at the periphery of the Golgi apparatus (GA) and migrate to, and fuse with (double arrows, lower right and inset) the apical plasma membrane. Simultaneously, apical plasma membrane is expended as globulets of milk fat (MFG) are released into the gland lumen (see Fig. 9). The membrane that envelops the milk fat globule is derived from, and in various aspects is similar to, the apical plasma membrane (Keenan et al., 1970). Reprinted from, Morré, Kartenbeck and Franke, 1979b, BBA, 559:71–152, Copyright 1979 with permission from Elsevier. Scale bar = 1 μm.

similar to that encountered in mammalian cells involved in steroid biogenesis. This common morphological pattern may be indicative more of a general involvement of endoplasmic reticulum in isoprenoid metabolism such as synthesis of steroid hormones in animals and essential oils in plants rather than the Golgi apparatus.

8.3.7. Protein Secretion

An involvement of the Golgi apparatus in the secretion of proteins by acinar cells of the pancreas and parotid gland have already been alluded to. However, numerous examples exist where this function seems to exist as a major Golgi apparatus activity. Well-studied examples include the secretion of serum albumin by the liver and of casein and other milk proteins by the mammary gland (Fig. 8.16). Based on findings from autoradiography, other examples include seminal vesicles of the prostate, proteins of the adenohypophysis, and various antibody forming cells. The pattern of secretion is much the same in all of these various cell types with initial synthesis and processing occurring in rough endoplasmic reticulum, followed by additional processing in the Golgi apparatus or secretory vesicles and final transport to the extracellular space via Golgi apparatus-derived secretory vesicles. In many situations, a peripheral route with direct transfer to the secretory vesicles from endoplasmic reticulum may be followed as deduced for lipoprotein particles in the liver but the level with which subcellular trafficking may be monitored in most cells does not permit a definitive understanding of the precise routes followed.

8.3.8. Simple Sugars, Ions and Other Small Molecules

In specific cell types, secretory vesicles derived from the Golgi apparatus are responsible for the discharge to the cell's exterior of a variety of different substances including many types of small and intermediate-sized molecules. Examples include milk sugar (lactose) and calcium ions secreted by Golgi apparatus of mammary epithelial cells and catecholamines secreted by adrenergic nerves and the adrenal medulla. In some types of salt water algae capable of also living in fresh water, some Golgi apparatus vesicles appear specialized primarily for the secretion of excess water to prevent the cells from bursting.

8.4. Processing of Large Molecules: An Integral Aspect of Golgi Apparatus Function in Secretion

Processing of secreted proteins and glycoproteins takes many forms (Table 8.1). Included are glycosylation reactions, removal of sugars already added, sulfations, phosphorylations and acylations, and the ubiquitous proteolytic processing of proproteins already alluded to in the brief discussion of albumin secretion above. Each of these processes occurs in Golgi apparatus to the extent that the posttranslational modification of proteins and glycoproteins emerges as a major Golgi apparatus function elucidated in the decade between 1970 and 1980.

Table 8.1.   Protein modification and cleavage.

Removal of signal peptide
Folding
Formation and rearrangement of disulfide bonds
Acetylation
α-Carboxyglutamination
Proline hydroxylation to form hydroxyproline
Phosphorylation and dephosphorylation of serine and tyrosine residues
Proteolytic cleavage and/or trimming
Glycosylation and modification of glycosyl additions
Sulfation
Glycation (addition of glucose)
Carbonylation (oxidation)
Prenylation
Oxidation
Aggregation

8.4.1. Glycosylation of Glycoproteins

Golgi apparatus are major sites in all cells for the attachment of terminal or capping sugars (terminal N-acetylglucosamine, galactose, fucose, and sialic acid) to the oligosaccharide chains of asparagine-linked (N-linked) to the nascent glycoprotein chains in the rough endoplasmic reticulum. The mannose-rich precursors synthesized from dolichol intermediates in the endoplasmic reticulum are first synthesized with extra glucose and mannose residues. These are trimmed subsequently, with removal of all of the glucose and six of the mannose residues. This must occur before the terminal hexoses, amino sugars and sialic acid are added. Not only are the glycosyltransferases of terminal addition of sugars located in the Golgi apparatus but, as well, the α-D-mannosidase capable of processing the asparagine-liked oligosaccharides.

Another example of the role of the Golgi apparatus in processing is in the formation of lysosomal enzymes. Lysosomal enzymes contain one or more mannose phosphates which function as sorting signals recognized by specific phosphomannosyl receptors associated also with the Golgi apparatus membranes. The addition of the mannose phosphate recognition signal involves transfer of an N-acetylglucosamine phosphate to appropriate mannose residues of the nascent lysosomal enzymes. These glucosamine residues are then removed to expose the mannose-6-phosphate recognition signal. Both the transferase activity (N-acetylglucosamine 1-phosphotransferase) and the trimming enzyme (α-N-acetylglucosaminyl phosphodiesterase) are concentrated in the Golgi apparatus of rat liver.

Sulfation. Sulfation, like terminal glycosylation, is a nearly exclusive Golgi apparatus function. The earliest evidence and still among the most conclusive came from autoradiographic investigations in which immediately after administration of radioactive sulfate, activity was most concentrated over Golgi apparatus cisternae. A variety of cell types have been surveyed that produce sulfated macromolecules and all appeared to have sulfotransferase activity concentrated in the Golgi apparatus. In sulfation, the sulfate is activated by binding to a nucleotide from which it is transferred to an appropriate acceptor molecule by a sulfotransferase. A number of sulfotransferase activities are localized in Golgi apparatus membranes. The first to be described was that of Fleischer and Zambrano (1973) and Fleischer (1977) where a cerebroside sulfotransferase, present in Golgi apparatus of rat kidney was shown to convert cerebroside to sulfocerebroside (a sulfated glycosphingolipid).

Phosphorylation. The best studied example of phosphate addition at the level of the Golgi apparatus is in the formation of the 6-phosphomannose residues that characterize lysosomal enzymes. However, a number of phosphoproteins exist with important regulatory functions influenced by the action of protein kinases of Golgi apparatus.

Acylation: Addition of Free Fatty Acids to Serine. A number of proteins, including the G protein of vesicular stomatitis virus, contain fatty acids covalently linked to hydroxy amino acids such as serine. This addition is yet another activity attributed to the Golgi apparatus in the post-translational modification of proteins.

Proteolytic Processing of Proteins. Most secretory and many membrane proteins are synthesized in pro-forms that undergo proteolytic cleavage prior to secretion or following insertion into the membrane. Many of the small peptide hormones secreted by the Golgi apparatus (proinsulin, proparathormone, proopicocortin) as well as secretory proteins (proalbumin) undergo processing of this type to yield their mature discharged form. Processing occurs not only in the Golgi apparatus proper but also in the secretory granules. It is a post endoplasmic reticulum step that requires transport to the Golgi apparatus complex and continues well after the secretory proteins are packaged into vesicles. The endogenous activity responsible for the proteolytic cleavage of both presecretory and prosecretory proteins is still not known with certainty. Both endopeptidases and exopeptidases may be involved and all proproteins may be processed by fundamentally similar proteolytic activities. However, the proteases of Golgi apparatus have not been well characterized.

Albumin, the most abundant plasma protein with a molecular weight of 69,000 has a primary structure of a single polypeptide chain of approximately 570–580 amino acids that lacks a prosthetic group. It exists for a time in the cell as an intermediate precursor termed proalbumin. Early studies showed that albumin, like other secretory proteins, was synthesized on membrane-bound polyribosomes and that its subsequent vectorial discharge into the lumen of the endoplasmic reticulum occurred without passage through the cytoplasm. In the interval of 15 to 20 min between synthesis and secretion, the albumin migrates to secretory vesicles of the Golgi apparatus via smooth endoplasmic reticulum. During its passage, the ratio of the proalbumin form to the fully processed albumin form was 95:1 in rough endoplasmic reticulum, 51:1 in smooth endoplasmic reticulum, 33:1 in the Golgi apparatus and approached 0:1 in the serum. Thus, both smooth endoplasmic reticulum and Golgi apparatus are implicated as sequential sites of albumin processing as well as being involved in its secretion. More recently it has been shown that, in the secretory vesicles coming from the Golgi apparatus, the majority of the albumin is of the serum type indicating that secretory vesicles are major sites of albumin processing within the Golgi apparatus. Finally, the microtubule-directed drug, colchicine inhibits both the secretion of albumin and the conversion of proalbumin into albumin. A primary effect of colchicine in causing these effects is to inhibit the export of existing secretory vesicles from the Golgi apparatus with a corresponding inability to form new ones. The conversion of proalbumin to albumin is catalyzed by a cathepsin B-like enzyme present in the secretory vesicles.

8.5. Signal Hypothesis

Secreted and integral membrane proteins have several unique properties which Blobel and colleagues (Blobel and Dobberstein, 1975; Walter and Lingappa, 1986) have used to formulate a theory on the mechanisms by which their locations are established (Fig. 8.17). This theory, known as the signal hypothesis, states that secreted proteins are translated initially bearing an amino-terminal extension of predominantly hydrophobic amino acid residues, called the signal sequence. The theory states that it is the purpose of this region to cause ribosomes bearing nascent protein to be bound to the membrane and thus to vectorially discharge the nascent protein into or across the signal sequence, transfer of the nascent chain, and in eukaryotes, to direct ribosome binding to the membrane.

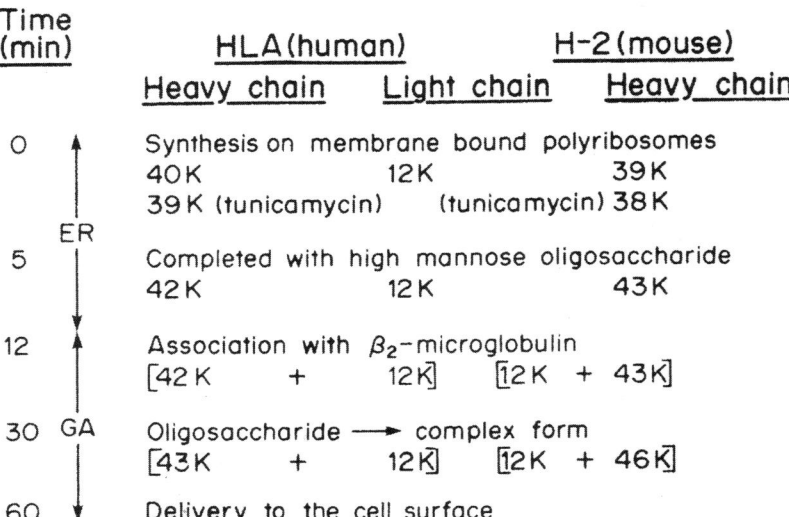

Fig. 8.17. Summary of some of the events that occur during assembly and maturation of histocompatibility antigens in mouse and man. Adapted from Kangel et al., 1979, Dobberstein et al., 1979, and Croze and Morré, 1981.

The presence of a signal sequence on both secreted and membrane-bound proteins, however, raises a question as to the mechanism used by the cell to distinguish these species. According to the signal hypothesis, this distinction must lie in the structure of the protein itself. Although it is possible that subtle differences in the signal sequence are responsible, none are apparent. For this reason "stop transfer" sequences have been hypothesized for membrane proteins. The hydrophobic nature of the signal sequence supposedly facilitates insertion of the peptide into the membrane. The traversing of the membrane occurs as synthesis of the peptide continues and is thought to be facilitated by proteins already in the membrane which may form a protein-lined tunnel spanning the bilayer as a transport site. The signal sequence consisting of the first 30 or so amino acids polymerized is subsequently cleaved off by a membrane protease which recognizes this sequence for secreted proteins that eventually become free in the lumen of the endoplasmic reticulum. It has been suggested by some that there is a specific internal sequence near the carboxyl terminus which is differentiated in secreted and transmembrane proteins and which signals the release of the glycoprotein from the membrane.

A signal recognition particle has now been identified that acts as a transient third unit of a ribosome in linking protein translation with vectorial discharge into the endoplasmic reticulum. The complex is made up of six polypeptides with molecular weights of 72,000, 68,000, 54,000, 19,000, 14,000 and 9,000 plus 7s RNA (Walter and Lingappa, 1986). When a messenger RNA encoding a secretory protein engages with the two subunits of the ribosome and the other components of translation, the first part of the polypeptide sequence to emerge is the signal sequence. The signal recognition particle immediately binds with the ribosome, presumably interacting with the signal sequence so that both translation and vectorial discharge may proceed simultaneously (Walter and Lingappa, 1986).

Figure 8.17 summarizes the events associated with processing of one class of intrinsic membrane proteins, the H-2 alloantigens of the mouse (Dobberstein et al., 1979) and HLA alloantigens in man (Krangel et al., 1979). The protein portion, like the many secretory counterparts, appears to be synthesized first as a precursor in the endoplasmic reticulum. Translation is on membrane-bound polyribosomes with cotranslational insertion into the membranes of the endoplasmic reticulum as they are proximally glycosylated. After about 10 min, the H-2 antigens appear in the Golgi apparatus where terminal glycosylation takes place. This is followed by their appearance at the plasma membrane after about 20 min. This same general pattern of activity is followed by the several intrinsic membrane proteins studies so far. Apparently the same pathway is followed by viral coat glycoproteins (Green et al., 1981; Bergmann et al., 1981).

8.6. Control of Secretion

Little is known of the types of controls that might operate to control Golgi apparatus function. The primary Golgi apparatus output is a secretory vesicle normally destined to fuse with the plasma membrane. Presumably, Golgi apparatus function will be subject to induction–repression mechanisms in terms of specific enzymes, e.g., glycosyltransferases, changes in which may accompany differentiation or cell transformation. A second type of control might be expected in terms of feedback regulation. In pollen tubes and fungal hyphae, when the accumulation of secretory vesicles in the apical zone is monitored in living cells with the light microscope, a relationship is seen between the size of the apical zone of vesicles and growth rate. The relationship observed suggests that vesicle production, vesicle coalescence at the growing tip, and growth rate are closely coupled. Growth rate is necessarily tied to the rate of surface increase, so a rapid incorporation of vesicles accompanies rapid growth rate and vesicles do not accumulate extensively. With declining growth rate, the increase in extent of the apical zone may reflect a decreased frequency of vesicle fusion. Yet, it is not known whether frequency of vesicle incorporation and/or production control growth rate or whether growth rate influences the rates of vesicle production and/or incorporation. In spite of this uncertainty, environmental factors that alter growth rates are expressed ultimately in an altered rate of vesicle production. Communication between cytoplasm and cell surface is achieved rapidly in tip-growing cells and indicates some form of feedback regulation which modulates vesicle production to keep pace with the requirements for surface growth. This suggests a possible function for membrane recycling in terms of control of Golgi apparatus function. If even a small fraction of the cell surface is internalized at regular intervals and returned to the Golgi apparatus, this may provide a means to transmit information from the cell surface back to the cell's interior as a means of regulating trafficking of materials back to the surface of the cell.

While a number of studies have accumulated over the years in which Golgi apparatus have been reported to respond in one way or another to various metabolic inhibitors, drugs, effectors, or environmental stimuli (Chapter 11), the most impressive feature of these studies is the resistance of the Golgi apparatus to morphological alteration. With few exceptions, the Golgi apparatus persists in some form although frequently extensively modified. When secretion is inhibited, for example, the extent of the cisternae may increase or one or more additional cisternae per stack may be added. Similarly, when secretion is stimulated, the Golgi apparatus may become slightly smaller with, on the average, one less cisterna per stack. In either situation, the Golgi apparatus quickly adapts to the new steady state. Beyond suggesting a requirement of Golgi apparatus functioning for energy and an ultimate dependency on RNA and protein synthesis, inhibitor studies have added little additional information to our knowledge of Golgi apparatus regulation. Most appear to affect either input (most metabolic inhibitors) or output (e.g., colchicine, vinca alkaloids, cytochalasins). Few are known to exert a direct action on the Golgi apparatus per se although direct inhibition of Golgi apparatus galactosyl transferase by puromycin has been reported.

One inhibitor which appears to affect the Golgi apparatus directly is the sodium-selective ionophore, monensin. While not selective in its action (mitochondria, autophagia vacuoles, plasma membranes, and other organelles may also be affected), this compound has a rather striking effect on the Golgi apparatus of a wide range of plant and animal cells (Figs. 6.15–6.17). Action is exerted near the point of exit of secretory vesicles from the stacked cisternae, secretion is blocked and transport units (secretory vesicles) may sometimes accumulate near the Golgi apparatus. In detailed studies using embryogenic cultures of carrot cells, an early response to $10\,\mu$M monensin was an increase in the number of cisternae per stack. An average of one additional cisterna per stack was formed within the first 2 to 4 min of monensin treatment; a second cisterna was formed within about 8 min. Thereafter, large vacuoles began to appear in the cytoplasm adjacent to the Golgi apparatus with a return of the number of cisternae per dictyosomal stack to the control number of about 5. By 1 h of monensin treatment, the regions of the cells containing Golgi apparatus stacks were populated by large numbers of vacuoles (up to 20 or more per electron microscope section). These vacuoles have been interpreted as swollen dictyosome cisternae that separated from the stack but had not migrated from the Golgi apparatus zone in the monensin-treated cells. The results permitted an estimation of the average time of formation of a new cisterna as being 2 to 4 min. This range of values agreed with estimates for mammalian cells from short time labeling and turnover experiments of 3 to 4 min assuming a dynamic model for Golgi apparatus function in which cisternae are released from a maturing face and new cisternae are built up at an opposite for forming face (Chapter 6). The monensin response is reversible and within 15 min after removal of the drug, normal secretory vesicles are again produced. These results indicate that the formation of discrete secretory vesicles can be uncoupled from the buildup and release of cisternal units. The latter continues in the presence of monensin while the former is inhibited completely. The observations are reminiscent of the situation with the Chrysophycean algae where, instead of forming discrete secretory vesicles, entire cisternae containing pre-formed wall units are released from one face of the Golgi apparatus with approximately the same rapidity as observed in the carrot cells blocked with monensin.

8.7. Segregation of Lysosomal Enzymes

Alex Novikoff (1964), introduced the acronym GERL to denote a specialized region of the endoplasmic reticulum involved in lysosome formation and as a reminder of a possible relationship to the Golgi apparatus. In most cells, a morphological relationship between Golgi apparatus, endoplasmic reticulum and lysosomes exists especially at the trans face of the Golgi apparatus. Based primarily on cytochemical localization of acid phosphatase (Fig. 8.18), the morphologists marker for lysosomal elements, the prevailing view among most cell biologists is the "the hydrolases are synthesized on ribosomes bound to the rough endoplasmic reticulum, and transferred to the interior of the endoplasmic reticulum, and transported to the Golgi apparatus. In the Golgi apparatus, primary lysosomes are believed to be formed by budding from the margins of Golgi cisternae." Yet, it was difficult to understand how, on the one hand, trans face cisternae of the Golgi apparatus can be involved in the production and release of secretory vesicles for discharge to the cell surface and, on the other hand, carry out the sequestration of acid hydrolyses for discharge into the interiors of primary lysosomes.

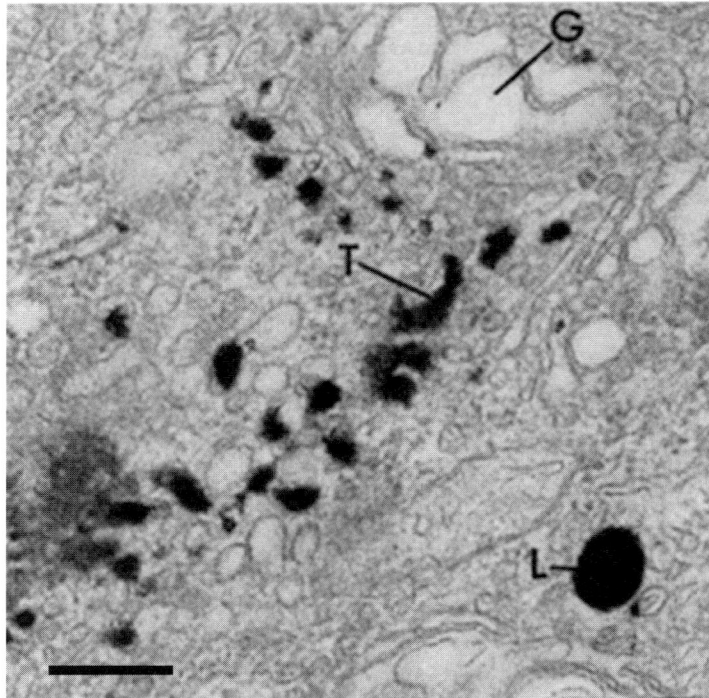

Fig. 8.18. Cytochemical detection of acid phosphatase in a portion of a neuron in thin section of fetal rat ganglion illustrating Novikoff's GERL. The Golgi apparatus stacks (G) are unlabeled. Acid phosphatase-positive tubular portions of GERL (T) are very much in evidence as is a strongly acid phosphatase-positive dense body of lysosomal origin (L). Reprinted from Novikoff et al., 1971, Journal of Cell Biology, 50:859–886. Copyright 1971 The Rockefeller University Press. Scale bar = 0.2 μm.

Subsequently, a role for the phosphomannosyl recognition marker of lysosomal enzymes, unmasked at the level of the Golgi apparatus, has been suggested to play a role in this process. The receptors have been shown both by direct analysis in isolated fractions and by ultrastructural immunocytochemical localization to be present in both endoplasmic reticulum and Golgi apparatus as well as on the plasma membrane. The plasma membrane form of the receptor is presumed to play a well established role in the adsorptive pinocytosis of lysosomal enzymes present in the extracellular environment. But what of the receptor present on endoplasmic reticulum and within the Golgi apparatus? If lysosometropic amines are added to displace the enzymes internally from their receptors, most of the hydrolyses are now secreted into the medium. However, left undisturbed, enzymes bearing the recognition marker are sequestered into lysosomes.

Thus, it is becoming increasingly clear that traffic patterns with the mammalian Golgi apparatus must be quite complex (Farquhar and Palade, 1981). Membrane differentiation, processing of lysosomal enzymes are perhaps just a few of the many processes that may take place simultaneously. The mannose-

6-phosphate recognition marker and its receptor protein affored, an early opportunity to begin to understand how traffic control was achieved.

8.8. NSF, SNAPS, and SNARES in Membrane Fusion and the Regulation of Membrane Traffic

Vesicular transfer of membranes among compartments within the cell's endomembrane system is an integral aspect of Golgi apparatus form and function. Vesicles bud from compartment one and fuse with an appropriate acceptor compartment. To ensure the fidelity and vectoriality of transport, each transport step is coupled to nucleotide hydrolysis via an ordered and regulated biochemical pathway involving specific recognition and docking proteins. In addition, fusion between transport vesicle and target membrane requires recruitment of an array of membrane-associated and cytosolic proteins to the point of vesicle docking.

Among the soluble factors are NSF and SNAPs proteins required by many of the membrane fusion events within the cell. They interact with a class of type II integral membrane proteins termed SNAP receptors, or SNAREs (Fig. 8.19). Interaction between cognate SNAREs on opposing membranes is a prerequisite for NSF dependent membrane fusion. NSF is an ATPase which will disrupt complexes composed of different SNAREs. Among the best characterized SNAREs in the Golgi apparatus are Sed5p in yeast and its mammalian homolog syntaxin 5, both of which localized predominantly to the cis Golgi. For example, Sed5p and syntaxin 5 mediate membrane flow from the ER into the cis Golgi.

The role of NSF in vesicular traffic was discovered through the germinal work of Rothman and colleagues who constructed a biochemical assay for Golgi apparatus transport (Rothman, 1994; reviewed by Nichols and Pelham, 1998). Golgi apparatus fractions from VSV infected, *N*-acetylglucosamine transferase deficient cells were incubated with Golgi apparatus from wild-type cells. Upon the addition of cytosol and ATP there was a progressive appearance of *N*-acetylglucosamine containing endo-H resistant VSV-G protein. In order for the VSV-G to become exposed to *N*-acetylglucosamine transferase, some membrane fusion event must have

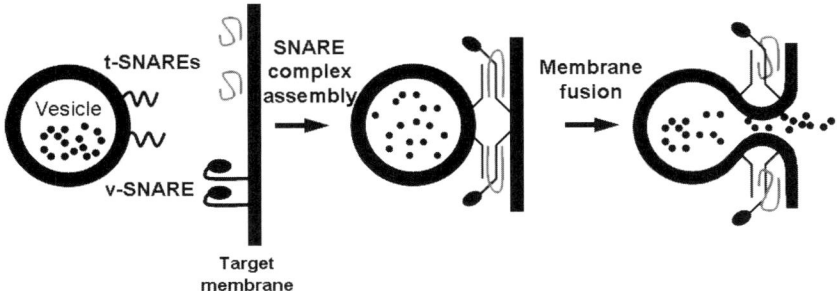

Fig. 8.19. Diagram representing a generalized SNARE (soluble *N*-ethylmaleimide-sensitive-factor attachment protein receptor)-mediated fusion reaction. Adapted from Ungar and Hughson, 2003.

occurred. It was subsequently shown that VSV-G can be packaged into coated vesicles, and that these vesicles can fuse with acceptor membranes (Ostermann et al., 1993). Transport is blocked by low concentrations of a cysteine-alkylating agent, N-ethylmaleimide (NEM), as well as by non-hydrolyzable analogs of GTP.

Purification of the target for NEM inhibition led to the identification of NEM sensitive fusion protein, or NSF (Block et al., 1988). The sequence of NSF is homologous to that of Sec18p in yeast (Wilson et al., 1989). Sec18 was originally identified in a genetic screen for mutants deficient in secretion (Novick et al., 1980). The functional equivalence of Sec18p and NSF is demonstrated by the fact that NSF complements Sec18 mutations in yeast cells, and Sec18p restores transport activity to NEM-treated Golgi membranes in the in vitro assay (Wilson et al., 1989). It is now apparent that NSF-dependent membrane fusion occurs at all stages in the secretory pathway from the ER to the plasma membrane, as well as during endocytosis. NSF is an ATPase, and hydrolysis of Mg^{2+}-ATP is critical in NSF function (Wilson et al., 1992; Whiteheart et al., 1992). NSF will not bind to Golgi membranes in the absence of cytosol, an observation which provided the assay for purification of soluble NSF attachment proteins, or SNAPs (Clary and Rothman, 1990; Clary et al., 1990). SNAP activity is required in the Golgi transport assay. Three species of SNAP, termed α, β and γ, have been identified in mammalian cells, though β-SNAP appears to be restricted to neural tissues (reviewed by Woodman, 1997). Sec17p is the yeast equivalent of α-SNAP. As evidenced by the ability of Sec17p to mediate NSF binding to Golgi membranes and of α-SNAP to restore protein transport in yeast cells containing *sec17* mutations (Clary et al., 1990; Griff et al., 1992).

The super family of SNARE proteins has been studied intensively since their discovery as essential elements of intracellular trafficking. SNARE (soluble N-ethylmaleimide-sensitive-factor attachment protein receptor) proteins facilitate complex formation between two membranes to facilitate membrane fusion by formation of parallel four-helix bundles that bridge the opposing membranes (Fig. 8.19). Tethering proteins together with SM (Sec 1/Munc 18 family) and other proteins with control by Rab GTPases may mediate both initial attachment and SNARE assembly also appear to participate. Early localization studies had suggested two types of SNAREs, V (vesicle) SNAREs and t (target membrane) SNAREs (Söllner et al., 1993). The v/t classification was subsequently replaced by a terminology based on the identity of a key residue, either arginine (R-SNARE) or glutamine (Q-SNARE) (Fasshauer et al., 2002). A fusion of two membranes generally requires four SNAREs, at least one of the two opposed membranes must contribute multiple SNAREs. Most fusions require one R-SNARE contributed by the vesicle and three Q-SNAREs contributed by the target membrane. Complexes once formed require a chaperone for disassembly. The chaperone NSF and the co-chaperone SNAP were discovered prior to SNARE identification to utilize the energy of ATP hydrolysis to disassembly SNARES that have already mediated one round of membrane fusion to be recycled (May et al., 2001).

A second family of proteins, the Rab GTPases, have also been implicated in regulation of SNARE complex formation. The general role of Rab proteins is discussed in the review by B. Goud (2002). Different Rab proteins are involved in various transport steps within the cell (Novick and Zerial, 1997; Lazar et al.,

1997). The Rab family member Ypt1p is required for ER to Golgi transport both in vivo and in vitro (Rexach and Schekman, 1991; Segev et al., 1988; Schmitt et al., 1988; Baker et al., 1990).

Evidence implying a role on several different classes of vesicle for putative v-SNAREs like Sec22p and Vti1p means that the targeting of such vesicles must require additional interactions that provide specificity. Key components already identified have paved the way for several detailed mechanisms whereby specific NSF-SNAP-SNARE proteins interact to precisely direct membrane traffic to specific sites within the cell. Phosphatidylinositol-4-phosphate (PI4P), localized to the trans-Golgi network, appears to exert a similar targeting role in the recruitment of trafficking proteins to this region of the Golgi apparatus (Behnia and Munro, 2005; Carlton and Cullen, 2005).

8.9. Summary

The Golgi apparatus is integral to the modification, sorting, segregation, condensation and packaging of substances synthesized primarily in the endoplasmic reticulum for secretion (exocytosis) to the cells' exterior for use by other cells of the organism. It is also involved in the formation of lysosomes. Golgi apparatus enzymes modify the secreted products as well as concentrate them as they are sequestered into transport vesicles. The most common modifications are through glycosylation and phosphorylation. Proteins that require a stimulus or trigger to elicit discharge from the cell are delivered through a regulated secretory pathway as typified by the pancreas or parotid gland and are stored in preformed granules in the cytoplasm prior to discharge by fusion of the granule membranes with the plasma membrane. A constituitive pathway allows for secretion of products by fusion of vesicle membranes with the plasma membrane that does not require special stimuli to elicit vesicle discharge. Many secretory proteins are labeled with a signal sequence which determines their final destination. The Golgi apparatus adds a mannose-6-phosphate to proteins destined for lysosomes. The Golgi apparatus also plays an important role in the synthesis of matrix polysaccharides of proteoglycans and glycosaminoglycans as well as in the synthesis of matrix polysaccharides of the cell walls of plants.

9

Replication

Golgi apparatus somehow replicate during the normal course of cytokinesis and in gamete formation so that their continuity is assured. Division of the Golgi apparatus or its parts was assumed for many years but usually without adequate documentation. The concept of fusion of endoplasmic reticulum - or nuclear envelope-derived vesicles within a zone of exclusion (Golgi apparatus matrix) or organizational field of a pre-existing stack or prestage (Morré et al., 1971c; Fig. 9.1) has subsequently been revisited and has emerged as a fundamental feature of Golgi apparatus ontogeny (Glick, 2002; Shorter and Warren, 2002).

Golgi apparatus stacks must multiply, since the original Golgi apparatus compliment is reproducibly maintained as cells divide. For example, at cytokinesis, two pools of Golgi membranes normally appear on opposite sides of the nucleus (Shima et al., 1998; Seemann et al., 2002). A functional Golgi apparatus is required for the membrane-fusion events either between the inner leaflets of the plasma membrane or between vesicles and the plasma membrane that are necessary to seal the intercellular bridge when cells divide (Robinson and Spudich, 2000). This requirement is especially evident in plant cells where a prominent cell plate is formed by fusion of Golgi apparatus vesicles (Whaley et al., 1966; Fig. 9.2) as well as for furrow formation at the end of cytokinesis in non-walled cells. Reorientation of the centrosome/Golgi at cytokinesis may reflect the need to direct delivery of vesicles toward the intercellular bridge and ensure plasma membrane sealing during cell separation.

9.1. A Mechanism of Golgi Apparatus Multiplication

A mechanism for replication of Golgi apparatus stacks as diagrammed in Fig. 9.3 is supported by experimental observations (Bracker et al., 1996). Each frame (A through O) represents the turnover of a single cisterna, so that between each frame one cisterna has been lost to vesiculation at the trans face, a new cisterna has been formed at the cis face, and as a consequence of this cisternal turnover, each cisterna in the stack has been advanced one position toward the trans pole. Several distinct stages are represented.

9.1.1. Extension of Forming Face Regions

An early stage in the replication of Golgi apparatus stacks is represented as extensions of the forming face regions of the nuclear envelope or endoplasmic

D. James Morré and Hilton H. Mollenhauer, *The Golgi Apparatus.*
© Springer 2009

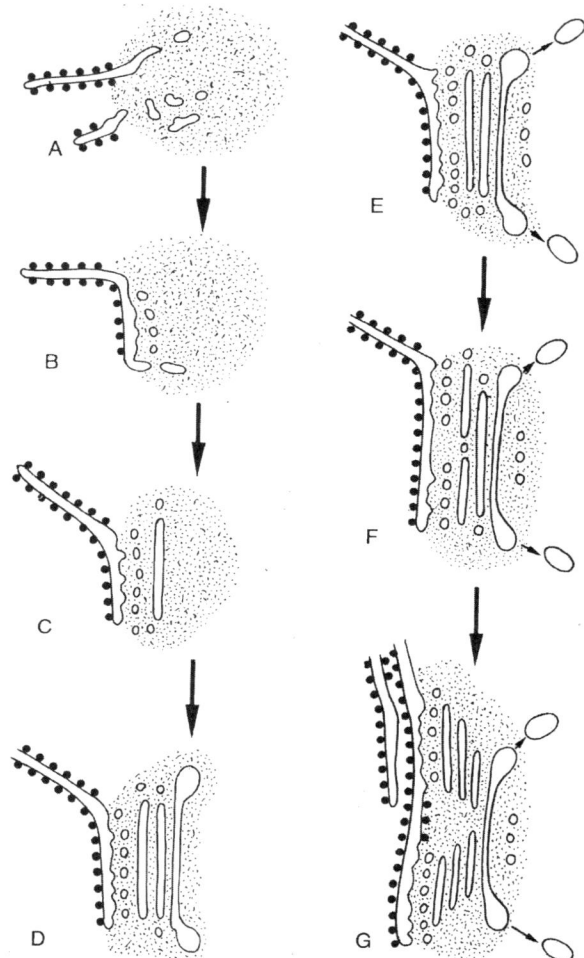

Fig. 9.1.   Diagram relating observations concerning the origin and continuity of Golgi apparatus. In the absence of pre-existing stacks, cisternae are presumed to arise in differentiated regions of the cytoplasm (or zones of exclusion) via prestages consisting of groups of small vesicules or tubular elements associated with endoplasmic reticulum or nuclear envelope (A). The elementary units, which may be tubules or vesicles (B), further differentiate to give rise to flattened plate-like portions of cisternae (C). Additional cisternae are then formed by repetition of this basic generative event with endoplasmic reticulum or nuclear envelope as one source of new membrane material (D). Multiplication of existing stacks and formation of complex Golgi apparatus is explained on the basis of multiplication of the forming regions associated with endoplasmic reticulum and associated zones of exclusion. One possibility for multiplication during periods of active secretory vesicle production (E-G) is that two sets of somewhat shorter cisternae are formed where a longer cisterna appeared previously. Continued separation of the two daughter forming regions with simultaneous production of new cisternae (F) would result in somewhat skewed stacks (G). Replication is completed when the last common cisterna is lost through formation of secretory vesicles. Ultimately, normal cisternal dimensions might be restored by growth in extent of the forming region. Extension of this scheme for Golgi apparatus replication in three dimensions would result in the eventual formation of complex Golgi apparatus consisting of numerous interassociated stacks (Golgi apparatus ribbon). Reproduced from Morré et al., 1971c with permission from Springer Science + Business Media.

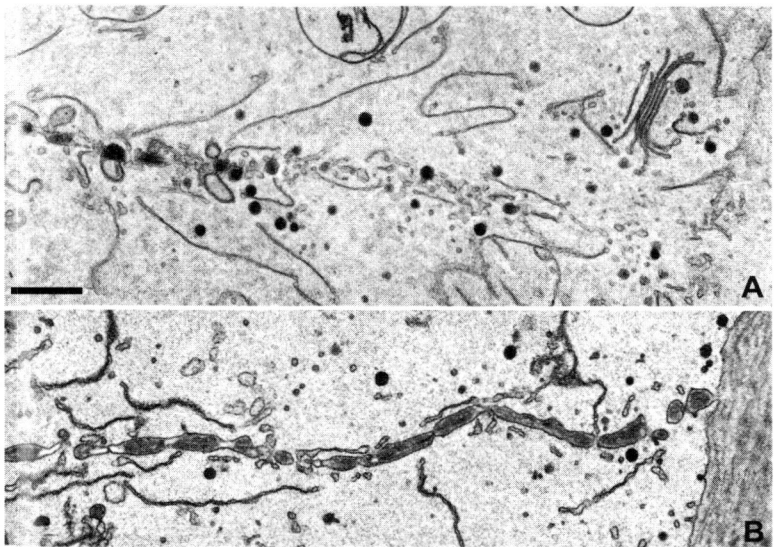

Fig. 9.2.   Golgi apparatus vesicles forming the cell plate in a dividing plant cell of the maize root tip. Reproduced from Mollenhauer and Mollenhauer, 1978 with permission from Springer-Wien. Scale bar = 0.5 μm.

reticulum where stacks lie adjacent to ribosome-free regions (A–C). This involves an increase in size of the ribosome-free portions of the nuclear envelope or endoplasmic reticulum adjacent to the stack (B) and by an increase in budding activity of the expanded smooth portions of nuclear envelope or endoplasmic reticulum especially in regions near, but not directly opposite an existing stack (C).

9.1.2. Appearance of Cisternae with Twice Normal Diameters at the Forming Face

Cisternae with diameters twice the normal size are found at positions corresponding to their derivation from the coalescence of the small vesicles from the membranes of the "extended" forming face regions (D). Stacks with 1, 2, 3 or 4 cisternae having twice normal diameters may be observed as well as complete stacks with double the normal diameter. The complete extended cisternae proceed from the cis face to the trans face following the polarity axis of the stack.

9.1.3. Replicating Forms having Two Stacks of Cisternae with Normal Dimensions on Top of a Single Stack of Cisternae with Twice Normal Dimensions

These forms have been observed for Golgi apparatus of a variety of organisms (Morré et al., 1971c) and are assumed to result in the formation of two stacks of normal dimensions where one existed previously (O). The process, however, seems to be initiated first by a "division" of the extended forming face regions (regions of smooth nuclear envelope or endoplasmic reticulum showing budding

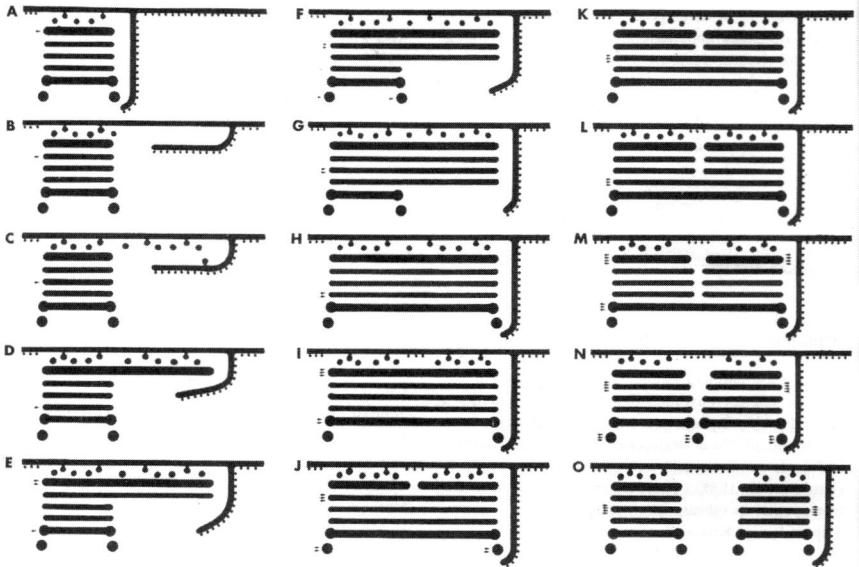

Fig. 9.3. Diagrammatic interpretation of a sequence of events to explain how Golgi apparatus stacks multiply. This scheme takes into account images of Golgi apparatus stacks in the subapical region of hyphae of a fungus, *Pythium*, during Golgi apparatus replication. Each sequential step in the diagram represents the turnover of a single cisterna in a system that replenishes a new cisterna at the proximal pole as a mature cisterna is lost to vesiculation at the distal pole. Throughout the sequence, the normal pattern of turnover continues. The forming-face area is first expanded to five a stack of larger-than-normal diameter, and then separated to yield two "normal" size stacks where the transitional "large" stack existed. Multiplication results from a pattern of physically separated formation of new cisternae rather than a true fission. The single, double, triple, and quadruple arrows mark the positions of selected cisternae during the stepwise progression of cisternal turnover. Reproduced from Bracker, Morré and Grove, 1996 with permission from Springer-Wien.

activity) into two such regions of "normal" dimensions. This is sometimes indicated by the appearance of a cluster of ribosomes attached to the membrane of the nuclear envelope or endoplasmic reticulum at the region between the two separated forming face regions (I, J). Thus, coalescence of transition vesicles derived from endoplasmic reticulum or nuclear envelope would direct not only the formation of the elongated cisternae at the proximal pole but also the eventual separation of the stacks each having two normal dimensions. In elongated stacks where complete cisternae extend across the stack, the appearance of cisternae with "normal" dimensions at the proximal pole is accompanied by a separation or cleft beginning at the cisterna closest to the subtending nuclear envelope or endoplasmic reticulum. This gives rise to the so-called "dictyokinesis" images of the early literature. Each cisterna of a stack normally has similar overall size, shape, and orientation as the cisterna next to it in the stack, even though the morphology otherwise differs according to the normal pattern of polarity. But in the images interpreted as stages of Golgi apparatus multiplication, the pattern of cisternal formation changes as the enlarged forming face areas become separated on the nuclear envelope or endoplasmic reticulum, and new cisternae no longer conform to the size and shape of underlying cisternae (J). This lack of conformity in the final stages (J–N) results in separation into two "normal" size stacks. With failure of new Golgi apparatus elements to coalesce completely during formation of a new cisterna at the cis pole, a separation appears to develop gradually and to segregate the new cisternae into two adjacent stacks (N, O).

9.1.4. Separation into Two Stacks having Normal Dimensions

As the stacks separate following duplication, the region of the nuclear envelope or endoplasmic reticulum between the two adjacent forming face regions appears to increase in extent (M–O). This has the effect of pushing the paired stacks apart. Eventually, stacks may become widely separated, presumably as a result of the migration of the forming face regions in different directions and/or membrane growth in the region separating adjacent forming faces on the nuclear envelope or endoplasmic reticulum rather than migration of the Golgi apparatus stack per se.

9.1.5. Control of Multiplication of Golgi Apparatus Stacks

A major corollary of the scheme in Fig. 9.3 is that Golgi apparatus multiplication is not necessarily controlled at the level of the Golgi apparatus stack but rather by events at the forming face on the nuclear envelope or endoplasmic reticulum. The first morphological indications of division involve extensions of the regions of budding activity of the nuclear envelope or endoplasmic reticulum. Attachment and detachment of polyribosomes at critical sites appear to be involved as well.

In organisms where the Golgi apparatus is paranuclear, the timing of stack multiplication appears to be co-ordinated with nuclear divisions to assure that each daughter nucleus winds up with a complement of associated Golgi apparatus. Dividing nuclei regularly show Golgi apparatus associated with each pole. It is not always clear when in the nuclear cycle the stacks multiply, but the process is completed by the time a mitotic spindle has formed within the nucleus. Thus, Golgi apparatus multiplication most likely occurs in late interphase or during the earliest stages of mitosis, before centrioles migrate to opposite poles.

9.2. Precisternal Stages of Golgi Apparatus Ontogeny

For the most part each cell at its birth inherits endoplasmic reticulum and already formed Golgi apparatus from its mother. Golgi apparatus are inherited maternally. In the few examples where Golgi apparatus prestages have been described (Morré et al., 1971c), new stacks arise in a ribosome-free region of the cytoplasm or zone of exclusion (Fig. 9.4). In early stages of oogenesis, for example (Ward and Ward, 1968), this region is surrounded by endoplasmic reticulum and smooth-surfaced tubular membranous elements. In dormant seeds of higher plants, the embryos contain zones of exclusion surrounding clusters of vesicular profiles prior to germination when normal stacks are not seen (Fig. 9.4). Stacks appear rapidly with the onset of germination, and their appearance coincides with the disappearance of the clusters of vesicles. Also, in young oocytes of *Oryzias* (Yamamoto,

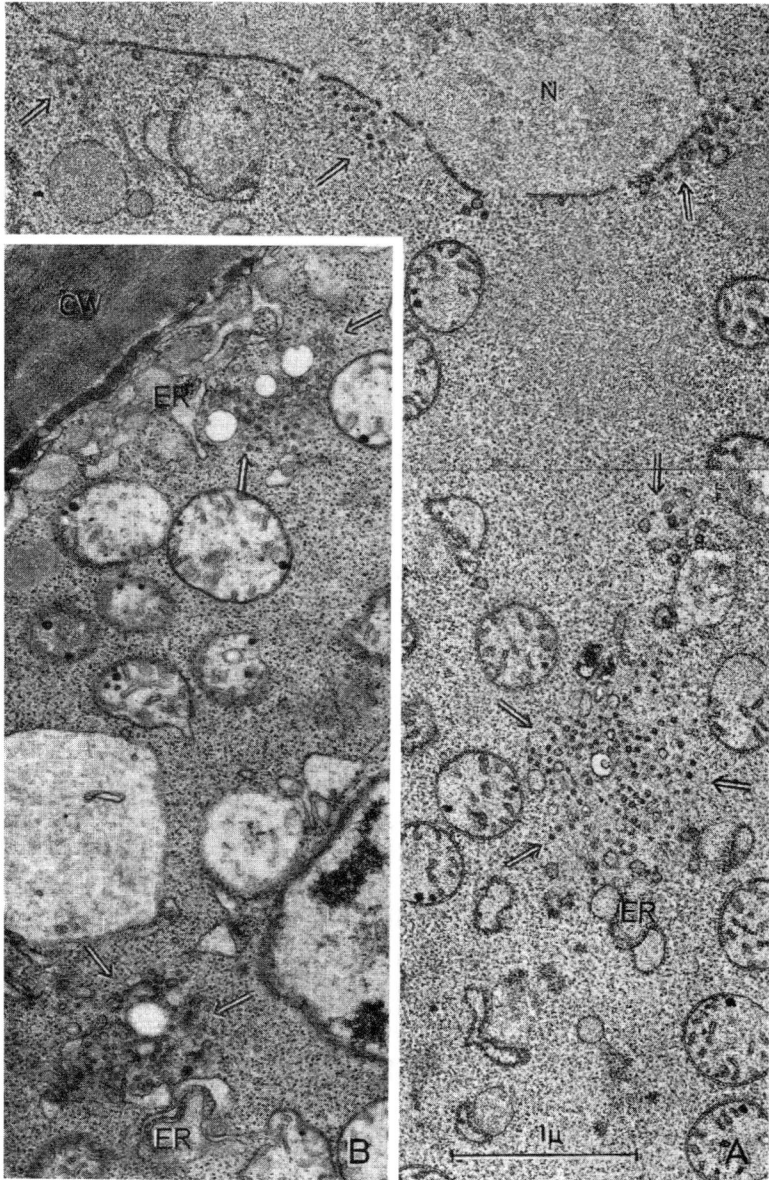

Fig. 9.4. Portions of endosperm cells of mature, dry maize seeds. These seeds do not contain normal appearing Golgi apparatus with stacked cisternae at this stage of development. Instead, zones of exclusion (double arrows) containing numerous small vesicular profiles are scattered throughout the cytoplasm or are located adjacent to the nucleus (N) and near the cell surface. These structures are suggested to represent Golgi apparatus prestages. Associations with fragments of rough-surfaced endoplasmic reticulum (ER) are frequently observed at the periphery of the zones of exclusion. CW = cell wall. (A) Glutaraldehyde fixation followed by brief exposure to aqueous KMnO$_4$.B. Glutaraldehyde-OsO$_4$ fixation. Reproduced from Morré et al., 1971c with permission from Springer Science+Business Media.

1964) and the guinea pig (Adams and Hertig, 1964), the Golgi apparatus develops from clusters of small vesicles within a zone of exclusion.

Where precisternal stages have been described, they are encountered in generative or resting cells and are associated with a regional differentiation of the cytoplasm. If differentiation of the cytoplasm precedes the appearance of Golgi apparatus, what determines the sites of this differentiation? Observations with *Acetabularia* support the notion that these prestages are sites of previous stacks. In this organism, a decrease in the numbers of stacks during encystment is accompanied by an increase in the number of cytoplasmic regions resembling prestages (G. Werz, personal communication, cited in Morré et al., 1971c).

The appearance and disappearance of stacked cisternae in cytoplasmic regions previously occupied by single cisternae also has been described for *Tetrahymena pyriformis* (Elliott and Zieg, 1968). In logarithmic growth, normal vegetative cells of this organism did not contain stacked cisternae. Rather, numerous, separate and smooth-surfaced cisternae were present with tubules between them. During starvation, before mixing the cells for mating, stacks of lamellae were formed in the region previously occupied by the smooth-surfaced cisternae and tubules. During late stages of conjugation, these structures were modified to resemble classical Golgi apparatus. When the cells underwent fission following a period of feeding, the stacks disappeared and the situation characteristic of cells in logarithmic growth was restored. In the yeast *Schizosaccharomyces pombe*, cells often do not normally contain Golgi apparatus stacks, but stacks appear in protoplasts during wall regeneration, for example.

Related to the problem of dictyosome multiplication is the question of the extent to which Golgi apparatus are self-perpetuating or control their own formation. In certain protists such as *Trypanosoma* and *Toxoplasma*, the Golgi apparatus stack is a single-copy organelle (Pelletier et al., 2002) and so must be replicated prior to cell division. In experiments of Flickinger (1969a,1970,1974), enucleated ameba cells were held for periods up to 5 days, until most of the normal-appearing Golgi apparatus stack disappeared. The cells subsequently regenerated normal-appearing Golgi apparatus stacks when supplied with a transplanted nucleus from another ameba. Although these observations establish a high degree of nuclear control they do not prove a de novo origin of the Golgi apparatus.

That the Golgi apparatus matrix can partition even in the absence of Golgi apparatus membranes adds to the complexity of the process. However, the organizing factors that determine Golgi apparatus formation or multiplication may arrive in progeny Golgi apparatus via the ER or with the Golgi matrix fraction, provided there is accurate duplication and/or inheritance of the matrix (Seemann et al., 2002). It may even be that as long as the Golgi matrix is inherited the Golgi apparatus can be rapidly regenerated. It may be that the organizing factors are mostly carried by the matrix fraction into the daughter cells during normal cell division. Such matrix structures are found in proximity to, but distinct from, ER exit sites. This argues that the unit of Golgi apparatus inheritance is a Golgi apparatus template that can specify the rebuilding of a functional Golgi apparatus at any time even in the absence of an existing Golgi apparatus (Morré et al., 1971c; Shorter and Warren, 2002). Early structures are marked by both peripheral (GM130) and membrane (p27) proteins that appear in some manner to nucleate subsequent assembly of the cisternal stacks (Kasap et al., 2004). Yet cytoplasts containing abundant endoplasmic reticulum but lacking a previous Golgi apparatus region fail to make new Golgi apparatus structures (Pelletier et al., 2000) as if at least some component of a preassembled Golgi apparatus were required.

9.3. Golgi Apparatus Fragmentation and Reformation during Mitosis

During mitosis in most mammalian cells, the Golgi apparatus fragments along with the nuclear envelope only to reassembly post-mitosis. This is in sharp contrast to plants where the Golgi apparatus do not fragment in mitosis.

Isolated Golgi apparatus from rat liver when incubated with prophase cytosol unstack (Fig. 9.5, *left panel*). Incubation with interphase cytosol induces their reassembly into Golgi apparatus stacks, which mimics the morphological events that occur during the initial assembly of Golgi apparatus stacks at telophase in vivo. Thus, both the fragmentation and the reassembly of Golgi apparatus fragments has been reconstituted in a cell-free system.

The requirement for interphase cytosol for the reassembly of Golgi cisternae from mitotic Golgi apparatus fragments can be replaced with the purified components NSF, α-SNAP, γ-SNAP, and p115 (Rabouille et al., 1995). Incubation with the purified proteins generates stacks of Golgi cisternae having individual cisternae shorter than normal, suggesting that cytosolic factors promoting lateral growth are still lacking (Shorter and Warren, 2002).

9.4. Experimental Golgi Apparatus Fragmentation and Reformation

In addition to mitotic disassembly, experimental fragmentation of the Golgi apparatus may be achieved in a number of ways, some reversible and some irreversible, including the now classic method of treatment with brefeldin A

Rat Liver Golgi Apparatus stacks	Mitotic Golgi Apparatus Fragments	Reassembled Golgi Apparatus stacks

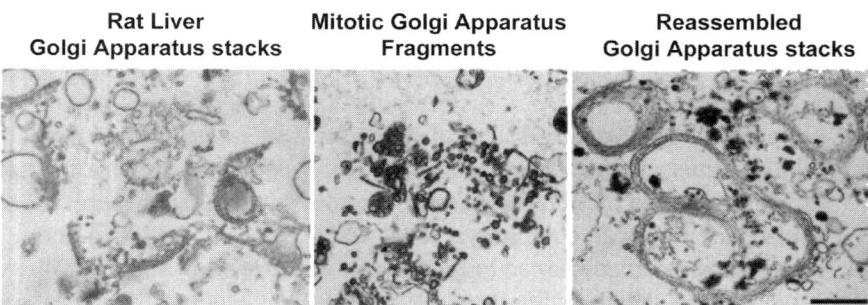

Fig. 9.5.   Cell-free mitotic Golgi fragmentation and reassembly. Rat liver Golgi apparatus stacks (left panel) were incubated with mitotic cytosol to generate a population of mitotic Golgi apparatus fragments (middle panel). Incubation of mitotic Golgi apparatus fragments with interphase cytosol regenerates stacked Golgi apparatus (right panel). Reproduced from Shorter and Warren, 2002 with permission from Annual Reviews. Bar = 0.5 μm.

(Chapter 11), or treatment with nocodazole (Storrie et al., 1998; Pecot and Malhotra, 2006) or overexpression of a dominant-negative form of sar1, a GTPase that recruits the COPII protein to ER exit sites (Barlowe et al., 1994; Miles et al., 2001). Irreversible fragmentation of the Golgi apparatus also is observed during apoptosis. Here disassembly results at least in part from caspase-mediated cleavage of several coiled-coil golgin proteins which normally function to maintain Golgi apparatus assembly (Mukherjee et al., 2007).

Since the brefeldin A response is both reversible and extensively investigated, some insight into possible de novo origins of Golgi apparatus stacks may be gained from the pattern of reassembly. Brefeldin A interferes with ARF1 activation of COPI-coated vesicles and loss of the COPI coat (Chapter 11). Triggered is a phenotypic response typified by rapid collapse of the Golgi apparatus and the generation of a hybrid endoplasmic reticulum–Golgi apparatus compartment (Lippincott-Schwartz, 1989. Brefeldin A removal results in Golgi apparatus reassembly within 30 to 60 min (Lippincott-Schwartz, 1989). During recovery, tubular–vesicular elements mature into concentric lamellae without obvious polarity. These subsequently convert into stacks in a Sar1 and Rab1-dependent process (Bannykh et al., 2005) suggestive of self assembly. Assembly is still likely from preexisting Golgi apparatus constituents not unlike the general mitotic assembly model (Fig. 9.6) except with mitotic assembly the majority of Golgi apparatus constituents do not return to the endoplasmic reticulum upon fragmentation. Even with brefeldin A treatment, the peripheral Golgi apparatus proteins GRASP65 (Golgi reassembly and stack protein 65) and GM130 (Golgi matrix protein 130)

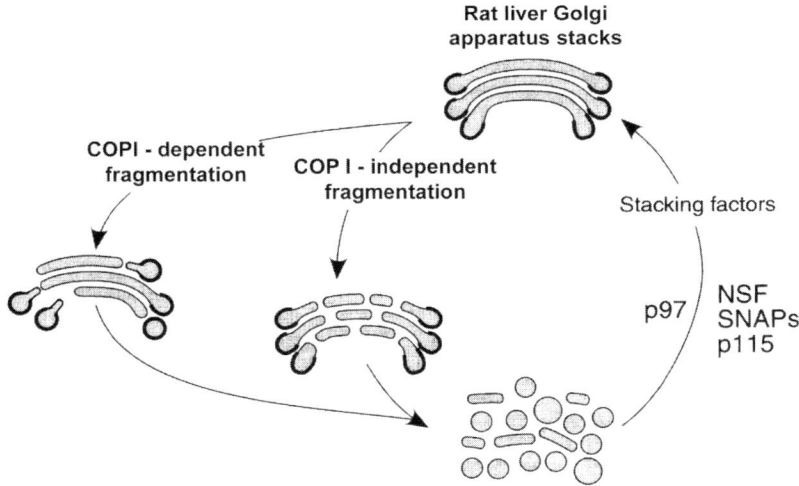

Fig. 9.6. A model for the disassembly and the reassembly of Golgi stacks. The stacks are disassembled by the continued budding of COPI vesicles and by COPI-independent fragmentation. Mitotic Golgi fragments are the substrate for cisternal regrowth by two fusion strategies, NSF/SNAPs/p115, and p97. To stack the cisternae, additional factors are required. Modified from Rabouille and Warren, 1997 with permission from Birkhauser.

apparently do not completely redistribute to the endoplasmic reticulum inferring that these two proteins are part of a permanent matrix that maintains Golgi apparatus identity (Pelletier et al., 2000).

9.5. Summary

The overall pattern of morphological change associated with Golgi apparatus origin and continuity suggests that multiplication follows the same pattern of hierarchy as organization. Golgi apparatus seem to be formed through the interassociation of stacks of cisternae while the cisternal stacks arise one cisterna at a time. The origin of cisternae in the absence of an existing Golgi apparatus appears to involve prestages of the Golgi apparatus zone of exclusion (matrix) where single cisternae are formed in a manner suggestive of fusion of tubular or vesicular elements within the differentiated cytoplasmic zone. Additional cisternae may arise, not from replication of the newly formed cisternae but by repetitive cycles of cisternal formation through fusion of vesicles in much the same manner as Golgi apparatus function in secretion and the accompanying flow differentiation of membranes across the polarity axis. Assembly likely results from features that self-assemble (self-organization) as well as features which may be template dependent.

Cell-Free Analysis

10.1. Cell-Free Systems Development

Cell-free systems have been instrumental in the elucidation of the roles of specific cytosolic factors on Golgi apparatus function including events in vesicle formation, transfer and docking (reviewed by Barlowe et al., 1994; Rothman and Warren, 1994). Both permeabilized cells and completely cell-free systems reconstituted from isolated membrane fractions have been used. Completely cell-free systems were applied initially to investigate intra Golgi apparatus trafficking (Fries and Rothman, 1980; Balch et al., 1984a, b; Hammond et al., 1994). Also investigated were transfers between the endoplasmic reticulum and the Golgi apparatus (Nowack et al., 1987; Paulik et al., 1988; Ruohola et al., 1988; Balch et al., 1987; Wuestehube and Schekman, 1992; Barlowe et al., 1994).

As already introduced in Chapters 4 and 8, one of the first contributions of cell-free systems was the identification of a cytosolic factor sensitive to N-ethylmaleimide by Rothman and collaborators (Block et al., 1988; Wilson et al., 1989), referred to as the N-ethylmaleimide-sensitive factor (NSF). A cytosolic protein with homology to the AAA-family of ATPases (Erdmann et al., 1991), NSF consists of a tetramer of four polypeptide chains each with a molecular mass of 76 kDa (Block et al., 1988; Wilson et al., 1989). Additional proteins, e.g., coat proteins, involved in vesicular membrane transfers and targeting also were identified. Clathrin-coated vesicles were shown to mediate transfers among the plasma membrane, endocytic and trans Golgi apparatus compartments (Pearse and Robinson, 1990). The coatomer proteins, COPI i.e. α-, β-, and γ-COPs (Ostermann et al., 1993; Peter et al., 1993; Schekman and Orci, 1996), together with a system of SNAPS and SNAREs (Rothman, 1994; Rothman and Warren, 1994; Subramanian et al., 1996) were subsequently identified and characterized.

Guanine nucleotide binding (G) proteins with major roles in cell-free transfers were first inferred from experiments where the nonhydrolyzable GTP analog GTP-γ-S was used to block cell-free transport in assays reconstituted both from isolated cell components and with permeabilized cells (Plutner et al., 1992; Rowe et al., 1996). Movement between sequential compartments required ATP and soluble proteins (cytosol) and was inhibited by non-hydrolyzable analogs of GTP and by an antibody toward NSF (Davidson and Balch, 1993). Subsequently both from genetic analyses and molecular studies, several related low molecular weight G proteins were identified as having roles in vesicle budding and targeting

D. James Morré and Hilton H. Mollenhauer, *The Golgi Apparatus.*
© Springer 2009

(Balch, 1989; Barlowe and Schekman, 1993). For example, the mammalian functional equivalent to Ypt1, the 21-kDa GTP-binding protein required for endoplasmic reticulum to Golgi apparatus transport in yeast extracts (Baker et al., 1990) based on antibody inhibition was identified (Balch, 1989, 1990). Also included were the small GTPase members of the p24 family, Rab1 (Pind et al., 1994a) and Rab2 (Chavrier et al., 1990) (Fig. 7.10).

10.2. Cell-Free Transfer Assay Development

Early cell-free assays for the study of molecular targeting relied on measurements of the completion of specific processing steps as evidence for transfer. Such assays had the advantage of measuring both vesicle attachment and vesicle fusion since the mixing of constituents within the fusing compartments was necessary for signal generation. Examples included the processing of vesicular stomatitis virus G proteins (Balch et al., 1987; Wattenberg, 1991) and the acquisition of specific α-1,6-mannose residues by [^{32}S]-labeled core-glycosylated pro-α-factor (gpαF), the precursor to a secreted mating pheromone in yeast (Salama et al., 1993).

A different approach to assay of cell-free transfer was introduced by Nowack et al. (1987). Highly purified unlabeled acceptor membranes immobilized on nitrocellulose were combined with radioisotopically labeled donor fractions. Transfer was initiated by immersing the immobilized donor into suspension of an acceptor membrane. Quantitation was by determination of radioactivity from labeled donor vesicles specifically attached to and/or fused with the unlabeled acceptor membranes. Metabolic labeling prior to cell fractionation was carried out in situ with tissues or tissue slices (e.g., amino acids for total proteins or acetate for lipids) (Nowack et al., 1987; Moreau et al., 1991). Labeling of fractions post isolation with [^{3}H]cholesterol or [^{3}H]inositol by exchange also was employed to examine transfer of sterols (Waits et al., 1990) or inositol lipids (Harryson et al., 1996). The choline head group of phosphatidyl choline was labeled enzymatically with CDP-[^{14}C]choline as a marker for phospholipid transfer (Moreau and Morré, 1991; Moreau et al., 1991). Developed initially for rat liver, such systems were subsequently applied to cells grown in culture (Section 5) and plants (Section 6). The approach had the advantage that the assays were sufficiently general to permit analysis of transfer of virtually any constituent of the donor compartment that could be labeled and analyzed (Morré, 1998). Also the methodology permitted heterologous transfers, i.e., between acceptor and donor compartment of different species (Morré et al., 1989b).

10.3. Reconstitution of Transitional Endoplasmic Reticulum to Golgi Apparatus Transfer

Transfer from transitional endoplasmic reticulum to the Golgi apparatus in these systems was efficient and exhibited considerable fidelity to the in vitro transfer process. Both donor and acceptor compartments consisted of isolated, homogeneous, and well characterized highly purified membrane fractions as described in Chapter 3.

Preparations enriched in part-smooth (lacking ribosomes), part-rough (with ribosomes) transitional elements of the endoplasmic reticulum were incubated with ATP plus a cytosol fraction. The preparations responded by the formation of blebbing profiles and ~60-nm vesicles (Fig. 10.1). The vesicles were indistinguishable from the transition vesicles formed in situ and considered to function in the transfer of membrane materials between the endoplasmic reticulum and the Golgi apparatus. Fractions of differing electrophoretic mobility were resolved by preparative free-flow electrophoresis. The main fraction contained the vesicles of the transitional membrane elements. A less mobile minor shoulder fraction was enriched in 60-nm transition (transfer) vesicles (Fig. 10.2). The vesicles when added to Golgi apparatus immobilized to nitrocellulose were active in the transfer of radioactivity to the Golgi apparatus membranes. The transfer was rapid ($T_{1/2}$ of about 5 min), efficient (10–30% of the total radioactivity of the transition vesicle preparations was transferred to Golgi apparatus), and independent of added ATP but facilitated by

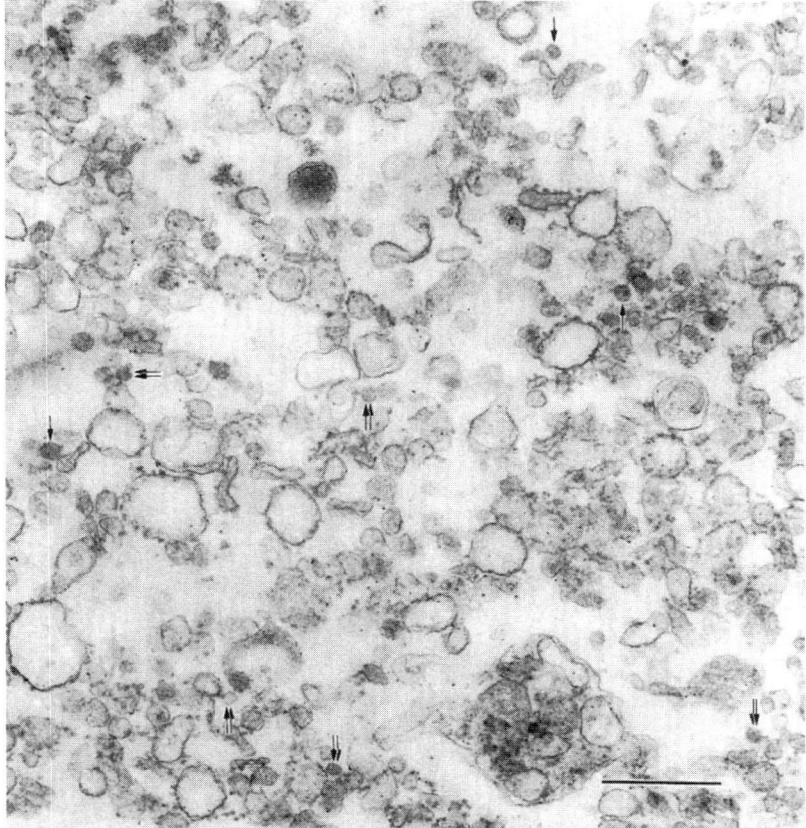

Fig. 10.1. Buds (arrows) induced in isolated transitional endoplasmic reticulum of rat liver by the addition of ATP plus cytosol. Many appear to remain attached to the endoplasmic reticulum fragments by short, narrow stalks (double arrows). Reproduced from Morré, 1998 with permission from Springer Science+Business Media, LLC. Scale bar = 1 μm.

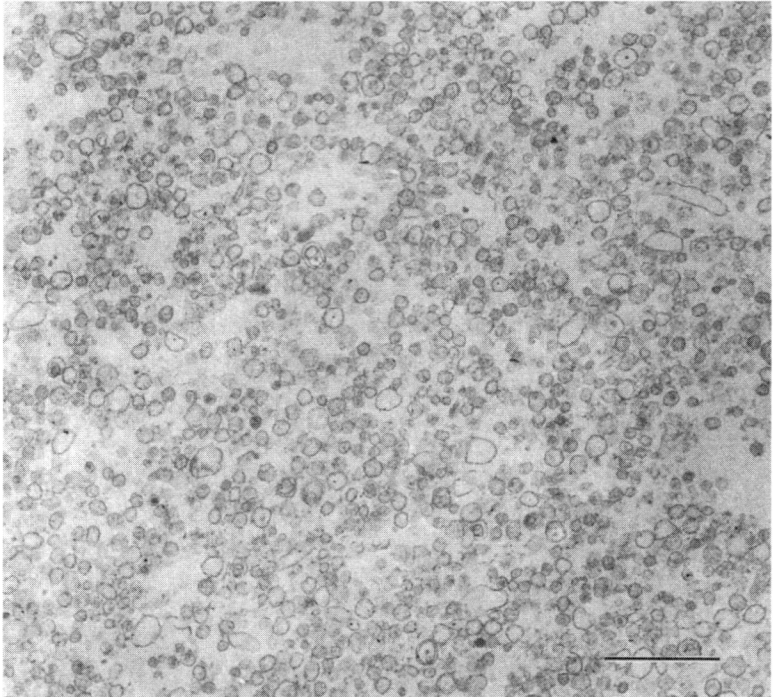

Fig. 10.2.   Enriched fractions from preparative free-flow electrophoresis according to Paulik et al. (1988). Numerous 50-70 nm vesicular profiles resembling the transition vesicles of the intact cell. The formation of the vesicles is time, ATP, and temperature dependent. The vesicles bind and fuse with *cis* Golgi apparatus acceptor membranes in the cell-free assay. Fusion is rapid and occurs in the absence of ATP. Reproduced from Morré, 1998 with permission from Springer Science+Business Media, LLC. Scale bar = 1 μm.

cytosol (Fig. 10.3). Transfer was specific and unidirectional in that Golgi apparatus membranes were ineffective as donor membranes and endoplasmic reticulum vesicles were ineffective as recipient membranes. Using a heterologous system with transition vesicles from rat liver and Golgi apparatus isolated from guinea pig liver, immunocytochemistry was utilized to verify coalescence of the small endoplasmic reticulum-derived vesicles with Golgi apparatus membranes.

The reconstituted transfer system for transitional endoplasmic reticulum (or isolated transition vesicles) to Golgi apparatus transfer consisted of a radiolabeled donor fraction, an unlabeled acceptor immobilized on nitrocellulose and an ATP + ATP regenerating system. The latter was patterned after that of Balch et al. (1984a). The cytosol fraction was the >10-kDa molecular mass fraction prepared by filtration of a microsome-free supernatant (90,000 g for 60 min) through a Centricon YM-10 filter (Amicon).

To immobilize the acceptor membranes on nitrocellulose, the purified plasma membranes were resuspended at a final concentration of 1–2 mg of protein/ml in 30 mM Hepes, 2.5 mM magnesium acetate and 30 mM KCl (Hepes/Mg(OAc)$_2$/ KCl). Strips (1 cm^2) of nitrocellulose (Nytran, S&S Scientific, Keene, NY) are

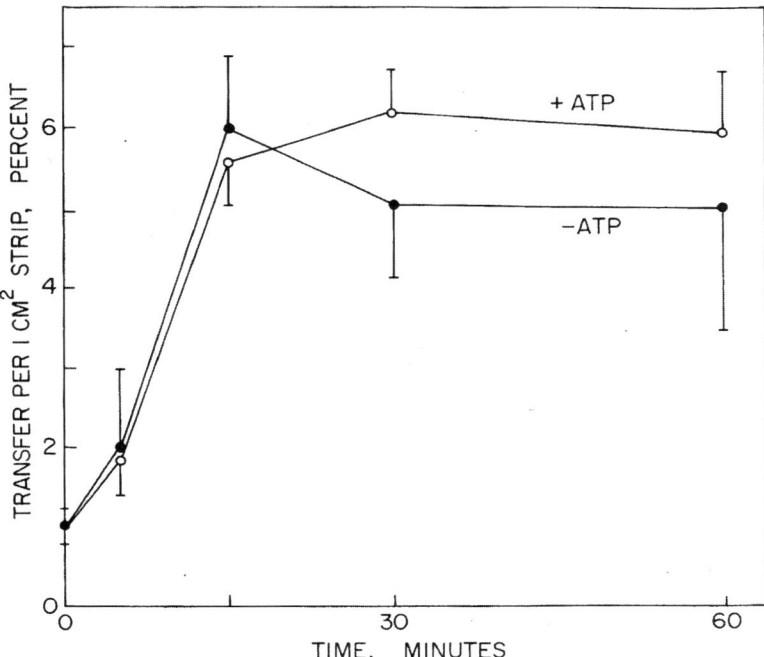

Fig. 10.3. Time course of transfer of radioactivity from donor transition vesicles prelabeled with [³H]leucine (10-30 μg protein, ca 300 cpm/strip for each of three strips/reaction) to acceptor Golgi apparatus (40 μg protein) immobilized on nitrocellulose strips, comparing transfer without ATP (●-●) to transfer with ATP (o-o) using a complete system (Paulik et al., 1988). ATP is required for transition vesicle formation. Fusion with acceptor Golgi apparatus is rapid and ATP independent. Reproduced from Morré, 1998 with permission from Springer Science+Business Media, LLC.

then added and incubated for 30 mm at 4°C with continuous shaking. To block unoccupied sites on the nitrocellulose, the strips next are transferred to 5% BSA in Hepes/Mg(OAc)₂/KCl followed by a further incubation at 4°C for 60 mm. The strips, each loaded with approximately 30 μg of protein, are rinsed through three changes of Hepes/Mg(OAc)₂/KCl and added to the reconstituted transfer system.

To label the donor fraction, livers were excised and slices were cut by hand with a razor blade as described (Morré et al., 1987a). The slices were incubated with 1 mCi [³H]acetate, [³H]leucine or [³⁵S]methionine in 5 mL phosphate-buffered saline (pH 6.2) for 1 h at 37°C, collected on a Miracloth (Chicopee Mills, NY) filter and rinsed to remove unincorporated radioactivity.

The complete reconstituted transfer system contained in addition to Hepes/(OAc)₂/KCL, donor membrane and acceptor strips 17 μM ATP, 83 μM UTP, and an ATP-regenerating system consisting of 0.5 μM creatine phosphate and 2.5 units creatine phosphokinase per milliliter. The final pH was 7.0. Usually two identical membrane systems were prepared and incubated in parallel. A complete mixture was maintained at 4°C as a control and the other complete mixture was incubated at 37°C (Fig. 10.4).

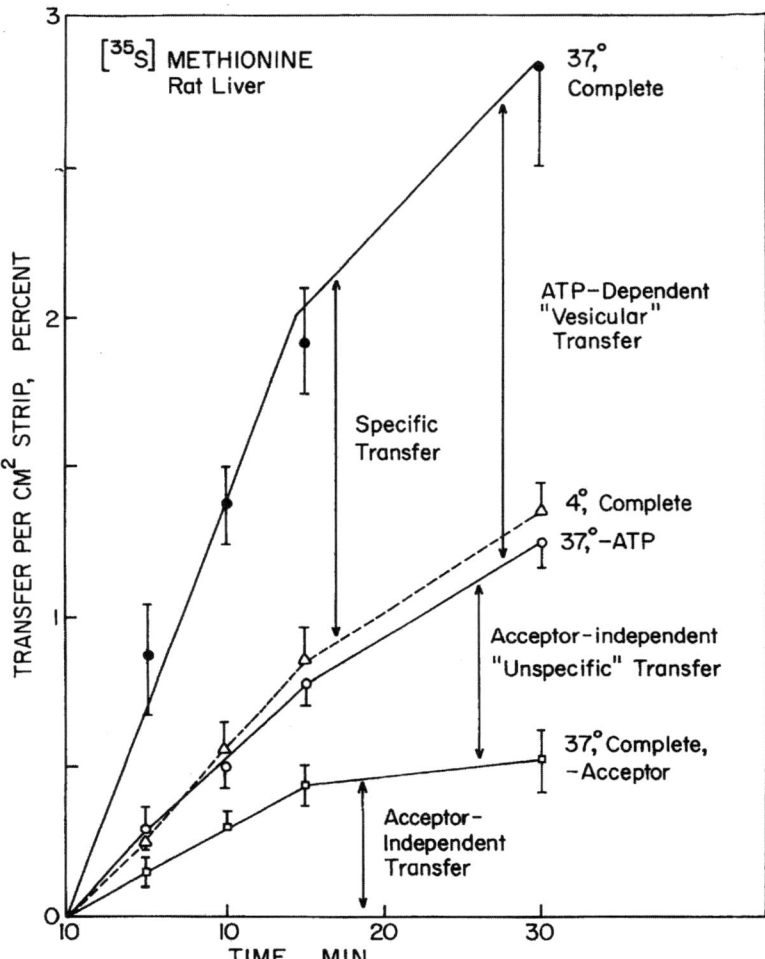

Fig. 10.4. Kinetics of endoplasmic reticulum to Golgi apparatus transfer in rat liver (Morré et al., 1993). Illustrated are the temperature- and ATP-dependent and temperature- and ATP-independent components of transfer. The ATP-dependent transfer components are inhibited by *N*-ethylmaleimide and correlate with vesicle-mediated transfers. Reproduced from Morré, 1998 with permission from Springer Science+Business Media, LLC.

Transfer in the presence of cytosol monitored over 30 mm exhibited both ATP dependent and independent components. Specific transfer, that occurring in the complete (+cytosol, +ATP) system at 37°C was ATP dependent (Fig. 10.4).

Inter Golgi apparatus transfer also was reconstituted using rat liver fractions but failed to result in significant transfer of radioactivity from radiolabeled donor Golgi apparatus to unlabeled acceptor Golgi apparatus (Nowack et al., 1987). Part of the difficulty was traced to an apparent lack of intermediate vesicle formation by rat liver Golgi apparatus donor in solution. The failure of the rat liver system to meet this requirement of inter Golgi apparatus transfer with Golgi apparatus acceptor immobilized on nitrocellulose has added to the argument for a tubular rather than a vesicular mechanism of inter Golgi apparatus exchange (Morré and Keenan, 1994; Wiedman, 1995; Morré and Keenan, 1997).

10.3.1. Fidelity and Efficiency of Cell-Free Transfer in Rat Liver: Comparison to Studies with Liver Slices and Tissues

Cell-free transfer from transitional endoplasmic reticulum to Golgi apparatus in rat liver exhibits acceptor-, temperature-, and pH-specificity comparable to that observed in vitro in addition to an efficiency approaching that observed in vivo. Specific cell-free transfer is proportional to the number of transition vesicles generated (Paulik et al., 1988). With transition-vesicle-enriched fractions, nearly 50% of the total radioactivity was transferred to cis Golgi apparatus. This corresponded to the proportion of vesicles present based on morphology. However, with transitional endoplasmic reticulum no more than 0.5 to 1% of the total membrane of the transitional endoplasmic reticulum was represented by blebbing profiles representing transition vesicles after 15 mm of incubation with ATP plus cytosol. Thus, specific transfers of 2.5 to 5% of the starting radioactivity represented a practical upper limit of transfer without first isolating and concentrating the transition vesicles.

The percentages of transfer observed in the cell-free system are comparable to those estimated in vivo or with liver slices (Dunkle et al., 1992). Estimates of 5.3% transfer over 10 to 20 min of radiolabel from the endoplasmic reticulum to the Golgi apparatus to the plasma membrane compares closely to transfers of between 2.5% and 5% estimated for isolated transitional endoplasmic reticulum in the cell-free system from rat liver (Morré, 1998).

10.3.2. Donor and Acceptor Specificity

The liver system is donor and acceptor specific (Nowack et al., 1987, Dunkle et al., 1992). When Golgi apparatus stacks were separated by preparative free-flow electrophoresis into subfractions enriched in cisternae derived from cis, medial, and trans portions of the stack (Chapter 3), efficient transfer of radioactive membrane proteins from radiolabeled transitional endoplasmic membranes was observed only with the cis elements (Dunkle et al., 1992). Trans elements were devoid of specific acceptor activity. Additionally, transfer between transitional endoplasmic reticulum and cis Golgi apparatus transfer exhibited saturation kinetics with respect to acceptor membrane suggesting a limited number of fusion sites (Dunkle et al.. 1992). Saturation kinetics with respect to acceptor Golgi apparatus (Dunkle et al., 1992), support the suggestion that the in vitro-generated vesicles utilize transition vesicle-specific docking sites of finite number associated with the cis Golgi apparatus cisternae (Rothman and Warren, 1994) as first predicted by Palade (1983).

10.3.3. Temperature Dependence and 16°C Temperature Block

Transition vesicle formation in vitro showed a temperature dependence in the cell-free system (Dunkle et al., 1992) similar to that of the intact cell. A low temperature block at 16°C was observed below which vesicle formation was prevented

(Morré et al., 1989a). When transfer was determined as a function of temperature, a transition was between $12°C$ and $18°C$ was encountered similar to that seen in vivo for formation of the so-called $16°C$ cis Golgi apparatus-located membrane compartment.

ATP-dependent transfer is largely blocked at $4°C$ (Morré and Paulik, 1993) and, in situ, remained blocked below a sharply defined temperature of transition (Tartakoff, 1986). This temperature transition is at about $16°C$ for endoplasmic reticulum to Golgi apparatus transport and at a somewhat high temperature of $23°C$ for Golgi apparatus to plasma membrane transport.

The temperature effect is primarily on vesicle attachment and fusion as transition vesicles do accumulate behind a temperature block but do not fuse with cis Golgi apparatus elements. By analyzing transition vesicle formed at low temperatures, they were found to lack the hexagonal II phase-forming phospholipids phosphatidylethanolamine and phosphatidylserine (Moreau et al., 1992). Apparently these lipids are not transferred to the transition vesicle upon budding at low temperature and the absence of these phospholipids may prevent the fusion of the low-temperature vesicles with the cis Golgi apparatus membranes. By adding lipid mixtures enriched in these lipids to vesicles, their ability to fuse with the cis Golgi apparatus was reconstituted (Moreau et al., 1992).

10.3.4. Lipid and Protein Cotransfer

Trafficking and sorting of lipids during transport from the endoplasmic reticulum to the Golgi apparatus also have been studied using the cell-free system from rat liver (Moreau and Morré, 1991; Moreau et al., 1991, 1992, 1993). Transitional elements of the endoplasmic reticulum prelabeled with $[^{14}C]$- or $[^3H]$acetate were prepared from liver slices as the donor fraction. Non-radioactive Golgi apparatus were immobilized on nitrocellulose as the acceptor. When reconstituted, the radiolabeled donor retained a capacity to transfer labeled lipids to the non-radioactive Golgi apparatus acceptor. Transfer exhibited two kinetically different components. One was stimulated by ATP, facilitated by cytosol and inhibited by N-ethylmaleimide. In parallel with protein transport, the ATP-dependent lipid transfer occurred with a temperature transition at about $20°C$. The other component of lipid transport was not stimulated by ATP, did not require cytosol, was acceptor unspecific, was unaffected by inhibitors and, while temperature dependent, did not exhibit a sharp temperature transition. The ATP-dependent transfer was vesicular.

Transition vesicles isolated by preparative free-flow electrophoresis, when used as the donor fraction, transferred lipids to Golgi apparatus acceptor with a 5–6-fold greater efficiency than that exhibited by the unfractionated transitional endoplasmic reticulum. The formation of these transition vesicles appeared to be the ATP-dependent step. Transferred lipids were chiefly phosphatidylcholine and cholesterol. Membrane triglycerides, major constituents of the transitional endoplasmic reticulum membranes, were both depleted in the transition vesicle enriched fractions and not transferred to Golgi apparatus suggestive of lipid sorting prior to or during transition vesicle formation.

The characteristics of the ATP plus cytosol-dependent transfer of lipids were similar to those for protein transfer mediated by transition vesicles. Thus, the 50–70 nm vesicles derived from transitional endoplasmic reticulum appear to function in the trafficking of both newly synthesized proteins and lipids from the endoplasmic reticulum to the Golgi apparatus.

10.3.5. Processing of Transferred Constituents: Evidence for Functional Fusion of Donor and Acceptor Compartments

10.3.5.1. Membrane Glyco (G) Protein of Vesicular Stomatitis Virus

One protein that has been followed extensively as a processing marker of widespread utility including the liver system is the membrane glycoprotein (G protein) of VSV (Rothman et al, 1984; Balch et al., 1987; Rothman, 1987; Wattenberg, 1991).

To validate ER to Golgi apparatus transfer, VSV G protein was radiolabeled and the conversion of the mannose$_{8-9}$ form present in transitional endoplasmic reticulum to the mannose$_5$ form of the Golgi apparatus was monitored (Paulik et al., 1999). Processing is the result of specific α-mannosidases located at the luminal surface of the cis Golgi compartment. Since man$_{8-9}$ to man$_5$ processing cannot take place without luminal continuity between donor and acceptor compartments, such processing is taken as an unequivocal demonstration of the functional transfer of VSV G.

The usual design of these experiments was to metabolically label virus-infected cells under conditions where synthesis and secretion of the virus coat was the dominant membrane biosynthetic function of the cell. The electrophoretic mobility on SDS-PAGE of the processed man$_5$ form of the protein being greater than that of the unprocessed man$_{8-9}$ form, permitted evaluation of the degree of processing from electrophoretic mobility of the radiolabeled VSV-G on SDS-PAGE. Temperature-sensitive virus strains were available where transfer of G protein from the endoplasmic reticulum could be delayed at the restrictive temperature and then induced synchronously at the permissive temperature (Younger et al., 1966). The temperature control was retained by purified fractions of transitional endoplasmic (Morré et al., 1993). The targeting and fusion mechanisms of ER-derived transition vesicles was conserved among widely divergent species (Paquet et al., 1986; Morré et al., 1989b) and was observed even in heterologous transfer systems (e.g., endoplasmic reticulum from CHO cells and cis Golgi apparatus from rat liver) (Fig. 10.5; Morré et al., 1993).

10.3.6. Lipid Processing

To investigate phospholipid processing, conversion of phosphatidylcholine to lysophosphatidylcholine by an enzyme located at the luminal surface of the cis Golgi apparatus compartment has been used as the indicator (Moreau and Morré, 1991).

A latent phospholipase A was shown to be concentrated in cis elements of rat liver Golgi apparatus, the presumed sites of fusion of the 50–70-nm transition vesicles formed from endoplasmic reticulum. As a result, conversion of

ATP-DEPENDENT
VSV G Transfer & Processing

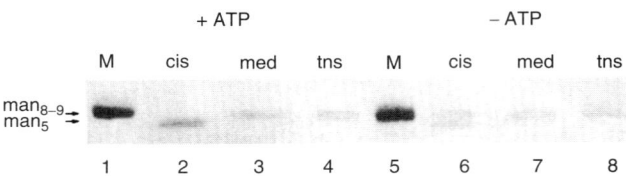

Fig. 10.5. Vesicular stomatitis virus G protein (VSV G) processing as a marker of functional transfer illustrating the requirement for ATP and *cis* Golgi apparatus acceptor specificity in a heterologous system from BHK microsomes to rat liver Golgi apparatus subfractions. *Cis* Golgi apparatus subfractions were isolated by preparative free-flow electrophoresis. The left four lanes show transfer and processing with ATP present. The right four lanes are transfer and processing in the absence of ATP 1 and 5. Starting microsomes incubated in parallel with the acceptor strips. With *cis* Golgi present as acceptor, processing was nearly complete in the presence of ATP (2). In the absence of ATP (6), processing also was observed with *cis* Golgi apparatus as acceptor but not to the same extent as in its presence. With electrophoretic fractions enriched in medial Golgi apparatus (3, 7), or *trans* Golgi apparatus (4, 8) elements, transfer was reduced and no evidence of processing was detected. To prepare the donor fraction, BHK cells were metabolically labeled at the restrictive temperature of 39.5° C with transfer for 30 min at the permissive temperature of 34° C using the temperature sensitive (ts045) form of the VSV. Reproduced from Paulik, Widnell and Whitaker-Dowling, 1999 with permission from Springer Science+Business Media, LLC.

transferred phospholipids to their corresponding lysoforms was utilized as an index of post transfer lipid processing in reconstituted membrane transfer in rat liver (Fig. 10.6) [^{14}C]CDP-choline and endogenous cytidyltransferases were used to label the phosphatidylcholine of transitional endoplasmic reticulum in vitro. In the reconstituted transfer system, the radiolabeled phosphatidylcholine was transferred via transition vesicles to Golgi apparatus immobilized on nitrocellulose strips in a time- and temperature-dependent process. Transfer was promoted by ATP and the ATP-dependent transfer and was specific for cis Golgi apparatus elements as acceptor. Trans Golgi apparatus elements were ineffective as acceptors. Median Golgi apparatus elements were intermediate. A portion of the transferred phosphatidylcholine was converted subsequently to lysophosphatidylcholine also in a time- and ATP-dependent manner. The phospholipase A activity of the Golgi apparatus was more than 90% latent (active site located on the lumens of the Golgi apparatus membranes). Therefore, the lipid-containing vesicles derived from endoplasmic reticulum must have combined with cis Golgi apparatus membranes as the basis for Golgi apparatus-dependent phospholipase A processing of endoplasmic reticulum-derived phosphatidylcholine. Since the lipids were processed by phospholipase A in approximately the same proportion as occurs in situ, the findings offer evidence both for the specificity of the ATP-dependent component of cell-free lipid transfer from endoplasmic reticulum to Golgi apparatus and its fidelity to lipid transfer observed in vivo.

    As an activity complementary to the latent phospholipase A, isolated Golgi apparatus, highly purified from rat liver, were found to contain an acyl transfer

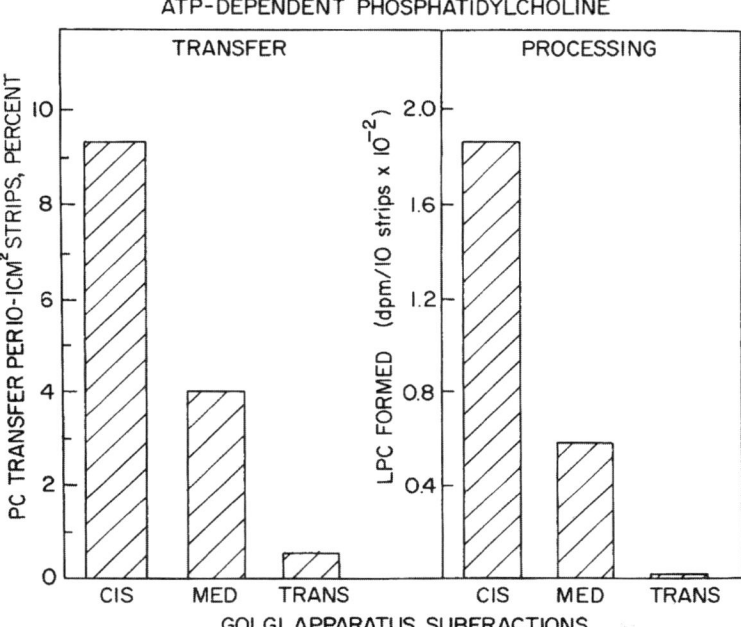

Fig. 10.6. ATP-dependent phosphatidylcholine (PC) transfer and processing in the endoplasmic reticulum to Golgi apparatus cell-free system from rat liver. Formation of lysophosphatidylcholine (LPC) from the transfused PC provides a measure of processing. Results are from Moreau and Morré, 1991.

activity capable of restoring the acyl chains of the lysophospholipid products of the action of phospholipase A_2 on phosphatidylcholine (Lawrence et al., 1994). The activity was located primarily in cis and medial Golgi apparatus fractions, had a pH optimum of 6.0 to 7.5 and was stimulated by various acyl-CoA derivatives but not by fatty acids plus ATP. The activity was determined from the conversion of [^{14}C]lysophosphatidylcholine to [^{14}C]phosphatidylcholine.

10.3.7. Glycoconjugate Processing

Incorporation of radiolabeled sugars from corresponding sugar nucleotides to endogenous membrane protein acceptors was utilized as a measure of glycoprotein processing with Golgi apparatus subfractions generated by preparative free-flow electrophoresis (Morré et al., 1983b). These measurements do not monitor glycosyltransferase distribution per se but serve as a measure of available acceptor. For example, with asparagine-linked oligosaccharides of glycoproteins, galactose acceptors decrease toward the mature face with a corresponding increase in availability, through galactose addition, of acceptors for sialyl transfer.

With cultured cells, glycolipid transport was measured as incorporation of labeled sialic acid into lactosylceramide (Wattenberg, 1990). Membranes of a

CMP-sialic acid transporter deficient mutant were used as donor and Golgi apparatus acceptor was from a cell line defective in UDP-galactose transport. In this system, the incorporation of [^{3}H]sialic acid into lactosylceramide to form GM$_3$ required both donor and acceptor membranes, cytosol, ATP and incubation at 37°C.

Fucosylation of dipeptidylaminopeptidase (DPP-IV) from rat liver endoplasmic reticulum by Golgi apparatus of guinea pig liver was utilized by Paulik et al. (1988) as a measure of functional protein transfer in the cell-free liver system. Unlabeled membranes were incubated in the complete transfer system containing GDP-1-[^{14}C](U)L-fucose. After 30 mm, the membranes were dissolved in detergent solution and DPP-IV was immunoprecipitated using rat specific antisera which did not cross react with guinea pig DPP-IV.

10.4. Nucleoside Triphosphate Dependence of Endoplasmic Reticulum to Golgi Apparatus Membrane Transfer

A characteristic of all cell-free systems described early was an energy dependence described initially as an ATP or nucleoside triphosphate (NTP) requirement (Balch et al., 1984a; Balch and Keller, 1986; Balch et al., 1987; Doms et al., 1987). NTPs were required for vesicle budding by rat liver transitional endoplasmic reticulum (Morre et al., 1986a; Nowack et al., 1987; Paulik et al., 1988) but vesicle attachment to or coalescence with cis Golgi apparatus was ATP independent (Paulik et al., 1988). The transitional endoplasmic reticulum is a part rough (with attached ribosomes), part smooth (lacking ribosomes) fraction of endoplasmic reticulum capable of forming the intermediate vesicles and membranes that transfer membrane to the cis Golgi apparatus compartment (Morré et al., 1986a).

With both transfer of membrane proteins (Nowack et al., 1987; Paulik et al., 1988) and of membrane lipids (Moreau et al., 1991) in the rat liver system, ATP-independent as well as ATP dependent components of transfer were observed (Fig. 10.4). While some transfer was observed to be ATP-independent there was a clear requirement for ATP to support sustained production of abundant vesicles from the transitional endoplasmic reticulum (Paulik et al., 1988).

When the cell-free formation of transition vesicles from isolated transitional endoplasmic reticulum was measured, budding was supported by either GTP or ATP although the yield was still greatest with ATP (Fig. 10.7). These measurements were at what was determined to be a saturating concentration of nucleotide triphosphate of 800 µM and in the absence of the ATP regenerating system. Transition vesicle formation was enhanced by the combination of ATP plus GTP (Fig. 10.8) compared to either ATP and GTP alone. However, an even greater enhancement was observed when UTP was substituted for GTP in the mixture (Fig. 10.8). This result was of interest since the transfer cocktail of Balch et al. (1984a) contained 250 µM UTP in addition to 50 µM ATP and an ATP regenerating system. The additional stimulation by UTP points, as well, to some unique characteristic of the budding step from endoplasmic reticulum in rat liver involving nucleotides other than GTP.

Based on antibody and cobalt inhibition data, an AAA-ATPase of the transitional endoplasmic reticulum was suggested to be involved in somehow

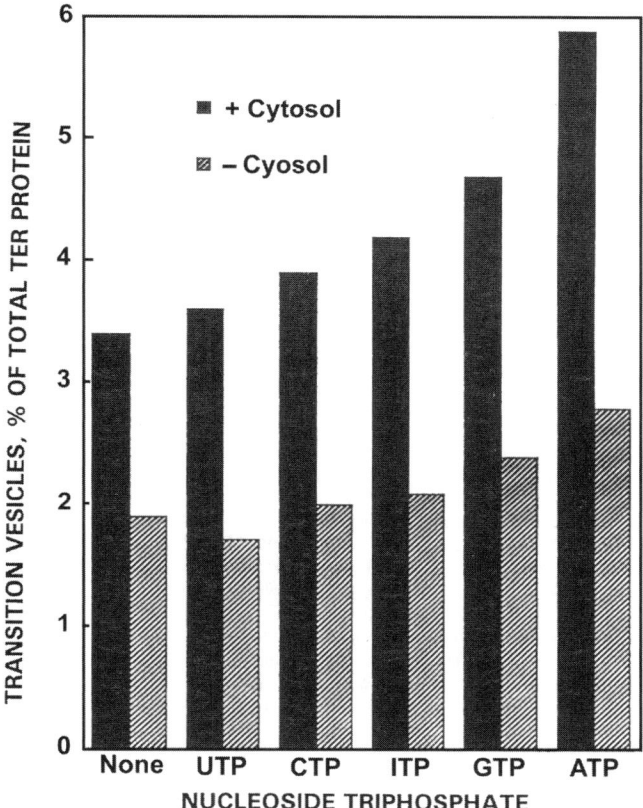

Fig. 10.7. Nucleotide specificity of nucleoside to phosphate-stimulated formation of transition vesicles from transitional endoplasmic reticulum of rat liver in the presence (solid bars) and absence (shaded bars) of cytosol. The isolated transition vesicle fractions were incubated for 10 min in the presence of 800 µM nucleotide but without an ATP regenerating system after which the transition vesicles formed were isolated by preparative free-flow electrophoresis. The protein content of the transition vesicles expressed as a percentage of the starting transitional endoplasmic reticulum (TER) served as the measure of transition vesicle formation. Both GTP and ATP supported vesicle formation but ATP was most effective. Reproduced from Morré, 1998 with permission from Springer Science+Business Media, LLC.

mediating the ATP requirement (Zhang et al., 1994). GTP might substitute for ATP at the active site of the ATPase to account for some degree ATP-independent transfers even in those systems driven by GTP but utilizing the AAA-ATPase. The p97 AAA-ATPase from *Xenopus* reported by Peters et al. (1990) did utilize GTP as substrate albeit at a lower specific activity than with ATP (1.5 µmol/mg/h for GTP compared to 5 µmol/mg/h for ATP).

A role for ATP in transport of membrane proteins from endoplasmic reticulum and the intermediate compartment to Golgi apparatus in situ has been deduced by Verde et al. (1995) from studies of VSV G protein transport in VSVtsO45-infected Vero cells. ATP depletion was achieved with dinitrophenol, carbonylcyanide chlorophenyl hydrozone (CCCP), or sodium azide (see also

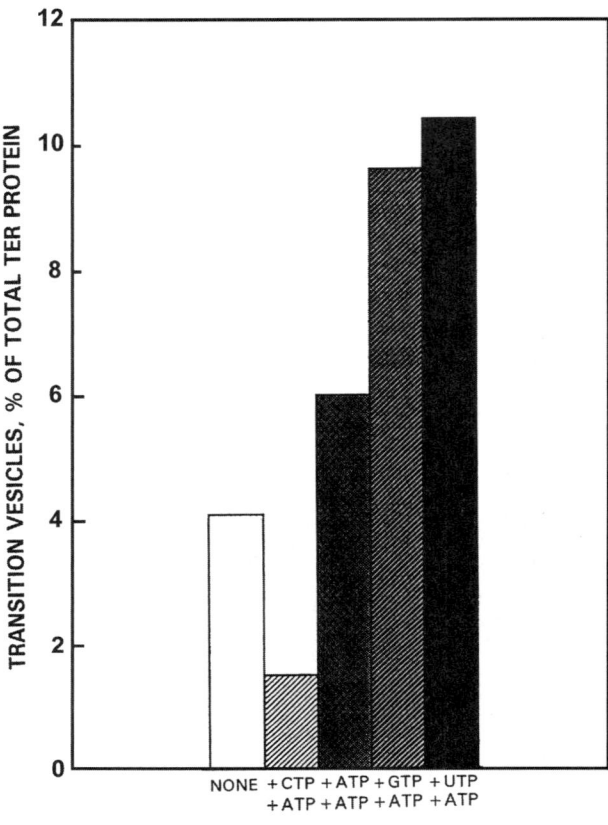

Fig. 10.8. As in Fig. 10-7 except nucleotides were added in equimolar combinations compared to ATP alone. The final concentration of total nucleotides, including ATP alone, was 1.6 mM. Cytosol was present. A mixture of ATP plus GTP was more effective in supporting transition vesicle formation than was ATP alone. However, a mixture of TP plus UTP was just as effective as ATP plus GTP. Reproduced from Morré, 1998 with permission from Springer Science+Business Media, LLC.

Burkhardt and Argon, 1989 for similar results with CCCP). ATP depletion in the cells was verified by luciferase assay of ATP levels. Maturation of newly synthesized CD8 protein, an O-glycosylated human membrane protein stably expressed by FRT-U1O cells after transfection, was blocked by ATP depletion in vivo in a manner approximately equivalent to a low temperature (15° C) block in G-glycoprotein transport in VSVtsO45-infected Vero cells observed in parallel.

The GTP-G protein requirement for cell-free transfer also has been studied widely following initial findings implicating GTP and G proteins (Beckers and Balch, 1989; Beckers et al., 1989; Baker et al., 1990) in trafficking. Transfer can be blocked by the inclusion of the non-hydrolyzable analog of GTP, GTPγS, in the assay (Plutner et al., 1992, Rowe et al., 1996).

Bourne (1988) emphasized the possible role of GTP hydrolysis in ensuring directionality. However, other roles also may be important. Baker et al. (1990) suggested that coupling of GTP hydrolysis, to some slow recycling steps, for example, could increase the rate of membrane budding. GTP is not likely to provide a major source of energy for the mechanical action necessary for vesicle budding (Hicke and Schekman, 1990). However, the energy carried by GTP in a high energy complex may serve to promote fidelity of vesicle budding and targeting in much the same manner as GTP-binding translation factors such as EfTu operate in protein synthesis (Hicke and Schekman, 1990).

10.5. Reconsititution of Golgi Apparatus to Plasma Membrane Transfer

Initial efforts to reconstitute cell-free trafficking of membrane constituents between the Golgi apparatus and the plasma membrane have been reported for rat liver (Rodriguez et al., 1992). Transfer was temperature dependent but not stimulated by ATP. In contrast to results with ATP, transfer was stimulated by NADH and especially by NADH in combination with ascorbate free radical. The results are consistent with an involvement of electron transfer reactions previously described for trans Golgi apparatus (Morré et al., 1987b) and for coated vesicles (Sun et al., 1984) in the dynamics of vesicular transport between the Golgi apparatus and the cell surface.

Generation of exocytic transfer vesicles in vitro from the trans Golgi apparatus network was reported by Salamero et al. (1990). Budding was dependent upon ATP, cytosol, and temperature and was maximal within 10 min.

10.5.1. Cell-Free Transfer in Cultured Cells

The cell-free transfer systems developed by Rothman, Balch and colleagues (Balch et al., 1983, 1984a, b; Balch and Rothman, 1985; Balch and Keller, 1986; Beckers et al., 1987) initially on VSV-G processing as a measure of transfer. An advantage of the VSV-G processing system has been the ability for comparisons with permeabilized cells.

Permeabilized cells have offered a system that combines some of the advantages of the completely cell-free system (access to impermeable reagents, opportunities to assay cytosolic factors) (Beckers et al., 1987, Simons and Virta, 1987, Warren et al., 1988; Gruenberg and Howell, 1989) without the complete loss of positional relationship characteristic of the intact cell. Especially advantageous has been the use of permeabilized cells for comparisons to completely cell-free transfer and to intact cells in the investigation of the steps and order of events involved in the processing of VSV-G-protein (Balch et al., 1987). Without serious exceptions, the steps and order of events are the same for all three.

Since the VSV-G protein processing assay is very specific, donor and acceptor compartments in cell-free assays need only to be isolated but do not necessarily need to be purified to homogeneity. One common form of the assay is that described earlier under processing of transferred constituent measured is the conversion of the mannose$_{8-9}$ form of VSV-G present in endoplasmic reticulum to the mannose$_5$ form in the Golgi apparatus (Beckers et al., 1987; Balch et al., 1983, 1984a, b; Balch and Rothman, 1985; Balch and Keller, 1986) with processing determined from molecular weight differences on SDS-PAGE (cf. Fig. 10.5). The VSV G protein is radiolabeled in these assays. Processing is the result of a luminal α-mannosidase of the Golgi apparatus. As a result Mann$_{8-9}$ to Mann$_5$ processing serves as an unequivocal demonstration of functional endoplasmic reticulum to Golgi apparatus transfer. Vesicle formation is ATP-dependent and vesicle fusion with the cis Golgi apparatus is blocked both by N-ethylmaleimide and GTP-γ-S and, presumably, requires hydrolysis of GTP (Melancon et al., 1987; Beckers et al., 1989).

From the work of Bannykh and Balch (1997), rather than to block transport, GTPγS permanently activated Sar1 and other GTPases to permit stable coat assembly and an accumulation of coated buds. Initially, semi-intact NRK cells were used to examine the potential role of these proteins known collectively as COPII components in the formation of both endoplasmic reticulum-derived buds and vesicle transfer complexes (VTCs) (Plutner et al., 1992; Peter et al., 1993; Balch et al., 1994; Pind et al., 1994b). To correlate morphological observations with previous in vitro biochemical studies, semi-intact cells were incubated for 30 mm at 32°C in the absence of cytosol. These preparations failed to generate detectable VTCs as expected. Any preexisting VTCs were apparently lost during permeabilization (Aridor et al., 1995). Essential components of the COPII machinery required for the export of cargo from the endoplasmic reticulum are contained in the cytosol (Barlowe et al., 1994; Kuge et al., 1994) such that VTCs generated in vitro in the presence of cytosol were nearly identical to those observed in vivo.

Thus export from the endoplasmic reticulum is mediated by COPII constituents (Balch, 1990; Harter, 1995; Schekman and Orci, 1996) and in the crude system vesicles can be formed in vitro by incubation of membranes with cytosol and ATP (Balch et al., 1984a, b). From studies with yeasts (see below), the COPII coat has been defined as containing five soluble components including the small GTPase Sar1 and two cytosolic coat complexes, Sec23p/24p and Sec13p/31p (Barlowe et al., 1994). Mammalian homologs to each of these proteins are known. Thus the coat machinery dictating export from the endoplasmic reticulum is highly conserved evolutionarily. Budding from the endoplasmic reticulum is considered to be initiated by activation of the Sar1 GTPase through the activity of Sec12p, a Sar1 specific guanine nucleotide exchange factor (GEF) (Barlowe and Schekman, 1993). Another resident membrane protein, Sec16p, could be involved in Sec23/24 recruitment (Espenshade et al., 1995; Gimeno et al., 1996). Members of the p24 gene family might be involved as well in contributing to the fidelity of these interactions (Elrod-Erickson and Kaiser, 1996; Fiedler et al., 1996; Schimmoller et al., 1995).

Another form of the assay used to evaluate inter-Golgi apparatus transfer has been to mark transport of G protein from the cis compartment of one set of Golgi apparatus membranes to one or more of the medial compartments in a second enzymatically distinct population (Rothman et al., 1984). One set of Chinese hamster ovary (CHO) cells was infected with VSV, lysed, and a crude Golgi apparatus fraction was prepared. The fractions used for the donor fraction were prepared from a glycosylation CHO defective mutant, clone 15B, which lacked the glycosyl transferase enzyme responsible for the addition of N-acetylglucosamine (GlcNAc) in the Golgi apparatus (Dunphy and Rothman, 1985). The acceptor fraction was prepared from uninfected wild-type CHO cells. The assay which consisted of two Golgi apparatus populations one containing a VSV G protein with an immature oligosaccharide, and lacking the enzyme necessary to process oligosaccharides to their mature form. The other has no G-protein, but does contain the transferase. Thus, when processing occurs as evidenced by transfer of $[^3H]$-N-acetylglucosamine to immature G protein oligosaccharide, evidence for fusion of donor and acceptor compartments is provided. Transport requires the presence of nucleotide triphosphates and magnesium as well as soluble cytosolic proteins (Balch et al., 1984a, Balch and Rothman, 1985; Doms et al., 1987; Davidson and Balch, 1992).

Inter Golgi apparatus transport has not been reproduced in the liver system with immobilized acceptor primarily due to the requirements of free vesicle formation to achieve transfer in this assay system. The absence of transfer and/or low transfer rates with immobilized acceptor have been traced not to the immobilized acceptor per se but to the failure of intermediate Golgi apparatus donor compartments to form free vesicles (Morré and Keenan, 1994).

10.5.2. Cell-Free Transfer in Yeast

Through a combined genetic and biochemical approach, a set of proteins (Sar 1p, Sec 23p complex, and Sec 13p complex) has been identified that in the presence of nucleoside triphosphate satisfies the requirement for cytosol in the cell-free production of budded vesicles from the yeast endoplasmic reticulum (Hicke and Schekman, 1990; d'Enfert et al., 1991; Salamero et al., 1993; Barlowe et al., 1994). The vesicles formed under these conditions are competent for fusion with Golgi apparatus membranes from yeast (Salama et al., 1993).

The yeast cell-free budding assay measures the production of slowly sedimenting (small) vesicle intermediates containing $[^{35}S]$-labeled core-glycosylated pro-α-factor (gpαF), the precursor to secreted mating pheromone in yeast (Rexach and Scheckman, 1991). The ability to target and fuse with the Golgi compartment is verified by the ability of the $[^{35}S]$ gpα F to acquire specific α-1,6-mannose modifications (Salama et al., 1993).

In the crude system, budding of yeast endoplasmic reticulum required membranes, cytosol, GTP, and ATP plus an ATP regenerating system (Rexach and Scheckman, 1991; Wuestehube and Scheckman, 1992) as do the CHO system and the ATP-dependent transfer component of the rat liver system. However, in more refined yeast systems, an ATP requirement was no longer observed. Not only was there no ATP requirement but the budding process was driven by non hydrolyzable analogs of GTP (Rexach and Scheckman, 1991; Barlowe and Scheckman, 1993; reviewed by Barlowe et al., 1994).

GTP-γ-S and GMP-PNP both promoted the formation of ER-derived transport vesicles. Initial rates of vesicle release were similar with different guanine nucleotides but vesicle budding may have ceased after about 10 mm (Barlowe et al., 1994). Since only GTP binding was required in the refined system, a certain amount of vesicle budding can occur in the absence of GTP hydrolysis.

10.5.3. ATP-Independent Vesicle Budding

Using purified components, vesicle budding from the endoplasmic reticulum has been reconstituted with washed membranes and three soluble proteins, the Sec 13 complex, the Sec 23 complex and the small GTPase Sar1p. In this reconstituted system, vesicle formation proceeds without ATP and, in the presence of GTP, these three proteins substitute for cytosol. What appears to be involved is a site-specific coating of membranes followed by pinching off of coated vesicles of uniform dimensions. Once vesicles are pinched off, the coat is disassembled exposing a set of membrane-bound components that promote vesicle fusion (Rothman, 1994). The membrane-bound coat of the ATP-independent vesicle is composed of COPII proteins and lacks clathrin (Barlowe et al., 1994).

The ATP-independent COPII vesicles resemble morphologically the ATP-dependent vesicles (Barlowe et al., 1994). By electron microscopy both types of vesicles are ca. 60 nm in diameter and are covered by a ca. 10-nm thick electron-dense, non-clathrin coat. Whereas the GTP-bound form appears to initiate budding, the role of GTP hydrolysis by Sar1p in the process is unclear but may be to serve as a resetting mechanism for additional cycles (Bourne et al., 1991).

In the presence of the non-hydrolyzable GTP analog GMP-PNP, COPII coat proteins Sec 23p/24p, Sec 13p/31p and the small GTPase Sar1p bind even to liposomes composed of pure lipids (Matsuoka et al., 1997). Incubation of COPII coat proteins with large multilamellar liposomes resulted in the formation of coated vesicles of about 80-nm diameter. ATP-dependent fusion of liposomes with the Golgi apparatus of perforated cells had been reported earlier by Kobayashi and Pagano (1988).

If budding could be initiated only by the association of COPII complex proteins with membranes or even with lipid bilayers, then budding within the cell and in cell-free systems with crude cytosol should be entirely promiscuous which it is not. ATP-dependent budding is restricted primarily to transitional endoplasmic reticulum and to specific regions within the transitional endoplasmic reticulum.

10.6. Cell-Free Membrane Transfer in Plants

In plants the donor and acceptor specificity of cell-free transfer of radiolabeled membrane constituents, chiefly lipids, was examined using purified fractions of endoplasmic reticulum, Golgi apparatus, nuclei, plasma membrane, tonoplast, mitochondria, and chloroplasts prepared from green leaves of spinach (Morré et al., 1991b, 1992a). Donor membranes were radiolabeled with [^{14}C]acetate.

Acceptor membranes were unlabeled and immobilized on nitrocellulose filters. The assay was designed to measure membrane transfer resulting from ATP- and temperature-dependent formation of transfer vesicles by the donor fraction in vesicles with the immobilized acceptor. Significant ATP-dependent transfer in the presence of cytosol was observed only with endoplasmic reticulum as donor and Golgi apparatus as acceptor. Transfer in the reverse direction, from Golgi apparatus to endoplasmic reticulum, was only 0.2 to 0.3 that from endoplasmic reticulum to Golgi apparatus. ATP-dependent transfers also were indicated between nuclei and Golgi apparatus. Specific transfer between Golgi apparatus and plasma membrane and, to a lesser extent, from plasma membrane to Golgi apparatus was observed at 25°C compared to 4°C but was not ATP plus cytosol-dependent. All other combinations of organelles and membranes exhibited no ATP plus cytosol-dependent transfer and only small increments of specific transfer comparing transfer at 37°C to transfer at 4°C. Thus, the only combinations of membranes capable of significant cell-free transfer in vitro were those observed by electron microscopy of cells and tissues to be involved in vesicular transport in vivo (endoplasmic reticulum, Golgi apparatus, plasma membrane, nuclear envelope). Of these, only with endoplasmic reticulum (or nuclear envelope) and Golgi apparatus, where transfer in situ is via 50 to 70 nm transition vesicles, was temperature- and ATP-dependent transfer of acetate-labeled membrane reproduced in vitro. Lipids transferred included phospholipids, mono- and diacylglycerols, and sterols but not triacylglycerols or steryl esters, raising the possibility of lipid sorting or processing to exclude transfer of triacylglycerols and steryl esters at the endoplasmic reticulum to Golgi apparatus step.

Other cell-free transfer steps using plant fractions include transfer between endoplasmic reticulum and Golgi apparatus in a green algae *Micrasterias americana* (Noguchi and Morré, 1991), ATP-induced budding of nuclear envelope (Hellgren and Morré, 1992) and ATP-dependent cell-free transfer of lipids from nuclear membranes to Golgi apparatus in germinating axes of garden pea (Morré et al., 1992), cell-free transfer of inositol phospholipids between membrane fractions of soybean (Harryson et al., 1996), cell-free transfer of lipids in leek (Sturbois et al., 1994) and cell-free transfer of envelope monogalactosylglycerides to thylakoids in developing chloroplasts (Morré et al., 1991a).

The endoplasmic reticulum to Golgi apparatus transfer systems from rat liver (Morré et al., 1989a) and from the green algae *Micrasterias* (Noguchi and Morré, 1991b) where temperature transitions have been studied, both exhibit a very precise 16°C block. At temperatures below 16°C, transfer is sharply reduced and at temperatures above 16°C, transfer proceeds normally.

10.7. Model for ATP-Dependent Vesicle Budding based on the Rat Liver System

Findings that reconcile both ATP-dependent and -independent steps of rough endoplasmic reticulum to Golgi apparatus conversion are provided by studies of J. Paiement and colleagues of the University of Montreal (Lavoie et al., 1996) using a cell-free system from rat liver. In their system represented diagrammatically in Fig. 10.9, rat liver microsomes, when incubated together with GTP, fuse to form large rough-surfaced ER sheets. In the presence of GTP plus 2 mM ATP, intermediate compartments composed of smooth ER tubules are induced to flow out from specific regions of the pre-existing endoplasmic reticulum complex. The latter appear to require both ATP and GTP hydrolysis. Their formation is blocked by both GTP-γ-S and ATP-γ-S and the resulting compartment has many of the characteristics ascribed to the ATP-dependent component of cell-free ER to Golgi apparatus transport described by us for rat liver. The formation of the smooth tubules is blocked by antisera, to the transitional endoplasmic reticulum AAA-ATPase (p97) of Zhang et al. (1994) and promoted by the addition of authentic AAA-ATPase (Roy et al., 2000). Total endoplasmic reticulum surface is unchanged during the ATP-dependent step as expected for the inter-conversion of rough sheets into smooth tubules. Upon addition of cytosol and GTP, the smooth tubules then convert into what appear as COPII-coated transition vesicles, thus duplicating both the ATP-dependent and ATP-independent steps in the conversion of rough endoplasmic reticulum into transition vesicles (Fig. 10.9).

The above described findings are consistent with a proposed model that is general to membrane displacement and relates not only to ATP-dependent smooth compartment formation but to cell motility and cell enlargement as well (Morré, 1994b, c). The premise followed in developing the model was to seek enzyme systems, cysteine residues or coiled-coil domains of proteins, that could interact with membrane proteins to incorporate or translocate specific membrane components and result in physical membrane displacement or cell enlargement. Criteria considered were dependence on ATP, sensitivity to thiol reagents and use of specific inhibitors or activators capable of blocking or enhancing vesicle budding.

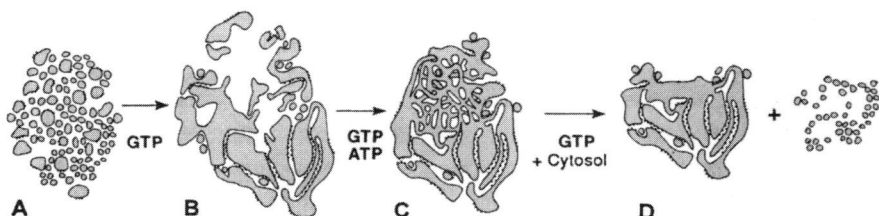

Fig. 10.9. Diagrammatic representation of the cell-free system of smooth compartment formation from isolated vesicles of endoplasmic reticulum in the system of Lavoie et al. (1996). Isolated rough microsomes from homogenates of rat liver (A), when incubated at 37° C for 2 h in the presence of 0.5 mM GTP, fuse to assemble large sheets of rough endoplasmic reticulum (B). Neither cytosol nor ATP are required for rough endoplasmic reticulum assembly. When both GTP and 2 mM ATP are present, as system of smooth tubules is assembled from the rough endoplasmic reticulum by displacement out from ribosome covered areas (C). The formation of these smooth tubules is ATP-dependent but does not require cytosol. Finally if GTP plus cytosol from rat liver is added, the smooth tubules convert to numerous small vesicles resembling transition vesicles (D). Although an ATP-dependent step cannot be ruled out in the formation of transition vesicles, vesiculation in this system requires cytosol and is promoted by either GTP or GTP-γ S and is partially inhibited by brefeldin A. Reproduced from Lavoie and Paiement, 1996 with permission from the American Society for Cell Biology.

The components of the model to achieve membrane displacement are fourfold. The AAA-ATPase is located on the cytosolic membrane surface and appears not to be membrane spanning. However a membrane spanning component (extensase) would appear to be necessary to transduce the energy of ATP hydrolysis into physical membrane displacement. A nucleotide exchange enzyme could facilitate the exchange of ADP for ATP in step D and the breakage and reformation of disulfide bonds would be the function of the protein disulfide–thiol interchange activity of the NOX protein at the luminal surface.

At the center of the proposed model is a still hypothetical nucleoside triphosphate (ATP) binding protein capable of undergoing displacement in space (e.g., Fig. 10.10). The condensed form of the protein was assumed to bind ATP (Fig. 10.10A). When the ATP was hydrolyzed by the associated transitional endoplasmic reticulum AAA-ATPase, the energy of hydrolysis would be dissipated as the protein extended. Displacement of ADP and addition of ATP is suggested to result in recoiling. The reverse could just as easily apply where ATP hydrolysis would result in contraction and rebinding would result in extension.

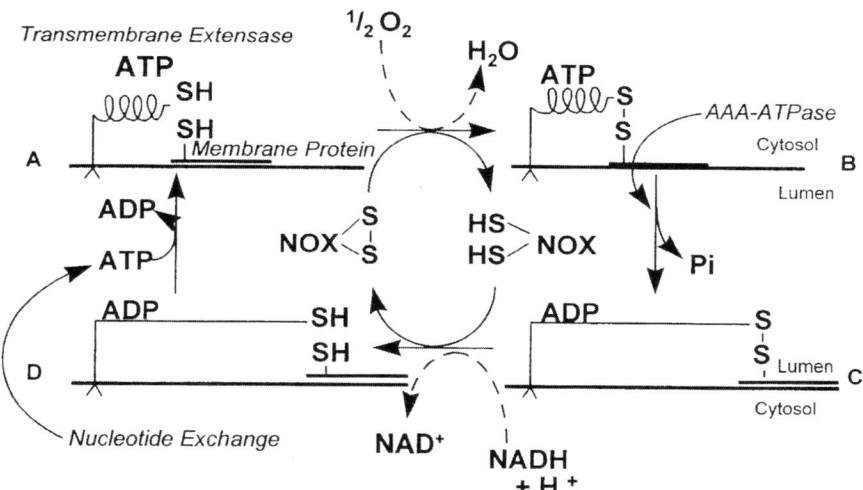

Fig. 10.10. A model for how the NADH:protein disulfide reductase (NADH oxidase = NOX) with protein thiol-disulfide interchange activity that is stimulated by retinol (Morré et al., 1998) and inhibited by brefeldin A (Morré et al., 1994b) could be important to physical membrane displacement and membrane budding. A. A hypothetical transmembrane protein (extensase) with a condensed region capable of extension with nucleoside triphosphate (ATP) bound at the cytosolic surface. B. A disulfide bond is formed between the extensase and a membrane protein in a NOX catalyzed oxidation at the luminal surface. The transmembrane extensase must somehow be permanently anchored (via the cytoskeleton) so as not to slip. C. The ATP is hydrolyzed by the transitional endoplasmic reticulum ATPase and the condensed region extends, displacing the bound protein within the membrane. D. The disulfide is then reduced by the associated NOX protein releasing the bound protein. The ADP is exchanged for another ATP, the condensed extensase reforms, and the system is ready to displace another membrane protein. Repeated cycles would result in the physical displacement of membranes. Reproduced from Morré, 1994b with permission from Springer Science + Business Media.

In order to couple the contraction–extension mechanism to displacement of membrane constituents, oxidation and reduction of thiols were included in the model. In step A prior to displacement, a disulfide bond would be formed between the protein being displaced and the displacing element (extensase). Following displacement in step B, the disulfide bond would be reduced in step C prior to subsequent contraction of the displaced element in step D.

The alternate formation and breakage of disulfide bonds would be accomplished in a number of ways, the simplest of which would involve a protein disulfide–thiol interchange (see next section on retinol stimulation of membrane budding). A final contributor to the mechanism would be a nucleotide exchange protein operative between steps D and A to release the ADP and rebind ATP.

The search for a requirement for ATP hydrolysis in the rat liver system led to the isolation and characterization of the AAA-ATPase (p97 protein) from transitional endoplasmic reticulum of rat liver (TER ATPase) as a candidate protein responsible for ATP hydrolysis (Zhang et al., 1994). Antisera to a 15-amino acid portion of p97 inhibited cell-free transfer between transitional endoplasmic reticulum and Golgi apparatus (Zhang et al., 1994). A member of the AAA-ATPase family, the ca. 100 kDa TER ATPase has duplicate copies of two different ATP binding regions. Cobalt chloride, an inhibitor of the ATPase, also was a very potent inhibitor of cell-free transition vesicle production. Other ATPase inhibitors such as vanadate, KNO_3, oubain, oligomycin or azide did not affect either the AAA-ATPase or transition vesicle production with the transitional endoplasmic reticulum fraction from rat liver (Zhang et al., 1994).

The derived amino acid sequence of cytosolic p97 was homologous with but not identical to the observed amino acid sequence of a cyanogen bromide fragment derived from the TER ATPase. The possibility, therefore, still exists that the TER ATPase represents a member of the AAA-ATPase family closely related to but not identical with the cytosolic form of p97.

10.7.1. Retinol Stimulation of Vesicle Budding in Rat Liver

Vitamin A (retinol) is concentrated in the Golgi apparatus of rat liver (Nyquist et al., 1971). Both vitamin A deficiency and excess influence Golgi apparatus architecture (Morré et al., 1981) and membrane flux through the Golgi apparatus (Morré et al., 1988). Among the ultrastructural alterations consistently associated with livers of rats receiving vitamin A in excess was an increased number of transition vesicles located in the vicinity of the cis Golgi apparatus (Morré and Morré, 1987). When experiments were carried out to determine if donor transition elements isolated from liver might be responsive to retinol, Nowack et al. (1990) showed that, at a near optimum concentration of 1 µg/ml, the rate and amount of transfer from endoplasmic reticulum donor to Golgi apparatus acceptor was approximately doubled. In the complete system (ATP, ATP-regenerating system, and cytosol) plus retinol, there were approximately twice the numbers of 50 to 70-nm vesicular profiles as there were without retinol (Nowack et al., 1990). Results were equivalent when the retinol response was monitored directly by electron microscopy or when transition vesicles were isolated from the incubated transition vesicle preparations by preparative free-flow electrophoresis. When bisected into vesicle formation and vesicle fusion steps, vesicle formation, but not fusion, was found to be retinal responsive both by electron microscopy and from quantitation of numbers of vesicles produced (Nowack et al., 1990).

A direct response of the Golgi apparatus to vitamin A was demonstrated for bovine mammary gland epithelial cells by video-enhanced optical microscopy (Morré et al., 1992b).

Two retinol-responsive enzymatic activities associated with transition elements of rat liver have been described, either or both of which might be involved with the production of transition vesicles. The first of these was a retinol inhibited GTPase of unknown function which might function as a guanine nucleotide exchange protein or as a GTPase (Zhao et al., 1990). The other was a retinol modulated NADH:protein disulfide reductase (NADH oxidase) protein with protein disulfide–thiol interchange activity (Morré et al., 1998). The latter activity is inhibited by brefeldin A (Morré et al., 1994b) has been implicated previously in physical membrane displacement (Morré, 1994a, b; Fig. 10.10). The activity has the ability to restore activity to scrambled, inactive ribonuclease A (Jacobs et al., 1996) and is stimulated or inhibited by retinol depending on the redox environment (Morré et al., 1998). Under reducing conditions, as might be encountered within the cytoplasm, the partial reaction stimulated by retinol appears to be the oxidation of membrane protein disulfides including those associated with the membranes of the transitional endoplasmic reticulum (Morre et al., 1998). This stimulation is not given by retinoic acid and implicates the operation of a retinol stimulated protein disulfide–thiol interchange activity as part of the vesicle budding process in the rat liver system as diagrammed in Fig. 10.10.

10.8. Summary

Advances in understanding the molecular mechanisms of membrane traffic to and through the Golgi apparatus have been predicated in large measure on the use of permeabilized animal cells, and on completely cell-free systems. These systems have included those addressing inter-Golgi apparatus membrane traffic, endoplasmic reticulum to Golgi apparatus traffic, and endocytotic events. Development of cell-free systems depends on the use of isolated fractions. Several cell-free systems have evolved which employ highly purified and well-characterized cell fractions. Central to their development has been the availability of isolated Golgi apparatus fractions in useful yield and fraction purity. The latter may be utilized in the absence of a compartment-specific assays for validation. However, when specificity is achieved by using a compartment-specific assay even relatively crude fractions can be employed. With either type of assay, the major advantage of cell-free systems is that they are most directly amenable to the investigation of molecular mechanisms of membrane trafficking.

11

Growth and Cell Enlargement

Cell growth (cell enlargement) is as fundamental for growth of organisms as is cell division. Implicit in cell growth is an increase in cell size, i.e., cell enlargement. Without cell enlargement, no organism can continue to grow. Cells unable to enlarge eventually are unable to divide such that growth will cease. Yet, compared to cell division, the enlargement phase of cell growth has been largely neglected by cell biologists. Moreover, it is this part of the growth process where the Golgi apparatus may be most closely involved.

Proliferating cells in culture tend to double both their size and their mass before each division (Mitchison, 1971). A dividing cell, if spherical, in order to double in volume, must increase the surface covered by the plasma membrane by a factor of 1.6 (Graham et al., 1973). In addition, they must produce membrane material to compensate for degradation (turnover). Included within this category are dividing cells which form large parts of their surface membrane de novo along the plane of division within a relative short time such as during cell plate formation in higher plants (for review see Whaley, 1975) and furrow formation (Albertson et al., 2005) in non-walled cells during cytokinesis at the end of mitosis.

The observation that continuously proliferating cells precisely double their size during each cycle to maintain constant volumes has resulted in the suggestion of the existence of an active cell size control mechanism in eukaryotic cells, some form of checkpoint control, that functions to prevent delayed or premature cell division at inadequate size or mass. Such a mechanism seems well defined in yeasts but remains an open issue with mammalian cells. Eukaryotic cells coordinate cell size with cell division by regulating the length of the G1 and G2 phases of the cell cycle (Sweiczer et al., 1996).

The coordination of cell growth (enlargement) and division is of fundamental importance in maintaining the size of a living cell stable within certain limits over time. Organisms have developed elaborate strategies to ensure that cell division does not occur until a particular minimum cell size is achieved to guarantee the production of viable progeny cells after each cell division. The pathways that regulate growth are often deranged in human diseases such as cancer (Chapter 12).

For rat hepatocytes, the absolute rate of plasma membrane protein synthesis has been estimated to be about 7 fg/cell per min (Franke et al., 1971b).

D. James Morré and Hilton H. Mollenhauer, *The Golgi Apparatus*.
© Springer 2009

11.1. Golgi Apparatus and Growth

Most investigations would agree intuitively that the Golgi apparatus is essential to sustained membrane growth as part of cell enlargement process. However, direct evidence for such a role is scarce.

Direct observations of a contribution of the Golgi apparatus to the growth process have largely been restricted to electron microscopic observations of stages where membrane formation is rapid either transiently such as during plate formation of dividing cells (Fig. 9.2) or sustained as in tip growth of neurites, pollen tubes and plant hairs (Fig. 8.12). See also Chapter 8.6 control of secretion.

As with most cause and effect relationships, correlative studies are usually a first line of evidence along with inhibitor studies. Both correlative and inhibitor studies support a role for the Golgi apparatus in cell enlargement and tissue growth.

11.1.1. Inhibitor Studies

There are surprisingly few Golgi apparatus-specific inhibitors. Most affect guide elements (microtubules and/or microfilaments), chiefly cytochalasin b and colchicine, such that vesicle migration is blocked or inhibited by disrupting ionic gradients such as the macrolide monensin. Among the best candidates for a Golgi apparatus-specific inhibitor is brefeldin A.

11.1.1.1. Brefeldin A

Brefeldin A (BFA) (Fig. 11.1) is a heterocyclic lactone (macrolide antibiotic) with potent inhibitory effects on trafficking to and through the Golgi apparatus (Klausner et al., 1992). Transport of G protein is inhibited in vesicular stomatitis virus-infected baby hamster kidney cells (Takatsuki and Tamura, 1985), and secretion of plasma proteins, such as albumin and α_1-protease inhibitor, are blocked in rat hepatocytes (Misumi et al., 1986; Oda et al., 1987). Also blocked is secretion of [^{15}S]SO$_4$-labeled secretogranin II from PCI2 cells, a marker of secretory granule budding from the trans-Golgi network (TGN) (Miller et al., 1992). Despite earlier indications that brefeldin A resulted in more or less complete disassembly of the Golgi apparatus and its relocation to the endoplasmic reticulum (Klausner et al., 1992; Takatsuki and Tamura, 1985; Misumi et al., 1986; Oda et al., 1987, 1990; Doms et al., 1989; Lippincott-Schwartz et al., 1989), it now appears that, in most instances, a modified Golgi apparatus remains (Hendricks et al., 1992; DeLemos-Chiaraudini et al., 1992).

Brefeldin causes rapid dissociation of a 110 kDa protein from Golgi apparatus membranes (Donaldson et al., 1990). This protein, β-COP, of the COPI complex, is a component of the non-clathrin-coated vesicles that accumulate as the Golgi apparatus-associated buds often attached to tubules (Chapter 4) when intercisternal transport is blocked by GTP-γ-S (Malhotra et al., 1989; Serafini et al., 1991). These findings suggest that brefeldin A exert similar effects on both pre- and post-Golgi apparatus compartments, blocking forward transport but not

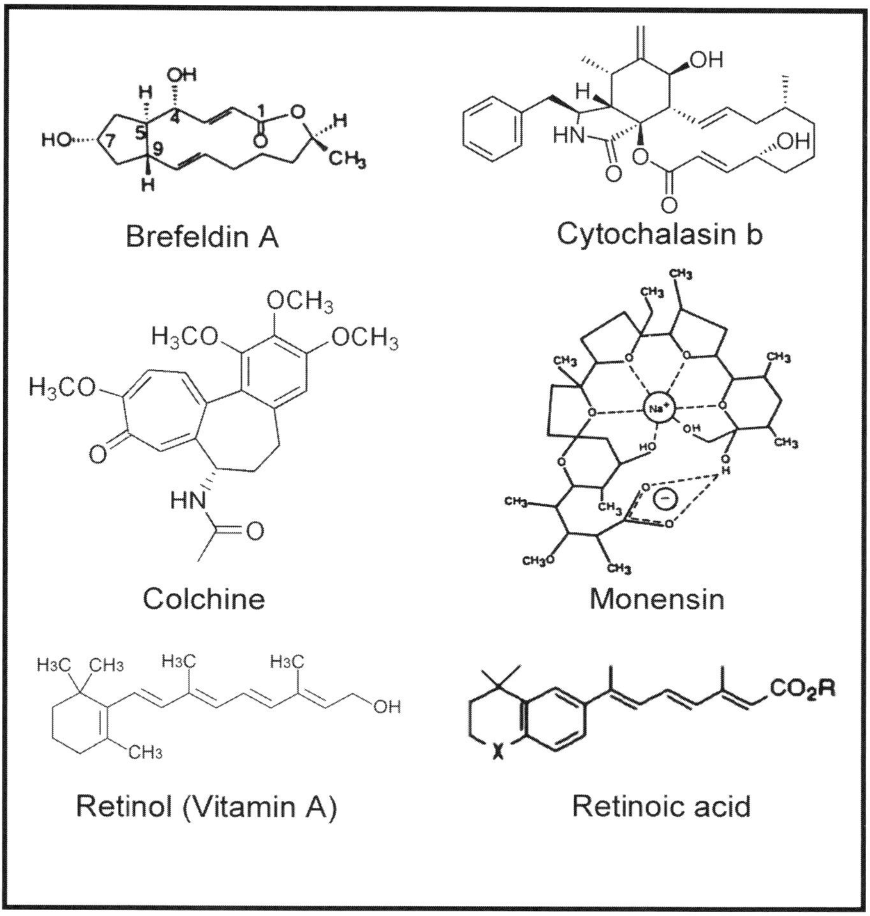

Brefeldin A

Cytochalasin b

Colchine

Monensin

Retinol (Vitamin A)

Retinoic acid

Fig. 11.1. Chemical structures of Golgi apparatus-disturbing agents.

the return pathway (Miller et al., 1992). The Golgi apparatus tubular system and its associated COPI-coated elements have been suggested to function as repositories of resident Golgi apparatus proteins during flow differentiation across the stacks of saccules (Chapter 4). Activity of COPI-coated elements is triggered by membrane binding of the GTP-bound form of ADP-ribosylation factors (ARFs). Brefeldin A blocks the process by inhibiting a guanine nucleotide exchange factor for ARF, GBF1, that is localized in the Golgi apparatus region (Kawamoto et al., 2002). Inactive ARF with GDP bound is converted to active ARF-GTP with the guanine nucleotide-exchange protein being required to accelerate GDP release (Cox et al., 2004). The inhibition of transport by brefeldin A was reported very early to be mediated through interactions with low molecular weight GTP-binding proteins (Nakano et al., 1988; Nakano and Muramatsu, 1980; d'Enfert et al., 1991). An enzyme that catalyzes guanine nucleotide exchange and promotes binding of ARF and other accessor proteins by the Golgi apparatus was initially described as the primary target for brefeldin A by Donaldson et al. (1992) and Helms and Rothman (1992). Dissociation of COPI from Golgi apparatus membranes occurs within 20 s after drug addition (Donaldson and Klausner, 1994).

Experiments with brefeldin A also have been instrumental in elucidating a retrograde (Golgi apparatus to endoplasmic reticulum) pathway involved in returning Golgi apparatus proteins to the endoplasmic reticulum (Pelham, 1990). When treated with brefeldin A, the stacks often are depleted or replaced by tubules (Lippincott-Schwartz et al., 1989; Ulmer and Palade, 1989). The remaining tubular system may then also dissipate as the associated resident Golgi apparatus markers are returned to the endoplasmic reticulum (Bannykh et al., 2005). Such observations are consistent with the proposed role for the Golgi apparatus tubules in the addition and removal of resident Golgi apparatus proteins (including glycosyl transferases) as part of their normal functioning during flow differentiation of membranes according to the membrane maturation model of Chapter 6.

Reduced pyridine nucleotide has been reported to enhance cell-free transfer of membrane material from a radiolabeled Golgi apparatus donor fraction from rat liver to an acceptor fraction consisting of inside-out plasma vesicles immobilized on nitrocellulose (Rodriguez et al., 1992). Highly purified fractions of Golgi apparatus from rat liver were tested for their ability to oxidize NADH and this oxidation was inhibited by brefeldin A (Morré et al., 1994b) with inhibition augmented by GDP (Fig. 11.2). At near optimal concentrations of $7\,\mu M$ brefeldin A and $1\,\mu M$ GDP, the activity was >90% inhibited. Brefeldin A inhibition of NADH oxidation by the Golgi apparatus was time-dependent and GDP appeared to accelerate the time-dependent inhibition by brefeldin A. Presumably, as a consequence, cell enlargement is inhibited by brefeldin A (Fig. 11.5; Section 3.11).

11.1.1.2. Monensin

Monensin is a monovalent polyether antibiotic in which the oxygen functions are concentrated at the center of the structure where they are available for the complexation of a suitable cation (in this case sodium). The alkyl groups are spread over the outer surface rendering the complex lipid soluble thus allowing the antibiotic to enter and diffuse through biological membranes. As such, monensin represents a Na^+ ionophore capable of collapsing Na^+ and H^+ gradients (Fig. 11.3). The effect of monensin as a secretion blocker was discovered by Tartakoff and Vassalli (1997) in a study of the biosynthesis and secretion of immunoglobulin in plasma cells. Monensin subsequently gained wide-spread acceptance as a biochemical and biological investigative tool for study of Golgi apparatus function and to localize and identify the molecular pathways of subcellular vesicular traffic. Monensin functions as an ionophore (ion-carrier) to transport monovalent ions across lipid barriers as complexes soluble in the lipid phase of the membranes. Among its advantages are the low concentrations of which inhibitions are produced ($0.01-1.0\,\mu M$), a minimum of troublesome side effects (e.g., little or no change of protein synthesis or ATP levels), and a reversible action (Mollenhauer et al., 1990).

Monensin exerts its most profound effects on the trans cisternae of the Golgi apparatus stacks in those regions of the apparatus primarily associated with the

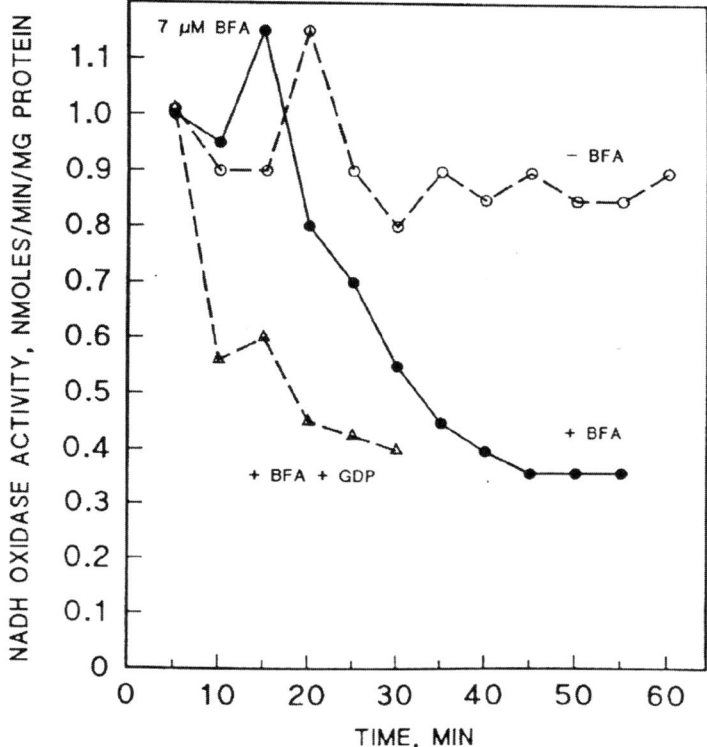

Fig. 11.2. Time-dependent inhibition of the NADH oxidase of rat liver Golgi apparatus by 7 μM brefeldin A (BFA): GDP may only serve to accelerate the rate of BFA inhibition of the oxidase. Even in the absence of added GDP, BFA inhibited the NADH oxidase in a time-dependent manner to a maximum at about 45 min. In the absence of BFA (equivalent amount of ethanol alone), the activity was not inhibited. Reproduced from Morré, Paulik, Laurence and Morré, 1994b with permission from FEBS.

final stages of secretory vesicle maturation and in post-Golgi apparatus structures primarily associated with endocytosis and membrane/product sorting (Figs. 6.15 to 6.17). Because of its relative specificity, monensin has been used extensively as an inhibitor of trans Golgi apparatus function. A detailed listing of examples of the use of monensin as a secretion blocker is provided by Dinter and Berger (1998).

In addition, monensin is a potent inhibitor of cell enlargement and growth (Section 3.1). Cultured cells, cells of tissue slices or explants, and plant organs that have received a topical exposure to monensin sufficient to inhibit growth usually also show deviations in Golgi apparatus structure and function (Mollenhauer et al., 1990).

11.1.1.3. Cytochalasin B

Cytochalasin B (Fig. 11.1) is a cell-permeable mycotoxin isolated from a fungus, *Helminthosporium dermatioideum*. It inhibits cytoplasmic division

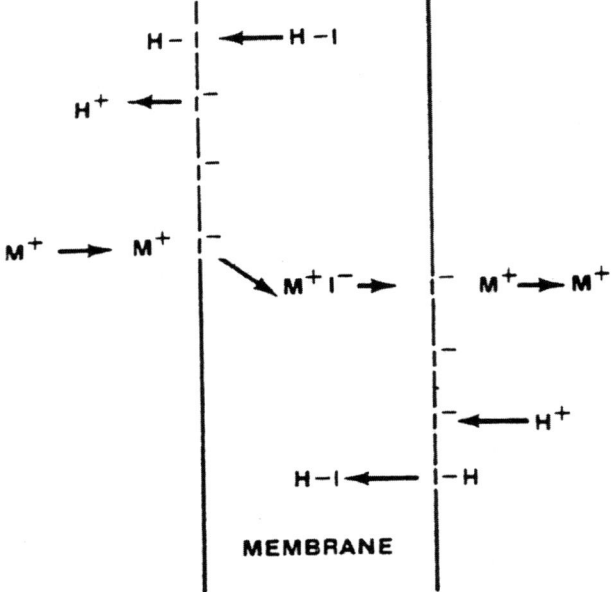

Fig. 11.3. Diagrammatic interpretation of carboxylic ionophore-mediated cation transfer across a bimolecular lipid membrane. M^+, metal cation; I, ionophore; H^+, proton; H-I, protonated ionophore; $M^+ I^-$, zwitterion of metal cation and anionic form of ionophore from: Bergan and Bates, (1984). The anionic form of the ionophore is stabilized by the polar environment characteristic to the surface of a membrane. The ionophore is capable of ion pairing with a metal cation either at the terminal carboxylic acid moiety or at other internal sites. The binding of a cation initiates the formation of a lipophilic, cyclic cation-ionophore complex that can diffuse through the interior of the bimolecular membrane structure. After traversing the membrane, the complex is again subjected to a polar environment where the electrostatic forces that had stabilized the complex are no longer greater than the unfavorable Gibbs free energy change of cyclization. The ionophore then releases its enclosed cation and reverts to the low energy acyclic conformation. Reprinted from BBA, 1031, Mollenhauer, Morré and Rowe, 225–246, Copyright 1990 with permission from Elsevier.

by blocking the formation of contractile microfilaments. It inhibits cell movement and induces nuclear extrusion. Cytochalasin B shortens actin filaments by blocking monomer addition at the fast-growing end of polymers. Cytochalasin B inhibits glucose transport and platelet aggregation. It blocks adenosine-induced apoptotic body formation without affecting activation of endogenous ADP-ribosylation in leukemia HL-60 cells.

As a Golgi apparatus inhibitor, it is most effective in blocking exit of secretory vesicles from the trans Golgi apparatus face in cells where vesicles are not dependent on microtubules as guide elements. The result is often massive accumulations of vesicles around each Golgi apparatus stack (Fig. 6.18) and inhibition of cell enlargement.

11.1.1.4. Colchicine

Colchicine (Fig. 11.1) is a highly poisonous alkaloid, originally extracted from plants of the genus *Colchicurn* (Autumn Crocus, also known as the

"Meadow Saffron"). Colchicine inhibits microtubule polymerization by binding to tubulin, one of the main constituents of microtubules. Availability of tubulin is essential to mitosis, and therefore colchicine effectively functions as a "mitotic poison" or spindle poison. Colchicine is most effective in blocking Golgi apparatus transport and cell enlargement in those cells where secretory vesicle migration is obligatorily coupled to microtubules.

11.1.1.5. FCCP

FCCP is an agent which lower the cellular ATP content by inhibiting oxidative phosphorylation (Antoine and Tonanne, 1986). FCCP has been shown to rapidly inhibit (in less than 5 min) the secretion of immunoglobulins (Ig) and to partially block the release of fucosylated Ig. This indicates that the drugs inhibit the transport of Ig from the Golgi apparatus (GA) (where fucose is added to Ig) to the plasma membrane. The degree of inhibition reached 70 to 80% with FCCP, whereas ATP stores were depleted by only 45 to 55%. These results are consistent with multiple effects of FCCP on the secretion pathway of Ig. FCCP, because of its protonophore properties may not only reduce cellular ATP levels but may also neutralize the Golgi or post-Golgi acidic compartments involved in the transport of plasma membrane and secretory proteins.

11.1.1.6. Retinoids

Retinol (Vitamin A) and retinoic acid (Fig. 11.1) both exert profound effects on Golgi apparatus morphology (Brown et al., 1985; Morré et al., 1981) and on cell enlargement (Dai et al., 1997). Effects of retinol on formation of transition vesicles in situ and in studies with cell-free systems are summarized in Chapter 10.

11.1.1.7. Metabolic Inhibitors

The Golgi apparatus is sensitive to growth inhibitory respiratory poisons (Jamieson and Palade, 1968b, 1971). Application of cyanide and other inhibitors of ATP formation identified two energy requiring steps: transport from the transitional elements of the endoplasmic reticulum to the proximal side of the Golgi apparatus and the other for the actual release of the granule to the cell's exterior (Jamieson and Palade, 1968b, 1971). In pea stems, cyanide inhibited polysaccharide secretion by Golgi apparatus but not cellulose synthesis (Robinson and Ray, 1977). The secretory process is less sensitive to classic inhibitors of protein or RNA synthesis (cycloheximide, fluorsphenylatanine, ethionine, puromycin, emetine and actinomycin b) than is protein synthesis, for example (Jamieson and Palade, 1968a). In addition to cyanide, inhibitions have been observed with dinitrophenol sodium fluoride and caffeine. Additional Golgi apparatus-disturbing agents are discussed by Dinter and Berger (1998).

11.1.1.8. Low temperature block

A non-pharmacological opportunity for correlative analysis is the low temperature block characteristic of Golgi apparatus. At temperatures below 16 to 18°C,

Golgi apparatus activity ceases. If secretory vesicle formation by Golgi apparatus is essential to cell growth, cell growth should show a corresponding abrupt low temperature transition below which cell growth should cease (Section 3.1). As such, the low temperature block provides a reversible means to block Golgi apparatus related growth.

In pancreatic ascinar (Tartakoff 1986) and other cells (Lagunoff and Wan, 1974; Holmes et al., 1981; Fries and Lindstrom, 1986; Saraste et al., 1986), at temperatures of 16 to 18°C or below, secretory proteins were blocked and accumulated in a pre-Golgi apparatus compartment. At 20°C a medial Golgi apparatus compartment was reached (Saraste and Kuismanen, 1984). Similar results were obtained with the intracellular transport and surface expression of proteins in several other cell systems (Matlin and Simons, 1983; Tooze et al., 1984; Tooze et al., 1988; Griffiths et al., 1985; Copeland et al., 1988).

11.2. Evidence from Tip-Growing Cells for a Role of Golgi Apparatus Activity in Cell Enlargement

There are no examples where a participation of the Golgi apparatus in cell enlargement is clearer than in tip growth. Nerve fibers (Pfenninger and Bunge, 1974), pollen tubes, rhizoids, fungal hyphae, and plant hairs (Roelofsen, 1959) elongate exclusively by tip growth. New plasma membrane is obtained predominantly, if not exclusively, from membranes of secretory vesicles coming from the Golgi apparatus (Grove et al., 1970; Pfenninger and Bunge, 1974). As such they have long periods of very high rates of plasma membrane production. Here rates of plasma membrane formation may vary from 30 to 300 μm^2 per min or more (Morré, 1975). In such cells it is likely that plasma membrane proteins are largely conserved with less turnover than in non-dividing cells (Warren and Glick, 1968; Roberts and Yuan, 1974). In pollen tubes of Easter lily it has been calculated that, for each single cell, Golgi apparatus produce and export in excess of 1000 secretory vesicles per min needed to generate 300 μm^2 of new plasma membrane per min during pollen tube elongation (Van Der Woude et al., 1971; Morré and Van Der Woude, 1974) (Table 11.1). Other examples of involvement of secretory vesicles in tip growing systems are given by Mollenhauer and Morré (1966a; 1976b), Sievers (1967) and Grove et al. (1970).

11.3. Physical Membrane Displacement

Vesicle formation or membrane budding is a well established manifestation of the vectorial displacement of membranes that occurs at the transitional endoplasmic reticulum to deliver membranes to the cis Golgi apparatus, at the trans Golgi apparatus to deliver materials to the plasma membrane and cell exterior along an exocytic pathway (Farquhar, 1985), and at the plasma membrane to internalize both membrane and substances from the external milieu along an endocytic pathway (Silverstein et al. 1977). In addition, displacements are observed to occur,

Table 11.1. Calculated rate of secretory vesicle and cell surface (membrane and cell wall) production for cultured pollen tube cells of Easter lily*.

Rate of pollen tube elongation	6 μm/min
Pollen tube diameter	16 μm
Thickness of cell wall matrix	0.05 μm
Secretory vesicle diameter	0.30 μm
Secretory vesicle volume	0.014 μm
Secretory vesicle surface area	0.28 μm²
Increase in volume of cell wall	15 μl/min
Rate of secretory vesicle production	1075 vesicles/min
Rate of membrane production by secretory vesicles	300 μm²/min
Rate of increase in plasma membrane due to growth	300 μm²/min

*From Morré and Van der Woude (1974).

especially in cultured cells, as part of normal cellular activities and cellular movements. These include ruffling, the formation and retraction of pseudopods and pleomorphic changes in cell shape (Hynes, 1979). These three manifestations of physical membrane displacement (budding, pleomorphic shape changes and lateral displacement of membrane sheets) are diagrammed in Fig. 11.4. Budding phenomena are central to both exocytic and endocytic processes (Farquhar, 1985).

For Golgi apparatus or endoplasmic reticulum membranes to form buds or vesicles, membranes must be displaced in time and space. As will be developed later in this chapter, physical membrane displacement can continue in a post Golgi apparatus process of membrane thinning leading to increases in the cell surface independent of addition of new membrane.

11.3.1. Membrane Budding

One of the properties of the fungal macrolide antibiotic inhibitor, brefeldin A, is to inhibit trafficking from the Golgi apparatus to the plasma membrane (Miller et al., 1992, Schindler et al., 1994). Additionally, brefeldin A specifically inhibits auxin-induced cell elongation in plants (Fig. 11.5) . Because the action of brefeldin A is normally thought to reside exclusively somewhere within the Golgi apparatus, such findings might be regarded as evidence for an obligatory role of secretory vesicles derived from the Golgi apparatus in cell enlargement at least in plants.

There are, however, several lines of evidence that argue against this interpretation. One of these is given in Figs. 11.6 and 11.7. In response to temperature, elongating segments of soybean show extensive accumulations of membranes at the trans Golgi network at temperatures of 18°C or less (Fig. 11.6). These accumulations are reminiscent of temperature blocks seen in other plant (Noguichi and Morré, 1991b) and animal cells (Tartakoff, 1986). On the other hand, cell enlargement induced by auxin shows no sharp transition in response to temperature over the entire range of 4–25°C (Fig. 11.7). This would argue that elongation growth in plants induced by auxin may continue for a time in a manner independent of the

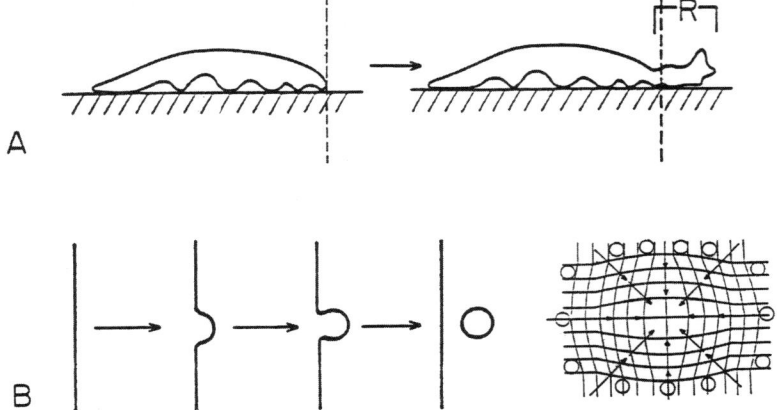

Fig. 11.4. Diagram depicting physical membrane displacement. A. Pleomorphic shape changes associated with mammalian cells during movement. B. Membrane budding. Planar physical membrane displacement would occur in membrane sheets during cell enlargement. Reproduced from Morré, 1994b with permission from Springer-Wien.

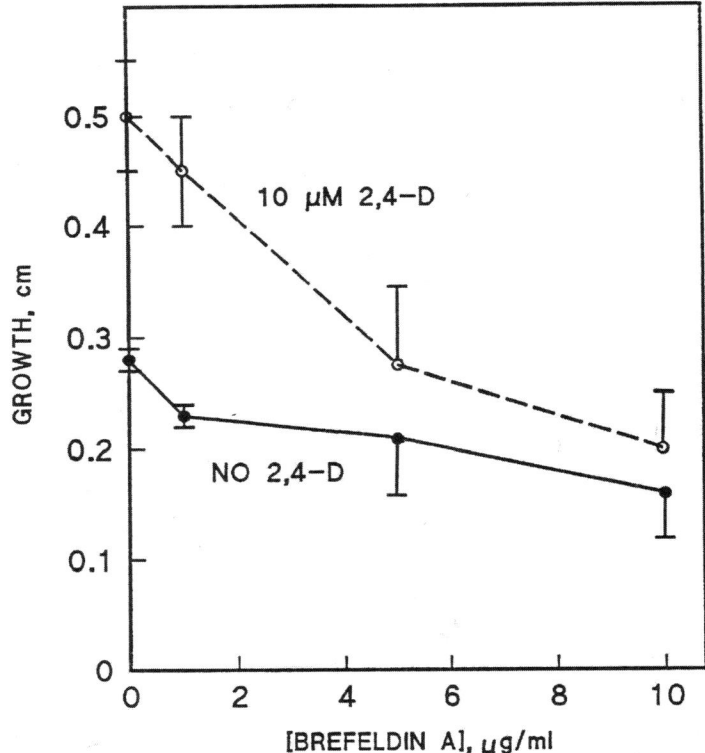

Fig. 11.5. Inhibition of elongation of 1 cm sections of soybean hypotcotyls by brefeldin A. The growth induced by auxin (indole-3-acetic acid, IAA, not shown; or 2,4-dichlorophenoxyacetic acid, 2,4-D) is specifically inhibited by brefeldin A. Reproduced from Morré, 1994b with permission from Springer-Wien.

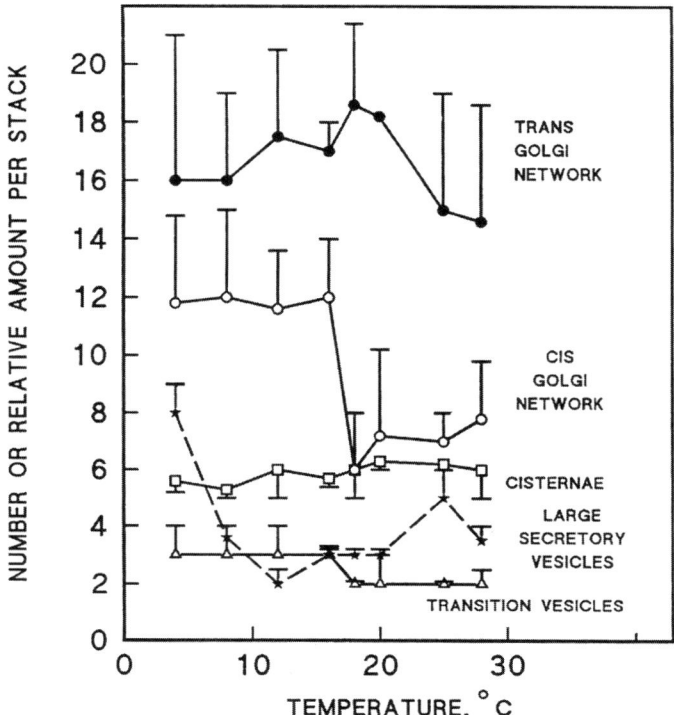

Fig. 11.6. Morphometric quantitation of trans Golgi apparatus membranes associated with stacked Golgi apparatus cisternae of 1 cm segments of etiolated soybean hypocotyls incubated for 30 min at each of the temperatures indicated. There is a temperature-dependent increase in trans Golgi apparatus membranes which disappears abruptly at about 18°C indicative of a low temperature block in membrane trafficking through the plant Golgi apparatus. Reproduced from Morré, 1994b with permission from Springer-Wien.

vesicular transport pathway. This is in sharp contrast to proliferation of mammalian cells where little or no growth occurs at temperatures below 28°C (Fig. 11.8; Wu et al., unpublished). The most temperature sensitive stage of the cell cycle is G_1, the cell cycle stage where cell enlargement takes place (Watanabe and Okada, 1967). Cell viability is unaffected. CHO cells paused for 3 days or more at 4°, 12°, or 24° resumed normal growth when returned to 37°C (Hunt et al., 2004). Effects of temperature on enlargement of mammalian cells per se have not been determined.

Similar conclusions that cell enlargement could occur for a time independently of delivery of new membrane to the cell surface were reached earlier based on results with monensin (Fig. 11.9) . Experiments were done with the narrow setae of the moss *Pellia*. The setae elongate rapidly in response to auxin (Schnepf et al., 1979). Yet they are thin enough that the monensin is able to penetrate all of the cells in the tissue segment. That the monensin is able to penetrate is evidenced by electron microscope observations of swollen trans elements associated with all Golgi apparatus stacks following fixation with glutaraldehyde (Fig. 11.10; Morré et al., 1986b). Despite the nearly complete inhibition by monensin of

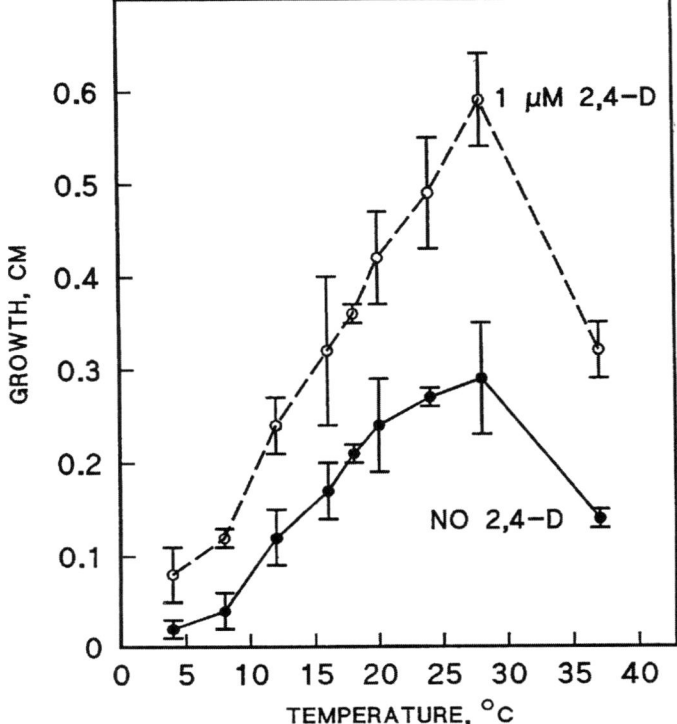

Fig. 11.7. Auxin-induced elongation of 1 cm segments of soybean hypocotyls exposed at different temperatures in parallel to the morphometric experiments of Fig. 8. The response to temperature shows no abrupt transition as would be expected for a typical low temperature block. Reproduced from Morré, 1994b with permission from Springer-Wien.

normal trans Golgi apparatus functioning, treated setae respond proportionately in a relatively normal fashion to auxin for times of 4 h or more after the onset of monensin inhibition. In contrast to results with brefeldin A, both auxin-induced and basal growth are inhibited to approximately the same degree by monensin (Fig. 11.9). As a result, the ratio of the rate of auxin growth to the rate of basal growth remains relatively constant over the range of monensin concentrations tested. Beyond 4 h, auxin-induced elongation appears to cease rather abruptly, possibly due to the depletion of essential Golgi apparatus-derived plasma membrane precursors required for sustained expansion.

    Du et al. (2006) have provided a mechanistic link between the SKERBP and PI3K/Akt pathways that may be growth related as synthesis of new membrane is an absolute requirement for cell growth and proliferation. Akt is a ubiquitous regulator of cell growth, proliferation, and survival that is activated by P13K (phosphatidylinositol-3-kinase) (Chapter 12). Processing of SREBP-2 (sterol regulatory element binding protein-2) is a master regulator of cholesterol homeostasis increased in response to various cholesterol depletion (including statin) treatments. This increase was blunted by LY294002, a potent and specific

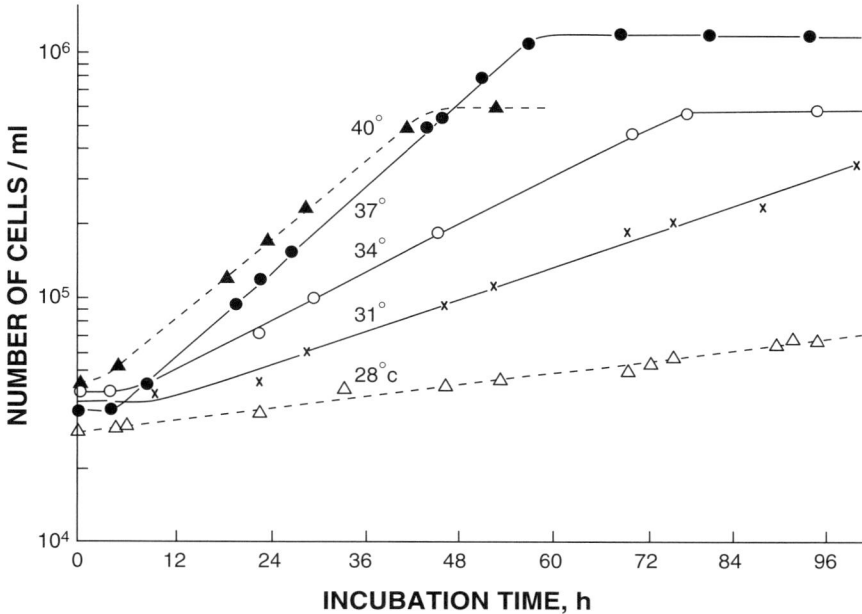

Fig. 11.8. Growth curves of L5178Y mouse leukemic cells at different temperatures. Reproduced from J. Cell Biol., 1967, 32, 309–323. Copyright 1967 The Rockefeller University Press.

inhibitor of PI3K. These several observations were linked through fluorescence microscopy where findings indicated that LY294002 disrupted transport of the SREBP escort protein, SCAP (SREBP cleavage-activating protein), from endoplasmic reticulum to the Golgi apparatus. Taken together, the findings indicate that the PI3K/Akt pathway is involved in SREBP-2 transport to the Golgi apparatus, thus contributing to the control of SREBP-2 activation and delivery of membrane cholesterol to the cell surface.

haCER2, a novel human ceramidase localized in the Golgi apparatus (Xu et al., 2006) regulates levels of both sphingosine (pro growth arrest) and sphingosine-1-phosphate (pro survival and pro cell proliferation) by controlling the hydrolysis of ceramides. High ectopic expression of haCER2 in HeLa cells caused fragmentation of the Golgi apparatus and growth arrest due to sphingosine accumulation. Low ectopic expression of haCER2 increased sphingosine-1-phosphate without sphingosine accumulation and promoted cell proliferation in serum free media.

11.4. Energy Requirements for Physical Membrane Displacement

One common requirement for physical membrane displacement appears to be an energy source. Budding of endoplasmic reticulum vesicles is driven by ATP whereas transfer between trans Golgi apparatus membranes and the plasma

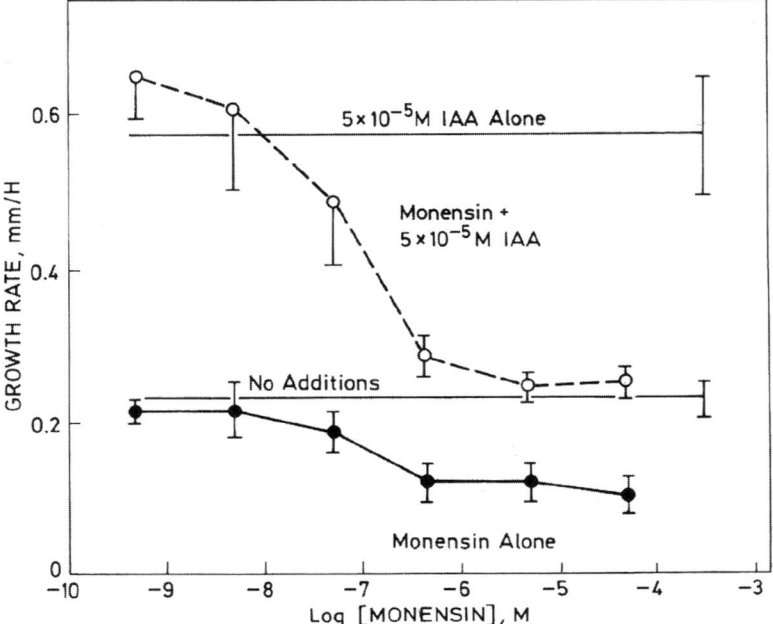

Fig. 11.9. Growth over 4 h of 5 mm long segments of *Pellia* setae as a function of monensin concentration in the presence and absence of 5×10^{-5} M indole-3-acetic acid (IAA). Values are means ± standard deviations from three experiments. Reproduced from Morré, Schnep and Deichgraber, 1986b with permission from The University of Chicago Press.

membrane is supported by reduced pyridine nucleotide. Using a reconstituted Golgi apparatus to plasma membrane transfer system, Rodriguez et al. (1992) found that transfer was promoted by reduced pyridine nucleotide (NADH) or reduced pyridine nucleotide plus ascorbate free radical. ATP was largely without effect in promoting this transfer. There are additional links between oxidation of NADH and Golgi apparatus. In the cell-free system described by Rodriguez et al. (1992), transfer of radiolabeled lipids or proteins from trans Golgi apparatus elements to inside-out plasma membrane vesicles immobilized on nitrocellulose was found to be promoted by NADH. In addition, trans elements of Golgi apparatus of rat liver (Morre et al., 1978b) were found to oxidize NADH as determined by reaction in the presence of a complex of copper and ferricyanide which, when reduced, produces electron dense deposits of insoluble copper ferrocyanide (Hatchett's brown) that are visible in the electron microscope (Fig. 7.3). The oxidation of NADH by isolated Golgi apparatus was shown as well to be stimulated by and reversibly regulated by the presence or absence of clathrin surface coats (Sun et al., 1983, 1984, Morré et al., 1983, 1987a). Cell enlargement in both plants and animals also emerges as having an energy requirement. For the endoplasmic reticulum to Golgi transfer step, several of the involved proteins have been identified and characterized. One of these is a 97 kDa ATPase (Zhang and Morré, 1993).

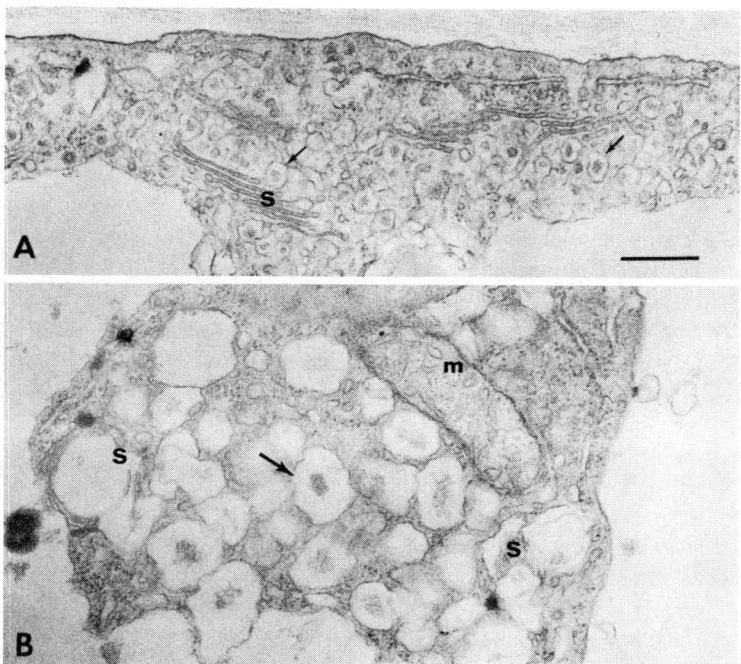

Fig. 11.10.   Portion of a setal cell of *Pellia* treated for 30 min without (A) or with (B) 5×10^{-7} M monensin. In B cisternal stacks (S) are replaced by groups of swollen vacuoles (arrows) often with centrally-located electron-dense inclusions similar to those seen in secretory vesicles prior to monension treatment (A). m = mitochondrion. Reproduced from Morré, Schnept and Deichgraber, 1986b with permission from The University of Chicago Press. Scale bar = 0.1 μm.

Despite much investigative effort, little or no information has emerged concerning mechanisms of physical membrane displacement. Even with vesicle budding, information is limited largely to the requirement for ATP, an involvement of low molecular weight G proteins and other cytosolic factors and inhibitions by thiol reagents such as *N*-ethylmaleimide (Wattenberg, 1991). Coat proteins appear to be involved (Donaldson et al., 1990) and the macrolide antibiotic brefeldin A inhibits (Donaldson et al., 1990; Miller et al., 1992). The response of the NADH oxidase activity of Golgi apparatus and plasma membranes to the latter first suggested to us a role of NADH oxidase in vesicle budding (Morre et al., 1994a, b). Polyclonal antisera to specific peptide portions of the ATPase block transfer in a cell-free system. With antisera to the transitional endoplasmic reticulum ATPase, the inhibition is primarily at the level of vesicle budding from the transitional endoplasmic reticulum. Since physical membrane displacement mechanisms are involved in the transfer of membrane from endoplasmic reticulum to Golgi apparatus and from Golgi apparatus to plasma membrane, these physical transfers are integral contributors to the physical membrane displacements associated with cell enlargement. Cell enlargement is especially relevant for plant cells where increases in size of the plant or of specific plant parts is sometimes due almost entirely to cell elongation rather than the result of a direct increase in the number of cells. Using excised plant stems floated on solution, the rate of elongation can often be accelerated several-fold by the addition of cell elongation promoters known collectively as auxins. This auxin-induced growth is primarily the result of cell elongation in which displacement of the cell surface membrane represents a significant end result. That both secretory vesicle formation and release from the trans Golgi and auxin-induced cell elongation could be inhibited in parallel by the fungal antimetabolite, brefeldin A (Schindler et al. 1994), suggested the possibility that plant cell elongation and vesicular trafficking were somehow correlated. This may be, but the correlation may be less direct than would be predicted from the parallel response to brefeldin A. In the first analysis, auxin-induced cell elongation emerges as being much less responsive to low temperature than would be expected were it strictly a vesicle-mediated process. Trafficking of materials through the Golgi apparatus emerges as a process with very precisely defined temperature limits. At temperatures below 16–18°C transfer appears to be largely blocked, whereas auxin-induced enlargement of plant cells is no more affected than would be predicted from the Arrhenius equation.

A dependency on metabolic energy (e.g., ATP) is often assumed for membrane displacements although proof is available only for vesicle budding where an absolute requirement for ATP has been established using permeabilized cells (Balch and Keller, 1986; Simons and Virta, 1987; Beckers et al., 1987, 1990) and cell-free systems (Morré et al., 1986; Balch et al., 1987; Paulik et al., 1988; Warren et al., 1988; Wattenberg et al., 1991). In plants, enlargement has been assumed to occur as cell wall stretching in response to turgor (Lockhart, 1965; Taiz, 1984). Yet, certain inhibitor studies have been interpreted by the author as indicative of strict metabolic requirements for plant cell enlargement as well (Morré and Eisinger, 1968). Transitional endoplasmic reticulum fractions from rat liver hydrolyze ATP. The principal ATPase has been characterized and found to have inhibitor characteristics dissimilar to those of standard ATPases from other membrane fractions such as the plasma membrane, mitochondria or vacuolar apparatus (Zhang and Morré, 1993). The ATPase of the transitional endoplasmic reticulum is not inhibited by ouabain, by oligomycin, nitrate or vanadate. The activity corresponding to a monomeric molecular weight of approximately 97 kDa was isolated from rat liver based upon activity stains on native gels and other methods. The activity was shown to have a native molecular weight of approximately 600 kDa and highly purified preparations by electron microscopy were shown to represent ring- like structures comprised of six apparently identical subunits. Partial amino acid sequence revealed homology with the valosin type of ATPases. This novel family of ATPases contains, in addition to valosin, CDC 58 and N-ethylmaleimide-sensitive factor (NSF). As a unique ATPase family these proteins are characterized by antisera raised against the 97 kDa monomer. Polyclonal antisera raised against the 97 kDa monomer inhibit both budding and transfer between transitional endoplasmic reticulum and cis Golgi apparatus in the cell-free system from liver. Should this membrane-associated ATPase be involved in the process of bud formation, the significance of a ring-like hexagonal structure to a membrane translocating ATPase presents interesting possibilities for future studies (Morré et al., 2007).

11.5. Summary

When Golgi apparatus activity is inhibited for a prolonged period growth will cease (both cell division and cell enlargement). This response appears to result indirectly from the need for preformed membrane constituents (proteins and lipids) at the cell surface to support cell size increases or to replace constituents lost to turnover. This is especially evident from work with plants (Section 3.1) when cell enlargement is investigated independently of cell division. Here, enlargement will continue for several hours even when Golgi apparatus activity is inhibited. As membrane thinning becomes excessive in the absence of membrane additions, growth ceases and protoplasts may actually rupture. Thus, cell enlargement and growth per se are not strictly vesicle-mediated processes but require provision of new membrane materials from the Golgi apparatus for sustained growth through both cell enlargement and formation of new cells by division.

12

Cancer

Beginning in the 1960s, considerable attention has been paid to alterations of cellular membranes associated with cell transformation and malignancy. Many such studies have been directed at a search for significant alterations particularly at the cell surface and, more specifically, in the plasma membrane. Focus shifted abruptly a decade later to oncogenes and elucidation of their participation in complex signaling pathways, mostly involving plasma membrane-associated receptors of transmembrane signaling. These pathways continue to be a principal focus of contemporary cancer research (Weinberg, 2007).

For membrane alterations to be expressed at the cell surface, it is most likely that they arise through biosynthetic or processing modifications directed through the ultimate action of altered genetic information but expressed through membrane-associated enzymes of the cell's internal endomembranes (Golgi apparatus, endoplasmic reticulum, and nuclear envelope). Characteristic transformation related changes in endomembranes were noted early but studies of more specific cancer-related endomembrane characteristics are still in their infancy.

12.1. The Ultrastructural Cancer Phenotype of the Endomembrane System

12.1.1. Rough (with attached ribosomes) Endoplasmic Reticulum

One of the hallmarks of the transformed cell cytoplasm is a decrease in the ratio of polyribosomes bound to the endoplasmic reticulum to the polyribosomes free in the cytoplasm (Dalton, 1964; Hruban et al., 1965). Within the range of Morris hepatomas, for example, rough endoplasmic reticulum may range from well-ordered and near-normal appearance in minimum deviation tumors to only occasional irregularly spaced cisternae in the more poorly differentiated invasive tumors. Rapid cellular division, reduction of rough-surfaced endoplasmic reticulum, and abundance of free polysomes in fast growing tumors are highly correlated (Becker, 1970). A well-developed rough endoplasmic reticulum with regularly arranged cisternae running parallel to one other (so-called stacked configuration) is primarily an expression of cells differentiated for protein export. Immature or undifferentiated cells such as stem, blast, or embryonal cells, or even cells in culture, have a much more sparse and less well-ordered complement of rough endoplasmic reticulum. By contrast, growing cells have a larger abundance

D. James Morré and Hilton H. Mollenhauer, *The Golgi Apparatus.*
© Springer 2009

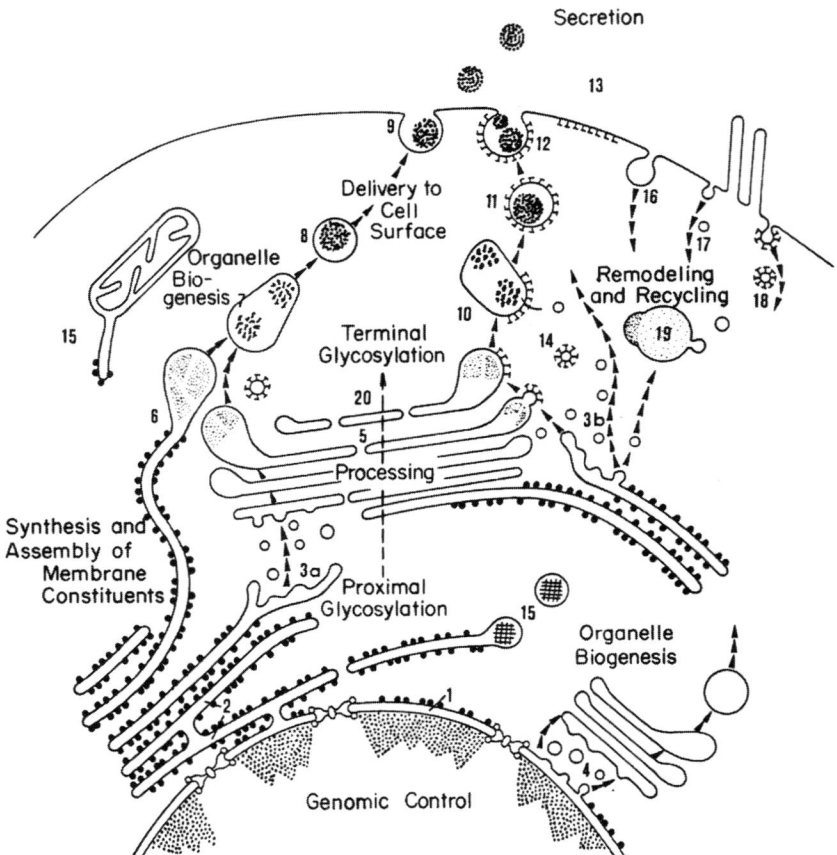

Fig. 12.1. Schematic representation of the role of endomembranes (and organelles) in membrane biogenesis and renewal essential for the expression of the neoplastic phenotype. The major processes indicated appear to occur normally in cancer cells but, through altered genomic control, become reprogrammed to generate an altered cell surface. Numbers refer to some of the many different compartments involved as follows: (1) nuclear envelope, which frequently exhibits continuity with (2) rough endoplasmic reticulum. Together with smooth endoplasmic reticulum (3), these membranes serve as major sites of membrane biogenesis as well as drug metabolism important to carcinogen activation. (3a) Transition vesicles bud from specialized endoplasmic reticulum regions to form new Golgi apparatus cisternae. This process emerges as a critical control point to regulate delivery of newly synthesized membrane quanta to the plasma membrane modified to keep pace with accelerated growth of transformed cells. (3b) Other transition vesicles may contribute in other routes of membrane trafficking. (4) In cells with Golgi apparatus adjacent to the nucleus, the nuclear envelope can replace endoplasmic reticulum in the formation of transition vesicles. (5) Golgi apparatus are major sites of processing of membrane and secretory proteins preparatory to their delivery to the cell surface. Modified in transformed cells, the Golgi apparatus emerges as a second critical control point, where tumor cell modification, i.e., altered glycosylation, is expressed. (6) Direct delivery of membrane material to the cell surface bypassing (the Golgi apparatus. Strong evidence for such a pathway is indicated from experiments where the Golgi apparatus route is blocked with monensin and incompletely processed proteins and glycoproteins are delivered to the cell surface. (7) Complex secretory vesicle (condensing vacuole) where materials for export are collected and perhaps further modified. (8) Mature secretory vesicle in transit to the cell surface. (9) Fusion of the secretory vesicle with the plasma membrane to complete exocytosis.

of free polyribosomes in the cytoplasm than do cells differentiated for export of proteins. The shift from membrane-bound to free polyribosomes in many types of tumors is consistent with the emphasis in tumors on synthesis of cellular proteins needed for cell growth and division.

The endoplasmic reticulum that remains in tumor cells appears to carry out functions similar to those of the endoplasmic reticulum in normal cells, including the co-translational addition of asparagine-linked high-mannose oligosaccharides to nascent polypeptide chains and their transport to Golgi apparatus for processing and delivery to the cell surface (Yeo et al., 1985).

12.1.2. Smooth Endoplasmic Reticulum

Smooth endoplasmic reticulum, endoplasmic reticulum lacking ribosomes, is most evident in such cell types as hepatocytes, in which it can be induced. Smooth endoplasmic reticulum is formed in large quantities through the administration of xenobiotic substances. In smooth endoplasmic reticulum xenobiotics are subjected to mixed-function oxidation as an initial step in the detoxification process or as occurs during the metabolism of chemical procarcinogens to their active forms. While such endoplasmic reticulum forms are encountered in preneoplastic livers following administration of high doses of many types of chemical carcinogens, they appear not to persist as a general characteristic of the cytoplasm of malignant hepatic tumors. In most malignant tumors, the smooth endoplasmic reticulum is restricted to transition regions in association with the Golgi apparatus or other organelles in which ribosomes may be naturally sparse or absent. However, in tumors producing steroid hormones, smooth endoplasmic reticulum is typically abundant.

12.1.3. Golgi Apparatus

Evidence for an altered Golgi apparatus in the transformed state has come both from morphologic and from biochemical investigations. A change in the dimensions of Golgi apparatus in hepatomas compared with host liver was noted by McCarthy et al. (1974). Similarly, in a series of 35 Morris hepatomas of differing growth rates, Hruban et al. (1972; Hruban, 1979) found an effect on the lengths or number of cisternae of the Golgi apparatus that was largely independent of hepatoma growth rate. Compared with normal liver, hepatomas have also lost to varying degrees the ability to elaborate specific secretory proteins into

---

Fig. 12.1. (continued) (10–13) As in (7) except secretory vesicles are partly or entirely covered by a clathrin coat. In most transformed cells and many normal cells, delivery to the cell surface does not involve a separate condensing vacuole as an obligatory step. Rather, the secretory vesicles bud directly from the Golgi apparatus periphery. Coated vesicles are found both at the Golgi apparatus (14) and at the cell surface (18). (15) Associations of rough endoplasmic reticulum with organelles including direct continuities. (16) Endocytosis of large material (phagocytosis) and (17) of small material (pinocytosis). (18) As in (16) and (17) except involving clathrin-coated membranes (or coated pits). (19) Secondary lysosomes. (20) Distal cisternae of the Golgi apparatus (trans-Golgi apparatus reticulum) sometimes separated from the stack as a cisternal fragment or as a thick cisternae. The latter may also be an important site of terminal glycosylation reactions although its relationship to cell surface formation remains to be investigated. Modified from Morré et al., 1979b with permission from Elsevier.

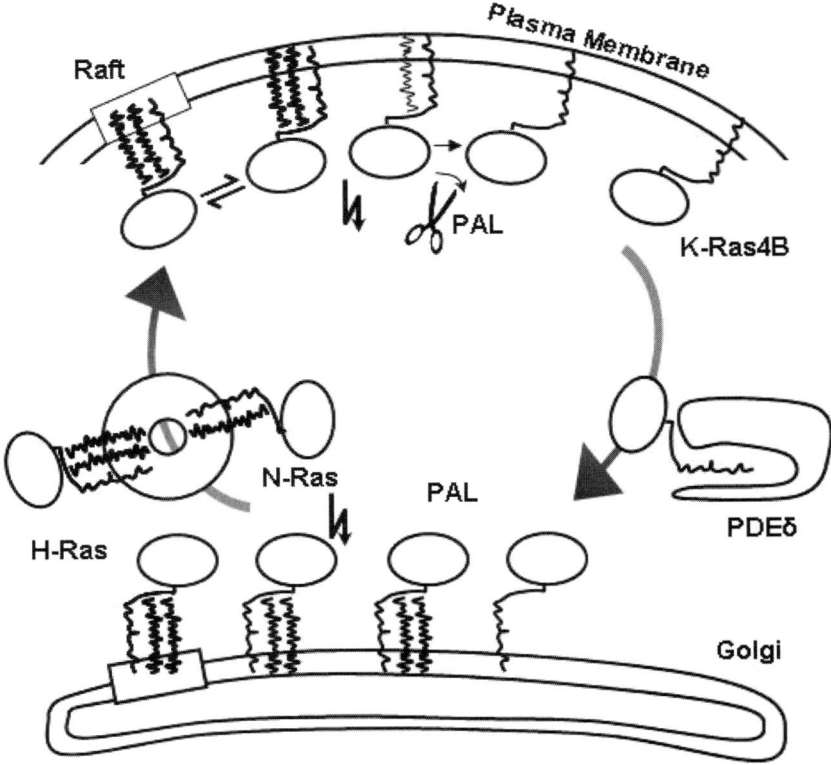

Fig. 12.2. Schematic of opportunities for Ras signaling via the Golgi apparatus where H-, N-, and K-Ras proteins are able to regulate separate signaling pathways because they localize to different cellular compartments. K-Ras4B is exclusively available for signaling at the plasma membrane, N-Ras and H-Ras appear to distribute between the plasma membrane and the Golgi apparatus depending on whether or not they carry a palmitoyl group. An increased rate of depalmitoylation leads to accumulation of N- and H-Ras in the Golgi apparatus. Increased palmitoylation favors accumulation at the plasma membrane. Modified from Meder and Simons, 2005 with permission from the AAAS.

serum (Redman et al., 1979). These, and other observations, led Reutter and Bauer (1978) to suggest that, in transformed cells, the Golgi apparatus may shift from a secretory to a membrane-generating mode of functioning [see also Hudgin et al. (1971) for a similar interpretation].

The Golgi apparatus of tumor cells does not appear grossly altered, although the stacks have been found to be of a smaller diameter relative to tissues of origin. For the most part, the tumor Golgi apparatus seems to acquire a morphology similar to that ascribed to Golgi apparatus forms in juvenile or dividing cells (Morré and Ovtracht, 1977; Fig. 2.2).

Many of the metabolic and biosynthetic processes that involve Golgi apparatus (Chapter 7) could be important to the progress of cancer development. These include sugar nucleotide transport, glycoprotein and glycolipid biosynthesis and sulfation reactions.

Biochemical evidence for a functionally altered Golgi apparatus in transformation has been more inferred from the known subcellular localization of glycoconjugate processing enzymes in various parts of the Golgi apparatus rather than from direct measurements either In situ or with isolated fractions. Glycoproteins and glycolipids that are altered in tumorigenesis frequently have L-fucose or *N*-acetylneuraminic acid as terminal sugars.

One approach to provide direct biochemical demonstrations of the distributions of glycosyltransferases across the polarity axis of the Golgi apparatus comparing cancer and non-cancer cells from the same tissue was that used by Hartel-Schenk et al. (1991). Here the technique of preparative free-flow electrophoresis described in Chapter 3 was employed to resolve cisternae of Golgi apparatus of rat liver and of rat hepatomas after unstacking into fractions enriched in cisternae from the cis, median or trans face of the Golgi apparatus (Fig. 12.3). In so doing, fractions from unstacked Golgi apparatus of rat liver and

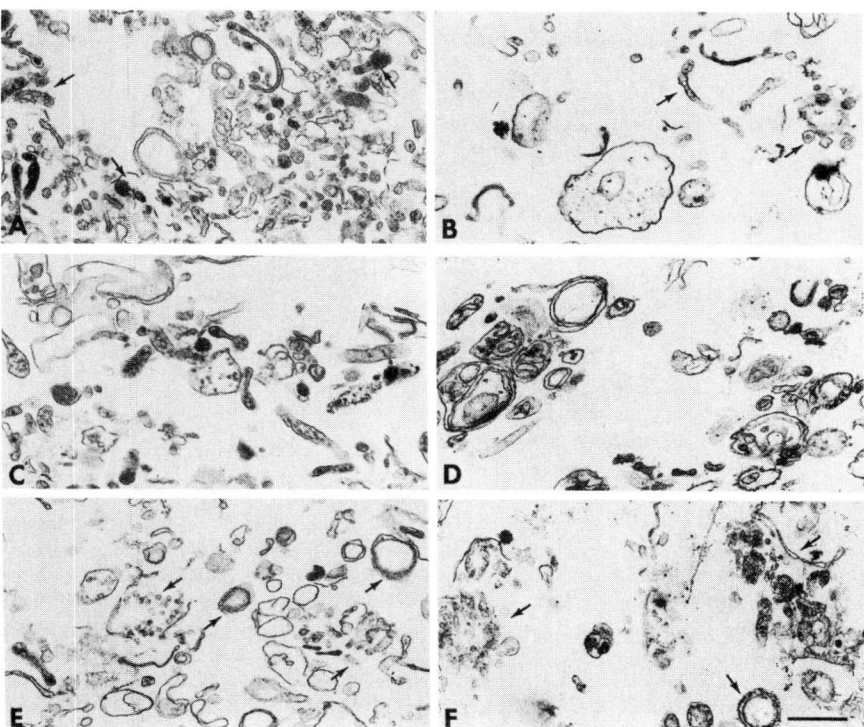

Fig. 12.3. Electron micrographs of free-flow electrophoresis fractions from the RLT-N hepatomas in comparison to rat liver. (A) Electropositive fraction (cis-most Golgi apparatus cisternae) number 58 from rat liver. (B) Electropositive fraction number 52 from RLT-N hepatomas. (C) Median Golgi apparatus compartment (fraction number 55) from rat liver. (D) Median Golgi apparatus compartment (fraction number 49) from RLT-N hepatomas. (E) Trans-most Golgi apparatus cisternae (fraction number 53) from rat liver. (F) Electropositive fraction number 46 from RLT-N hepatomas. In A and B, arrows mark elements containing large lipoprotein particles of the cis face. In E and F, arrows mark elements with morphological characteristics (thick membranes with many fenestrae) of the trans Golgi apparatus reticulum. Reprinted from Hartel-Schenk, Minnifield, Reutter, Hanski, Bauer and Morré, 1991, BBA 1115:108–122, Copyright 1991 with permission from Elsevier. Scale bar = 0.5 μm.

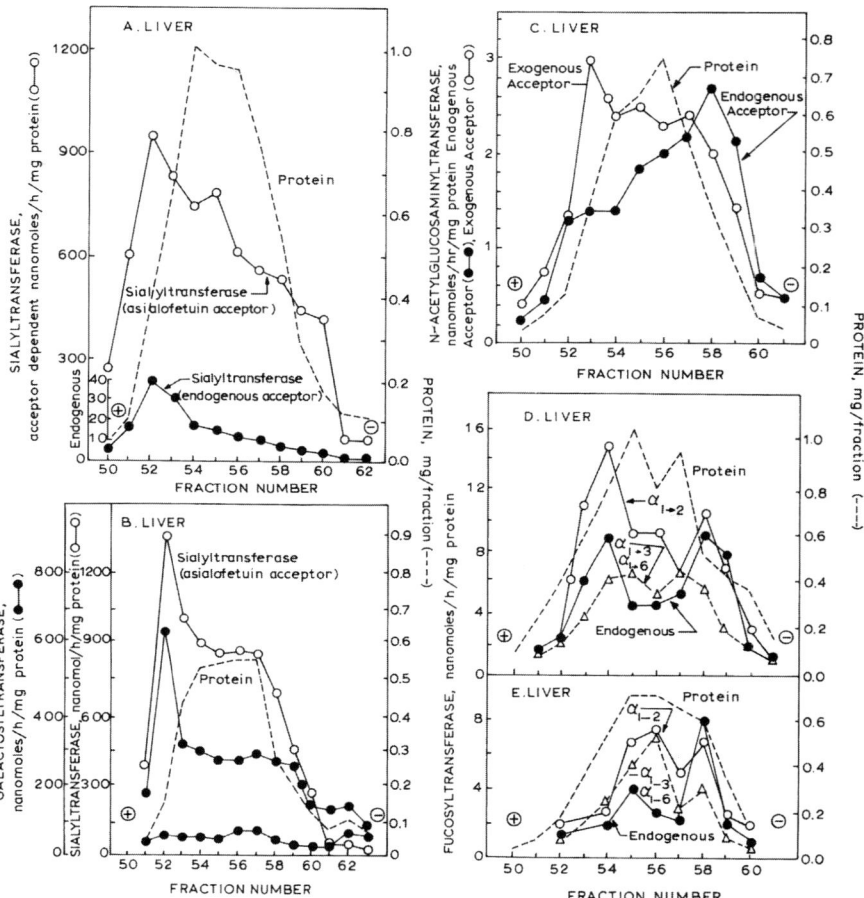

Fig. 12.4. Distribution of sugar transferase activities among Golgi apparatus subfractions prepared by free-flow electrophoresis from rat liver. (A) CMP-sialic acid:asialofetuin sialyltransferase activity. The specific activity was greatest in the most electronegative fractions containing Golgi apparatus (fractions 52-55) and increases steadily across the separation from the least electronegative to the most electronegative fractions. (B) UDP-galactose:ovomucoid galactosyltransferase and CMP-sialic acid:asialofetuin sialyltransferase. Galactosyl transfer to both endogenous and ovomucoid acceptors paralleled the sialyltransferase activity. (C) UDP-N-acetylglucosaminyl:endogenous acceptor N-acetylglucosaminyl:ovalbumin N-acetylglucosaminyltransferase. Reprinted from Hartel-Schenk, Minnifield, Reutter, Hanski, Bauer and Morré, 1991, BBA 1115:108–122, Copyright 1991 with permission from Elsevier.

of rat hepatomas were directly compared to reveal potentially important transformation-related differences in glycosyltransferase distributions.

Sialyltransferases, either with endogenous acceptors or with asialofetuin as acceptor (Fig. 12.4A), were found in the most electronegative fractions from rat liver (the lower numbered fractions) corresponding to cisternae of the trans face. Galactosyltransferase was also located in the trans half of the Golgi apparatus

(Fig. 12.4B). In contract, *N*-acetylglucosaminyltransferases were more cis located (Fig. 12.4C). Fucosyltransferae exhibited a bimodal distribution with both cis and trans locations (Fig. 12.4D). With rat hepatomas, surprisingly, both sialyl and galactosyltransferases (Fig. 12.5A and B) exhibited a predominantly cis location whereas both *N*-acetylglucoaminyl and fucosyltransferases gave bimodal distributions indicative of both cis and trans localizations (Fig. 12.5E and F).

These distributions, determined biochemically, are summarized in Fig. 12.6. With liver, cis to trans order of activities parallels the order of terminal sugar additions in N-linked glycoprotein biogenesis. For hepatomas, the order most closely parallel the order of terminal sugar additions for O-linked glycoprotein biogenesis. The findings would suggest that Golgi apparatus of hepatomas are adapted more for O-linked glycoprotein glycosylation than those of liver where N-linked glycoproteins are the dominant products.

Specific activities of GDP-fucose:glycoprotein fucosyltransferase were determined to be increased in a number of rat hepatomas at least two- to three-fold over that of normal liver (Reutter and Bauer, 1978), whereas the specific activities of sialyltransferase generally were decreased in these tumors (Reutter and Bauer, 1978). Galactosyltransferase-specific activity remained unchanged. During hepatocarcinogenesis induced in the rat by 2-acetylaminofluorene, CMP-sialic acid:glycoprotein sialyltransferase activity was found to be unchanged or decreased during tumor progression (Creek et al., 1984), while that of galactosyltransferase was unchanged (Elliott et al., 1984).

12.1.4. Plasma Membrane

Scanning electron microscopy (SEM) studies have suggested that the surface of malignant cells is generally more irregular than normal, with a relative increase in the number of surface microvilla and cytoplasmic lamellipodia suggestive of a retention in tumors of those features characteristic of rapidly dividing cells. Other changes may be more functionally related to the loss of contact inhibition or the failure of cancer cells to adhere and a tendency to round up (Table 12.1).

With cancer cells growing in organized tissues, important surface features have to do with cell junctions. Normal cells In vivo form cell-cell junctions that are ultrastructurally complex (Staehelin, 1974). Cells of tumors are often found to be deficient in the formation of junctional complexes. Despite some rather dramatic examples, Weinstein et al. (1976) concluded, after an extensive review of the literature, that no consistent pattern of junctional deficiencies existed in reference to malignancy as a whole.

The plasma membranes of malignant cells and/or that of transformed cells in culture exhibit a variety of biochemical and physical alterations that have been proposed to be either directly or indirectly related to an altered growth control or to invasion and metastasis. Modified protein, glycoprotein, phospholipids, and fatty acid patterns have been observed, but no consistent patterns of change have emerged. Plasma membranes of hepatomas are enriched in cholesterol compared with plasma membranes of normal hepatocytes. However, comparable cholesterol changes have not been observed in other neoplasms.

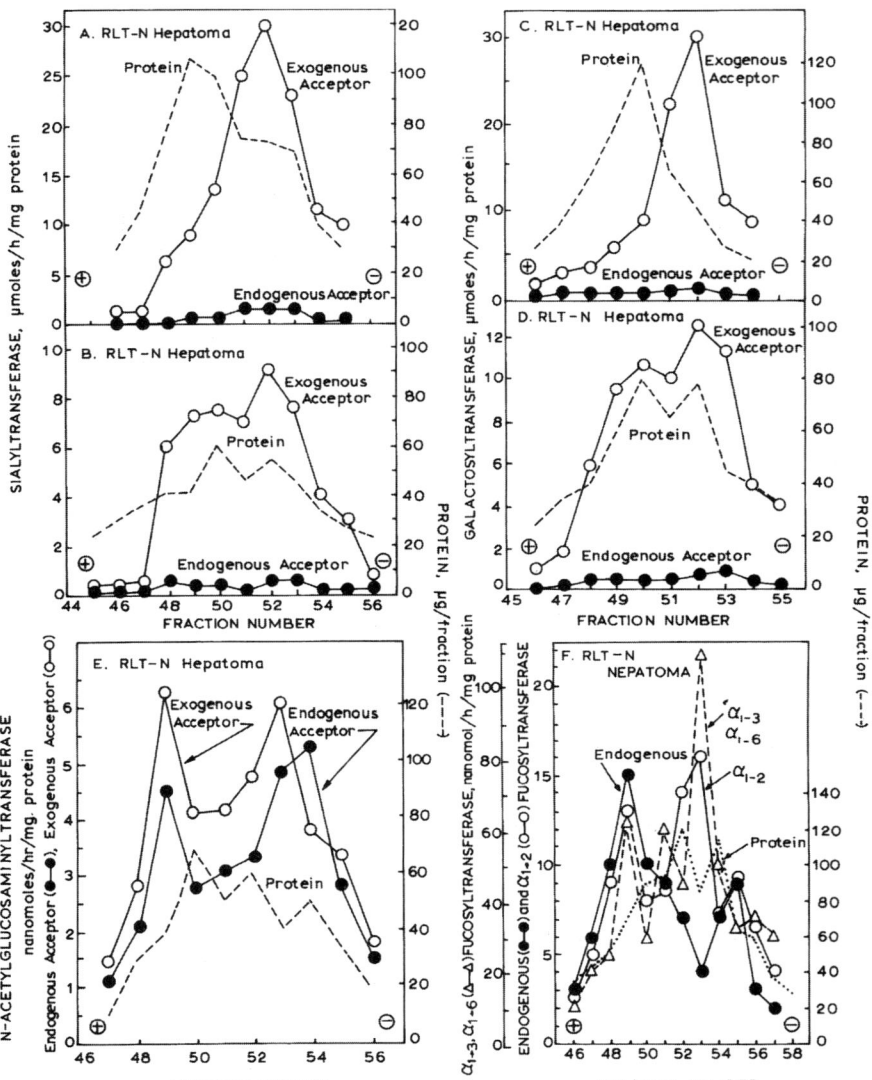

Fig. 12.5. Distribution of sugar transferase activities among Golgi apparatus subfractions of rat hepatomas prepared by free-flow electrophoresis. (A) CMP-sialic acid:asialofetuin sialyltransferase and CMP-sialic acid:endogenous glycoprotein sialyltransferase. (B) CMP-sialic acid:asialofetuin sialyltransferase and CMP-sialic acid:endogenous glycoprotein isalytransferase. (C) UDP-galactose: ovomucoid galactosyltransferase. (D) UDP-*N*-acetylglucosamine:endogenous acceptor. (E). UDP-*N*-acetylglucosamine:ovalbumin *N*-acetylglucosaminyltransferase. (F) GDP-fucose:asialofetuin (α1 → 2);GDP-fucose:asialogalactofetuin ((α1 → 6) and GDP-fucose:endogenous acceptor fucosyltransferase. Reprinted from Hartel-Schenk, Minnifield, Reutter, Hanski, Bauer and Morré, 1991, BBA1115:108–122, Copyright 1991 with permission from Elsevier.

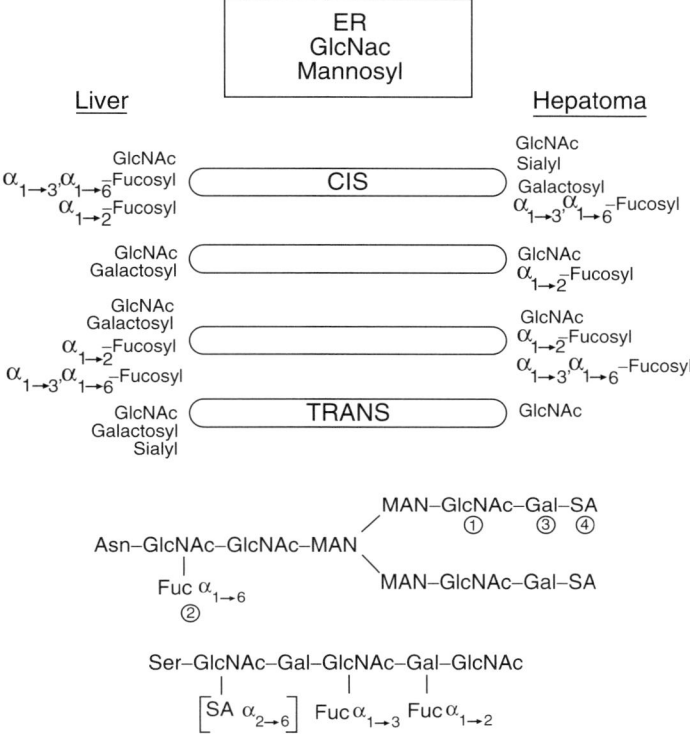

Fig. 12.6. Diagram summarizing the order and location of glycosyltransferase activities of rat liver and hepatoma Golgi apparatus as deduced from the free-flow electrophoretic separations. With liver the approximate location of transferase parallels the order of terminal sugar additions in *N*-linked glycoprotein biogenesis. For hepatomas, the location differs especially for sialyl- and galactosyltransferases and most closely parallels the order of terminal sugar additions in O-linked glycoprotein biogenesis. Reprinted from Hartel-Schenk, Minnifield, Reutter, Hanski, Bauer and Morré, 1991, BBA 1115:108–122, Copyright 1991 with permission from Elsevier.

Table 12.1. Properties of tumor cell surfaces.

Characteristic
Behavioral
Loss of contact inhibition of mitosis and movement
Loss of specific intracellular adhesion
Unregulated growth and division, invasiveness
Biochemical/Molecular
Altered glycoprotein and glycolipid chains
Changes in glycosyltransferases
Increased agglutinability by Con A
Reduction or loss of adhesion proteins
Increase in total sterol content
Altered transport properties
Oncogene expression and altered signal transduction pathways

Fluidity and other physical changes observed in plasma membranes of tumor cells emerge as highly dependent on tumor type and may differ widely.

Among the more consistent alterations in plasma membrane composition due to malignant changes have been to the glycolipids (Hakomori, 1981). These constituents represent a small fraction (about 1%) of the total lipids and may undergo extensive alterations upon cell transformation to include both early increases and late reductions and/or simplifications in carbohydrate chains (Morré et al., 1978a). These changes have been implicated as important to both metastasis (Matyas et al., 1986) and growth control (Spiegel and Fishman, 1987).

12.2. Role of Endomembranes in Signal Transduction and Oncogene Expression in Cancer

Despite the lack of emergence of a detailed understanding of the molecular lesions characteristic of a wide range of tumor types required to explain altered social behavior critical to the transformed phenotype, both loss of growth control and malignant behavior must involve fundamental alterations in the properties of the cell surface. Undoubtedly, the cell-surface alteration associated with cancer must somehow interact to affect one or more of the transducing mechanisms for generating mitogenic signals as well as surface properties affecting invasive ability and adhesion.

To the extent that oncogene products functioning at the cell surface are translated on membrane-associated polyribosomes and delivered to the plasma membrane via the Golgi apparatus and other endomembrane components, endomembranes play an essential role in oncogene expression. Examples include the v-*sis* oncogene of Simian sarcoma virus, which has strong sequence homology with platelet-derived growth factor (PDGF). Lacking an obvious stop-transfer sequence or membrane anchor, it might be secreted to stimulate growth in an autocrine fashion (Hunter, 1985). Another example is the receptor for epidermal growth factor (EGF), which has homology with the v-*erb*-B oncogene of avian erythroblastosis virus. The v-*erb*-B protein appears to be a truncated form of the EGF receptor, which lacks most of the external EGF binding domain, but which retains the proposed membrane anchor domain. It, too, is synthesized on polyribosomes of endoplasmic reticulum, is glycosylated in Golgi apparatus, and is delivered to the cell surface (Hayman and Beug, 1984; Schatzman et al., 1986). Many such examples have emerged in the interim although none specifically involve the Golgi apparatus as an obligatory contributor to the cancer phenotype.

For a Golgi apparatus protein to serve as a significant cancer-specific marker, the protein should be both cancer specific and Golgi apparatus localized. There are few, if any, such proteins. The tumor-specific ECTO-NOX protein, tNOX for example, is processed in the Golgi apparatus to its final mature molecular weight for delivery to the plasma membrane where it contributes to unregulated cell enlargement and growth characteristic of cancer (Morré and Morré, 2003).

Receptor tyrosine kinases (RTKs) are recognized widely as key regulators of normal cellular function but are implicated as well in the development and progression of human cancers. Oncogenic activations of RTKs often result from structural genomic rearrangements or genetic lesions such as point mutations or overexpression (Blume-Jensen and Hunter, 2001; Lamorte and Park, 2001). Several have been implicated in in-frame fusion events where unrelated sequences fused 5′ to the sequence coding for the tyrosine kinase domain produce constitutively activated kinases. The vast majority result from chromosomal translocations or inversions (Lamorte and Park, 2001). The resulting fusion proteins display constitutive tyrosine kinase activities attributable to dimerization domains found within the amino-terminal fusion partner. In these so-called forced dimers, the catalytic kinase domains are juxtaposed in a orientation optimal for transphosphorylation in a manner analogous to ligand-induced receptor dimers.

Charest et al. (2003) have reported a gene called FIG that encodes a protein that peripherally associates with the Golgi apparatus and likely plays a role in Golgi apparatus function. The transcript of this gene was reported initially as a 5′ fusion partner to the RTK ROS in a glioblastoma cell line (Sharma et al., 1989). Charest et al. (2003) found that the FIG-ROS fusion protein is activated constitutively. Unlike most other fusion proteins involving RTKs, the constitutive activation of FIG-ROS requires localization to the Golgi apparatus. The transformation capacity of FIG-ROS is eliminated by deletion of FIG sequences from the fusion protein crucial for Golgi apparatus localization.

The mitogen-activated protein kinase (MAPK) cascade is the major signaling cascade for cell proliferation and differentiation. Stimuli received from growth factor and hormones are transmitted to the cell through tyrosine kinase receptors. The guanosine triphosphatase (GTPase) Ras is the key regulator of this pathway. Mutations in Ras are found in many human cancers. Mammals have three different Ras genes that give rise to four highly homologous proteins that differ in their carboxyl termini through which they are anchored to cellular membranes. All Ras isoforms are farnesylated (modified with a farnesyl group). H-Ras and N-Ras localize to the Golgi apparatus where they acquire their palmitate groups enroute to the plasma membrane (Reviewed by Rocks et al., 2006). K-Ras isoforms bypass the Golgi apparatus and are mostly confined to the plasma membrane. H-, N- and K-ras isoforms are ocogenic and transform cultured cells.

The signaling pathways regulated by the different Ras isoforms have been attributed to differences in their subcellular locations (Fig. 12.2). K-Ras isoforms are confined to the plasma membrane, whereas H-Ras and N-Ras are found, as well, in the Golgi apparatus. At the Golgi apparatus, the bound Ras forms bind Raf to become stably activated in the GTP-bound state. Whereas Ras signaling was originally believed to occur exclusively at the plasma membrane, these recent data suggest that the Golgi apparatus may contribute to Ras signaling as well (Chu et al., 2002).

Guanine nucleotide exchange factors and GAPs are Golgi apparatus located. In T-lymphocytes, receptor stimulation leads to N-Ras activation at the Golgi apparatus via a Golgi apparatus resident Ras-GAP (Perez de Castro et al., 2004). Both a resident Ras affector (Mitin et al., 2004) and a MAPK scaffold

(Torii et al., 2004) have been reported as well for Golgi apparatus. One advantage of Ras signaling from the Golgi apparatus has been suggested to be that of close proximity to the nucleus, the ultimate destination for activated MAPK (Meder and Simons, 2005). Signaling from the Golgi apparatus would be closer to the target site and well suited for transcriptional regulation.

An important downstream affector of the Ras protein evoking a variety of cell responses derives from its ability to associate with and activate phosphatidylinositol 3-kinase (PI3K) with the resultant formation of phosphatidyl inositol (3,4,5)-triphosphate (PIP_3). This, in turn provides a docking site that leads to the tethering of various cytosolic proteins carrying PH (pleckstrin homology) domains (Weinberg, 2007). Among the most ubiquitous and potentially important of these is the Akt/PKB kinase which, once docked to the plasma membrane, becomes doubly phosphorylated and thereby activated. Activated Akt/PKB then proceeds to phosphorylate a variety of substances regulating proliferation, including endothelial nitric oxidase synthetase (ENOS) present and functionally coupled to nitric oxide (NO) production in the Golgi apparatus (Fulton et al., 2002). However, if active PTEN is also present, it dephosphorylates PIP_3 and converts it to PIP_2, thereby depriving Akt/PKB of a docking site at the plasma membrane (Wu et al., 2001). Some of the diverse biological functions assigned to activated Akt/PKB are summarized in Table 12.2.

Activated kt/PKB facilitates cell survival by both inhibiting pro-apoptotic (Bad, Caspase-9, FOX01 TF) pathways and activating anti-apoptotic proteins (IκB kinase and Mdm2) (Cory and Adams, 2002; Weinberg, 2007). Bad is an antagonist of Bcl-X. Bad and Bcl-X are both members of the Bcl-2 family of proteins regulating the channels with the mitochondrial membrane that release cytochrome c during the apoptotic cascade. Caspase-9 is a component of the protease cascade of apoptosis which is inactivated when phosphorylated by Akt/PKB kinase. AκB kinase (IKK) is phosphorylated and activated by Akt/PKB while phosphorylation of the forkhead (FOXO1) transcription factor (Brunet et al., 1999) prevents its nuclear translocation and subsequent activation of

Table 12.2.   Effects of Akt/PKB on survival, proliferation, and cell growth. Adapted from Weinberg (2007).

Biological effect	Substrate of Akt/PKB	Functional consequence
Anti-apoptotic	Bad (pro-apoptotic)	Inhibition
	Caspase-9 (pro-apoptotic)	Inhibition
	IκB kinase (anti-apoptotic)	Activation
	FOXO1 (pro-apoptotic)	Inibition
	Mdm2 (anti-apoptotic)	Activation
Proliferative	GSK-3β (anti-proliferative)	Inhibition
	FOXO4 (anti-proliferative)	Inhibition
	p21[Cip1] (anti-proliferative)	Inhibition
Growth	Tsc2 (anti-growth)	Inhibition
	mTOR (pro-growth)	Activation

pro-apoptotic genes. Mdm2, once phosphorylated by Akt/PKB, is activated and proceeds to trigger the destruction of p53 through ubiquitinylation and export of p53 from the nucleus. There are studies linking Golgi apparatus to apoptosis (Tsai et al., 2006) with irreversible Golgi apparatus fragmentation during apoptosis as one consistent feature (Mukherjie et al., 2007).

Activated Akt/PKB enhances cell proliferation through phosphorylation and inactivation of glycogen synthase kinase 3β (GSK-3β) activity. The latter is normally responsible for phosphorylating cyclin D1, causing its degradation (Doble and Woodgett, 2003). Once phosphorylated by Akt/PKB, FOXO4 is exported from the nucleus and induces expression of the CDK-inhibitor p27[Kip1] (Weinberg, 2007).

p21[Cip1] is a CDK inhibitor like p27[Kip1]. Its phosphorylation by Akt/PKB results in exit from the nucleus into the cytoplasm (Singerland and Pagano, 2000) where it acts as a caspase inhibitor.

The growth response from Akt/PKB activation is mediated through phosphorylation of Tsc2 causing dissociation of a Tsc1/Tsc2 complex and subsequent activation of mTOR which up-regulates protein synthesis and other contributors to an increase in cell size such as nutrient (glucose) availability (Vogt, 2001). Pancreatic islet cells selectively expressing constitutively active Akt/PKB kinase grow to a size more than twice the cross-sectional area of and almost four times the volume of their normal counterparts (Tuttle et al., 2001). This latter function of Akt/PKB kinase activation is manifested in the growth (increase in cell size) rather than actual cell cycle traverse (proliferation) of cells. The challenge now is to understand how such complex regulatory cascades are translated into the actual mechanics that drive cell enlargement (either by membrane displacement (Chapter 11) or delivery of new cell surface membrane via vesicles derived from the Golgi apparatus.

Activation of Akt by PI3K (phosphatidylinositol 3-kinase; Chapter 11) is associated with several human cancers (Osaki et al., 2004) as is overactive fatty acid and cholesterol synthesis (Freeman and Solomon, 2004). Du et al. (2006) have provided evidence of an Akt-dependent step in the transport of sterol regulatory element binding protein-2 (SREBP-2) from the endoplasmic reticulum to the Golgi apparatus mediated by the SREBP escort protein SCAP (SREBP cleavage-activating protein). A convergence of the two pathways involving the Golgi apparatus may be of fundamental regulatory importance in view of an absolute requirement of membrane synthesis for cell growth and proliferation.

12.3. Summary

Cancer cells exhibit characteristic membrane changes involving both the cell surface and intracellular membranes. Some, like decreases in the proportion of membrane-associated ribosomes to ribosomes free in the cytoplasm, are clearly linked to tumor progression and become most evident in either poorly differentiated or rapidly growing tumors, or both. Other changes observed with the Golgi apparatus and at the plasma membrane emerge as being potentially transformation linked, and appear to be basic to the expression of the malignant phenotype.

A major impetus for membrane research in cancer has been the need to determine mechanisms of how the transformed state is expressed phenotypically in terms of the regulation of cell proliferation and surface characteristics important to growth control. Here, contemporary research focuses not only on surface altera- tions but on various transducing and signaling mechanisms responsible for growth control and the generation of mitogenic stimuli as well. For membrane alterations to be expressed at the cell surface, biosynthesis and processing are directed by a complex system of internal endomembranes (e.g., nuclear envelope, endoplasmic reticulum, Golgi apparatus, vesicles) all ultimately under genetic control.

A major limitation to the design of effective strategies of cancer control is the overall lack of understanding of basic mechanisms of loss of growth control and adhesive properties manifest in most clinical malignancies. Knowledge of the types of changes that take place in cell-surface membranes and of how these changes result from coordinated activities of the endomembrane system may open the way to the rational design of new cancer control strategies as well as pro- vide a better understanding of the nature of the surface lesions responsible for the neoplastic phenotype and the ultimate underlying genomic alterations. However, as with other diseases (diabetes, atherosclerosis, aging, etc.) (Morré, 1991) the Golgi apparatus remains more as an important facilitator of disease-related events i.e., redistribution of glycosyltransferases, rather than causal to the disease.

Epilogue

The decade since the celebration in 1998 of the centenary year of Golgi apparatus discovery has seen the emergence of a new era of Golgi apparatus discovery with focus on cell and molecular biology. The electron microscopy era defined Golgi apparatus in terms of stacks and ribbons, COP-II-coated blebs and vesicles of the transitional endoplasmic reticulum and cis Golgi apparatus face, COP-I-coated tubule ends of the Golgi apparatus midregion, clathrin-coated vesicles and membrane surfaces at the trans Golgi apparatus network, and Golgi apparatus exit face where secretory vesicles and granules are released for eventual delivery to the cell surface. The new era has seen the beginnings of a definition of these same structures and attendant trafficking events in terms of specific membrane protein molecules and protein–protein interactions. However, despite their centrality for membrane trafficking and organization, the specific roles of many of the proteins involved in the multifaceted machinery they regulate are incompletely understood. Functional reconstitutions in cell-free systems still offer viable approaches to further resolution of details. Light microscopic observations using fluorescent constructs where vesicles and tubules are replaced by punctae and stacks and ribbons are seen as compact structures along with direct visualization of flow of membrane constituents across the stacked cisternae are unobtainable from static electron microscope pictures. However, occasional electron microscope validation to ensure that structures seen in light microscope images are, in fact, correctly identified may still be advisable. Historically, the early confusion about Golgi apparatus structure and even its reality were based up on light microscope images with final clarification coming only with the advent of the electron microscope and the precision and resolution of structure it provided. Golgi apparatus tubules, for example, are poorly resolved at the level of the light microscope and fragments derived from Golgi apparatus and fragments derived from endoplasmic reticulum may be difficult if not impossible to distinguish morphologically except through correlative comparisons from electron microscopy.

The electron microscope showed the Golgi apparatus to be a stack of cis, medial and trans cisternae. Autoradiographic and cell fractionation studies provided a time dimension to reveal passage of endoplasmic reticulum-synthesized products destined for secretion and of membrane proteins and lipids for delivery to the cell surface via passage through or around the Golgi apparatus stacks (Chapter 6). These findings were elaborated further in dynamic models where endoplasmic reticulum-derived transport vesicles coalesced at the cis face of the stack to generate new cisternae. The newly-generated cisternae then progressed through the stack eventually reaching the trans face where they were converted to a mixture of secretory vesicles as transport carriers to the plasma membrane and cisternal remnants. This cisternal maturation or flow-differentiation model (Morré et al., 1979b) implied that the Golgi apparatus was constantly being renewed by input from the endoplasmic reticulum and consumed at least in part as a source of new plasma membrane constituents to support growth and membrane renewal at the cell surface. However, membranes and export products do not pass through the Golgi apparatus stack unchanged. Proteins and lipids, for example, are glycosylated through the actions of specifically-located protein and lipid glycosylating enzymes which are concentrated in characteristic cisternal positions in the stack where glycosylations occur. From a biochemical perspective, one of the most defining features of the Golgi apparatus complex is this presence of carbohydrate-processing enzymes that are compartmentalized in the various cisternae comprising the stack and that display a polarized distribution in a cis to trans direction (e.g., Chapters 10 and 12). How the distinct compartmentalization and distribution of Golgi apparatus processing enzymes is achieved and maintained in the face of a continuous flux of membranes across the polarity axis is a challenging prospect.

Golgi apparatus are definitely not static compartments. Golgi apparatus collapse following brefeldin A involving the COP-I block (Lippincott-Schwartz et al., 1989) or following the use of Sar1 GTPase mutants locked in bound nucleotide conformations that act as dominant inhibitors of COP-II function (Barlowe et al., 1994; Kuge et al., 1994) provide strong evidence for complete turnover of all Golgi apparatus cisternae during normal Golgi apparatus functioning. How then are Golgi apparatus resident proteins recruited and maintained within the stack? This concern was largely avoided in models based on vesicular traffic which involved stationary cisternae (Farquhar and Palade, 1981) or in the distillation tower model of Rothman (1981). One possibility is that COPI-coated peripheral tubules of the Golgi apparatus serve as repositories and/or sources of such "resident" enzymes of the Golgi apparatus to add and remove required functional activities at specific positions within the stacked cisternae. This concept is supported as well by the classic experiments summarized above which show that when brefeldin A is added to block this process, the Golgi apparatus collapses and the resident Golgi apparatus enzymes are seen to remain associated with tubulo-reticular Golgi apparatus fragments or even to fold back on the endoplasmic reticulum per se. A role for Golgi apparatus-associated polyribosomes for addition and/or renewal of resident Golgi apparatus proteins remains an experimentally untested possibility as well.

To what extent do Golgi apparatus resident proteins constitutively cycle through the endoplasmic reticulum? This has not been determined rigorously for resident membrane proteins and/or proteins normally destined for export to the cell surface. Indication of recycling based on Golgi apparatus-associated proteins known to behave dynamically as a result of their functions in membrane trafficking (KDEL receptor, TGN-38) may not be representative of the recycling characteristics of Golgi apparatus resident proteins.

Delivery of membrane proteins to the Golgi apparatus presumably occurs directly to cis face-forming regions via endoplasmic reticulum-derived COP-II-coated vesicles. On the other hand, delivery of vesicle contents, especially as has been shown for lipoprotein particles of the liver and zymogens of the pancreas, may be seen to follow a peripheral route that bypasses the stacks. Luminal continuity is provided by a peripheral boulevard (boulevard périphériquie) of smooth Golgi apparatus tubules that connect the rough endoplasmic reticulum directly to secretory vesicles (see for example, Claude, 1970; Morré and Ovtracht, 1981; Chapter 8).

A logical extension of the cisternal maturation model of Golgi apparatus function is that under appropriate circumstances, entire Golgi apparatus stacks might be generated de novo from transitional endoplasmic reticulum exit sites. On the other hand, the general assumption that organelles proliferate by the growth and division of existing organelles (maternal inheritance of mitochondria, for example) support an alternative view where Golgi apparatus will always be assembled on a permanent Golgi apparatus template derived from pre-existing Golgi apparatus. Observations of reconstitutions of Golgi apparatus following disintegration during vertebrate mitosis or following brefeldin A treatment and the lack thereof in cytoplasts generated by microsurgery despite the presence of abundant rough endoplasmic reticulum favor the latter view. Most of the Golgi apparatus-derived material remains distinct from endoplasmic reticulum in mitotic vertebrate cells and it may do so as well in brefeldin A-treated cells. Thus while in one sense Golgi apparatus organization appears to be a product of self organization, some degree of template dependence is strongly indicated that may require either the presence of Golgi apparatus fragments or of matrix constituents to direct self-assembly and to maintain structure once assembled.

Assuming Golgi apparatus assembly within an existing Golgi apparatus region, the proteins and mechanisms involved in formation and maintenance of the stacked cisternae remain largely unknown. One possibility considered early was that the size of the central cisternal saccule at the juncture with the system of peripheral tubules is self-limiting or will not extend beyond a certain limit imposed by the transitional endoplasmic reticulum regions specialized for the production of transport vesicles (Fig. 9.3). When a newly formed cisterna reaches the determined size, a second would be formed. The final number of cisternae per stack might then be determined by membrane differentiation events that would serve as signals to initiate their eventual conversion into transport units for delivery to the cell surface. However, how a minimal spacing occupied by intercisternal elements in plants is formed and maintained in the absence of any discernable structure in vertebrates remains a mystery. On the other hand, molecular details of the organization and sorting and modification (differentiation) of proteins and lipids within the Golgi apparatus to establish and maintain its cis to trans polarization need no longer remain enigmatic. One possibility under investigation is that various Rab proteins along with their cognate effectors including tethers, scaffold proteins (COGs), and SNARE components operate as Rab-based hub complexes that direct dynamic domain organization within the Golgi apparatus stack (Bannykh, 2005).

The final challenge may be to extend our understanding of how forming membrane buds, secretory vesicles and tubules generate and direct the physical forces necessary to drive vesicle budding, tubule migration and other aspects of physical membrane displacement involved in normal Golgi apparatus functioning. Currently very little is known about what types of membrane motors might exist within the organization of the Golgi apparatus and how they might serve to convert chemical energy or membrane potential into the many physical displacements in time and space required to move membranes and cargo along the Golgi apparatus export route.

Appendix Tables

Appendix Table 1. History of the Golgi apparatus 1865–1965.

1865, 1867	La Valette St. George was one of the first to record an image in the light microscope of a structure later equated with the Golgi apparatus – the head cap of the acrosome "Samenkörper"
1889	Platner described "nebenkern" in snail spermatocytes
1898	Camillo Golgi published the famous "apparato reticolare interno"
1906	Golgi shares Nobel prize in medicine with Cajal
1908	Cajal accepts Golgi's discovery
1914–1923	Cajal refines Golgi's observations
1923, 1924	Nassonov concludes Golgi apparatus is an organelle of secretion
1929	Bowen publishes famous review on Golgi apparatus of mammalian germ cells and other secretory cells, "Cytology of Secretion"
1935	Beams and King demonstrated that the Golgi apparatus of plant cells and osmiophilic platelets were equivalent
1940	Nahm concludes that Golgi apparatus do not exist in plants
1943–1953	Baker and others. "The Great Golgi Controversy"
1953	Dalton and Felix publish first electron micrographs of the "Golgi substance" of epithelial cells of the epididymis and duodenum of the mouse
1954	Sjöstrand and Hanzon describe the ultrastructure of Golgi apparatus of the exocrine pancreas. Schneider and Kuff report initial studies of Golgi apparatus isolated from rat epididymis
1956	Hodge publishes an electron micrograph of a plant cell in which Golgi apparatus may be recognized
1957	Porter describes endoplasmic reticulum
1959	Grimstone deduces a role of endoplasmic reticulum in the formation of new Golgi apparatus cisternae in the flagellate *Trichonympha*
1959	Kuff and Dalton demonstrate that Golgi apparatus withstand rigors of cell homogenate preparation
1961	Schnepf provides first quantitation of Golgi apparatus activity and evidence for energy requirements
1962	Novikoff applies histochemistry to the study of Golgi apparatus enzymes
1962	Jarosch identifies and observes Golgi apparatus in situ in a living cell, the alga *Micrasterias*
1964	Morré and Mollenhauer isolate the Golgi apparatus in biochemically useful yield and fraction purity from plant cells
1964	Peterson and LeBlond employ radioautography to demonstrate synthesis of complex carbohydrates in the Golgi apparatus region

For a more detailed compilation and important research contributions between 1965 and 1982, see Table 1 of Berger, 1997.

Appendix Table 2.   Summary of Golgi apparatus terminology.

Golgi apparatus (*sing. or plural*) – The totality of Golgi apparatus stacks = dictyosomes in a cell that functions together. In popular usage, any system of stacked smooth surfaced cisternae or portion thereof involved in, or potentially involved in, the production of secretory vesicles

Golgi (*apparatus*) *complex* – Usually a somewhat broader term that includes the Golgi apparatus proper plus associated structures such as secretory and transition vesicles and cis and trans Golgi apparatus networks

Golgi (*apparatus*) *zone* – The Golgi apparatus complex plus the surrounding cytoplasm. May include transitional endoplasmic reticulum elements and Golgi apparatus-associated polyribosomes, lysosomes and microbodies

Dictyosome = *Golgi apparatus Stack* – The polarized stacks of smooth-surfaced cisternae that comprise the Golgi apparatus. Equivalent terms in the botanical literature include Golgi body and Golgi dictyosome. Probably equivalent to the osmiophilic platelet of the early literature

Zone of exclusion = *Golgi apparatus ground substance or Golgi apparatus matrix* – A differentiated region of cytoplasm surrounding the Golgi apparatus where organelles are scarce or absent, endoplasmic reticulum becomes smooth-surfaced (lacking ribosomes) and coated vesicles are common

Terms referring to polarity:

Cisternae with ER-like properties, also known as forming face, immature face, cis face, proximal pole (face)

Cisternae with PM-like properties, from which mature secretory vesicles exit, also known as maturing face, mature face, trans face, distal pole (face), secreting face (pole), exit face

Cisternae with intermediate properties are also known as intercalary cisternae

Trans Golgi network (*TGN*) – A morphologically distinct tubular–vesicular component at the trans-most aspect of the Golgi apparatus

Cis Golgi network – A subset of membranes at the cis Golgi apparatus pole formed from transitional endoplasmic reticulum by fusion of transition vesicles suggested to function as an intermediate compartment between the transitional endoplasmic reticulum and the cis Golgi apparatus

Cisterna = *saccule* = *lamella* (pl., cisternae saccules, lamellae) – A lumen or central cavity surrounded by a membrane lacking ribosomes which forms the individual units that when stacked form dictyosomes

Coatomer proteins (*COPs*) – COPI is a multimolecular complex composed of seven COPs: α, β, β', γ, δ, ε, and ξ. The seven COP polypeptide complex is generally called a coatomer (Waters *et al.* 1991, Schekman and Orci 1996). The eighth component of the COPI coat is ARF1, a small GTP-binding protein (Serafini *et al.* 1991). The mammalian COPII coat consists of small GTPase hSar1p and the heterodimeric protein complexes hSec23/24p and hSec13/31p (Barlowe *et al.* 1994).

Secretory vesicle = *Golgi apparatus vesicle* – A transport vehicle, surrounded by a membrane, formed from materials derived from the Golgi apparatus. The term is normally reserved for spherical or elongated vesicles that migrate to discharge their contents by fusion with another structure (usually the plasma membrane) shortly after release from the Golgi apparatus

Condensing vacuole = *immature secretion granule* – A transport vehicle tracing its origins to the Golgi apparatus but which remains for a time in the cytoplasm during which it accumulates, sequesters, concentrates or modifies its content (and membrane) and/or becomes altered morphologically

Secretory granules = *secretion granule* – Usually refers to a mature condensing vacuole (syn., mature secretion granule)

Transition vesicle – A small ca. 50-nm vesicle (with a nap-like or COP-II coat of the region between endoplasmic reticulum and the Golgi apparatus, usually within the Golgi apparatus zone of exclusion, that function in the transfer of membranes and other materials between the two structures (syn., primary vesicle)

Small Golgi apparatus vesicles – Refers to any of the small 50–60 nm vesicular profiles within the Golgi apparatus zone. Some may represent transition vesicles. Many small vesicular profiles associated with Golgi apparatus having electron transparent lumens represent cross sections through the numerous peripheral tubules continuous with Golgi apparatus cisternae

---

(continued)

Appendix Table 2. (continued)

Small clathrin-coated vesicles – Ca. 50–60 nm vesicles with spiny (clathrin-based coats that occur predominantly at the mature or trans Golgi apparatus face either attached to Golgi apparatus membranes or apparently free in the Golgiplasm. Clathrin-coated surfaces may exist as well on other portions attributed to the trans Golgi apparatus face including mature secretory vesicles

Intermediate compartment (*cis Golgi apparatus network*) – A subset of cisternal membranes, tubules, and vesicles located between the transitional endoplasmic reticulum and the cis-most Golgi apparatus cisternae observed with some but not all cell types containing Golgi apparatus

Golgi apparatus equivalent – Single cisternae that form secretory vesicles and constitute the secretory apparatus of certain fungal species that lack cisternal stacks (dictyosomes)

Terms to be avoided:

Golgi vesicle – In referring to an isolated fraction, a "vesicle" formed from Golgi apparatus-derived material during cell fractionation. Golgi apparatus fragments or Golgi apparatus membranes may be more descriptive

Golgi – As a single term referring to the apparatus of Golgi or its parts

Organelle – In referring to the Golgi apparatus, the term *cell component*, as suggested by deDuve (1964), seems more appropriate. Golgi apparatus functioning, in contrast to that of mitochondria or plastids, is highly dependent upon its associations as a part of the cell's internal endomembrane system

---

Appendix Table 3. Similar enzyme proteins and/or activities localized in endoplasmic reticulum, Golgi apparatus and plasma membrane.

Material	Activity detected in[a]	References
Enzymes concentrated in nuclear envelope and endoplasmic reticulum		
Glucose-6-phosphatase		
Liver, rat	NE; rough and smooth ER	Ericsson (1966); Farquhar et al. (1974); Wanson et al. (1975)
Liver (fetal and neonatal), rat	NE; rough and smooth ER; profiles of small vesicles at the periphery of the GA (depending on stage of development)	Leskes et al. (1971a, b)
Intestine (epithelium), mouse	NE; ER, especially one cisterna close to the GA	Hugon et al. (1971)
	NE; ER; GA; cisterna next to ER	Hugon et al. (1972, 1973)
Heart cells (cultured), chick	NE; rough ER; on cisterna of the GA	Schäfer and Hündgen (1971)
Vas deferens (epithelium), crayfish	ER (faint reaction); innermost cisterna of the GA	Kessel et al. (1969)
Arylsulfatase		
Bone marrow (leukocytes), rat and rabbit	NE; individual ER cisternae; inner most cisterna of the GA; immature granules (depending on leukocytic series and stage of development)	Bainton and Farquhar (1968, 1970)
Liver, rat	NE; ER; the GA and lysosomes; activity generally distributed	Wilkinson et al. (1972)
Root cells, cucumber	NE; ER; phagosomes (microbodies)	Poux (1966)
Nucleoside diphosphatase		
Liver, rat	IDP as substrate: NE (faint reaction); rough and smooth ER; innermost cisterna of the GA (faint reaction)	Goldfischer et al. (1971)
Adenohypophysis	IDP and ADP as substrates: as above; PM	Pelletier and Novikoff (1972)

(continued)

Appendix Table 3. (continued)

Material	Activity detected in[a]	References
Vas deferens (epithelium), crayfish	IDP as substrate: ER; outermost cisterna of the GA	Kessel et al. (1969)
Root cells, cress	IDP as substrate: ER; GA cisternae at the maturing face of dictyosome; profiles of small vesicles close to the GA	Zaar and Schnepf (1969)
Acetabularia	IDP, GDP, and UDP as substrates: increase in reaction product in successive dictyosome cisternae toward maturing face	Zerban and Werz (1975)
NADH ferricyanide oxidoreductase		
Liver, rat	NE; ER; GA; cisterna at forming face and peripheral GA tubules	Morré et al. (1974a, b)

Enzymes concentrated in plasma membrane[b]

Nucleoside monophos- phatase (5'nucelotidase)		
Liver, rat	AMP as substrate: secretory vesicles; PM; lysosomes	Farquhar et al. (1974); Morré et al. (1974a, b)
Bone marrow and blood (leukocytes), rabbit	AMP as substrate: some immature azurophil granules	Bainton and Farquhar (1968)
	CMP as substrate: innermost cisterna of the GA; small, coated vesicles; multivesicular bodies	Friend and Farquhar (1967)
Alkaline phosphatase		
Thyroid gland, rat	GA cisternae; PM	Calvert (1973, 1974)
Bone marrow and blood (leukocytes), rabbit	GA cisternae (gradient of activity within dictysomes); immature specific granules (depending on leukocytic series and stage of development)	Wetzel et al. (1967); Bainton and Farquhar (1968)
Intestine, rat	GA cisternae; PM	Hugon et al. (1973)
Various cells, mouse	GA cisternae; PM	Thiéry (1974)
NADH oxidase	Mature GA cisternae; membranes of mature secretory vesicles; PM	Morré et al. (1974a, b, unpublished)
Liver, rat Polymorphonuclear leukocytes, human	PM and endocytotic vesicles	Briggs et al. (1975)
Adenylate cyclase		
Liver, rat	ER; PM (including evidence for fluoride and glucagon stimulation)	Morré et al. (1974a, b); Howell and Whitfield (1972); Wagner et al. (1972); see, however, Lemay and Jarett (1975)
	This enzyme normally thought to be restricted to PM	
Na[+]-, K[+]-, MG[2+]-ATPase		
Red cell ghosts	Normally restricted to PM	Marchesi and Palade (1967); Benedetti and Delbauffe (1971)

(continued)

Appendix Table 3. (continued)

Material	Activity detected in[a]	References
Enzymes concentrated in Golgi apparatus		
Thiamine pyrophosphatase		
Liver, rat	Rough and smooth ER (faint reaction); GA cisternae (activity restricted to one or a few cisternae at the maturing face of dictyosomes; when present in several cisternae, the reaction is heaviest at the maturing face of the dictyosome)	Cheetham et al. (1971); Goldfischer et al. (1971); Farquhar et al. (1974)
Pancreas (acinar cells) and male genital tract, rat	GA cisternae (heaviest activity at the maturing face of the dictyosome). No activity in the GERL complex	Friend and Farquhar (1967); Cheetham et al. (1971)
Adenohypophysis, rat	As above	Pelletier and Novikoff (1972)
Neurons (dorsal root ganglia), mouse	As above	Novikoff et al. (1971); Boutry and Novikoff (1975)
Intestine (epithelium), mouse	NE; ER (faint reaction); GA cisternae at the maturing face of the dictyo some (in goblet cells, no reaction in NE and ER)	Hugon (1970)
Embryo (first-cleavage stage), toad	GA cisternae; GA-derived vesicles; PM (depending on the developmental stage)	Sanders and Singal (1975)
Amebas	ER; GA	Wise and Flickinger (1971)
Acetabularia	ER; GA	Zerban and Werz (1975)
Root tips (various cells), maize	Mostly localized in the GA	Dauwalder et al. (1969)
Acyltransferase		Benes et al. (1973)
Schwann cells, rat	GA cisternae; GA-derived vesicles which fuse with the PM	
Liver, rat	ER; proximal GA cisternae	Higgins and Barrnett (1972)
Intestine, rat	ER: proximal GA cisternae	Higgins and Barrnett (1971)
Salt gland, duck	GA cisternae; vesicles; PM (depend ing on the stage of development)	Levine et al. (1972)
Enzymes concentrated in lysosomes		
Acid phosphatase		
Neurons (ganglion nodo sum, dorsal root, ganglia), rat	One specialized ER cisternae associated with the inner (concave, trans) face of the GA stack and related to lysosome formation, that is, the GERL complex; lysosomes	Holtzman et al. (1967); Novikoff et al. (1971)
Neuron (dorsal root ganglia), mouse	Large neurons: activity restricted to the GERL complex; lysosomes. small neurons: all GA cisternae show activity (no obvious polarity); lysosomes	Boutry and Novikoff (1975)
Hepatoma, rat	GERL complex; lysosomes	Essner and Novikoff (1962); compare with Bennett and Leblond (1971)
Vas deferens, rat	GA cisternae (at the inner, concave face); GA-derived vesicles; lysosomes	Friend and Farquhar (1967)

(continued)

Appendix Table 3.   (continued)

Material	Activity detected in[a]	References
Bone marrow and blood (leukocytes), rat and rabbit	ER; GA cisternae (with heavier reaction on inner or outer cisternae, depending on stage of development); immature granules	Wetzel et al. (1967); Bainton and Farquhar (1968); Compare with Bainton and Farquhar (1966)
Adenohypophysis, fish	CA cisternae; GA-derived vesicles which fuse with lytic bodies (depending on functional state)	Hopkins (1969)
Root cells, cress, cucumber, and maize	Provacuoles	Berjak (1972)
	NE; ER (occasionally); GA-related vesicles; provacuoles giving heaviest reaction	Poux (1970)
	GERL complex (depending on stage of development); lysosomes?	Dauwalder et al. (1969); Coulomb et al. (1972)
β-Glucuronidase		
Liver, rat and mouse	NE; within cisternae of ER and GA; lysosomes	Smith and Fishman (1969)

Other enzymes

Peroxidases

Liver (Kupffer cells), rat	NE; ER; GA cisternae; dense bodies; multivesicular bodies	Fahimi (1970)
Parotid and lacrymal exorbital glands, rat	NE; ER; transition elements (including transition vesicles); GA cisternae at maturing face of dictyosomes; condensing vacuoles; lysosomes	Herzog and Miller (1970, 1972); Essner (1971)
Thyroid gland, rat	NE; ER: some inner GA cisternae; dense bodies (depending on stage of development)	Strum and Karnovsky (1970); Strum et al. (1971); Novikoff et al. (1974)
Lacrymal gland, rat	ER; GA; secretory vesicles from the GA	Herzog et al. (1976)
Bone marrow (leukocytes), rat, rabbit, cat, and human	NE; ER; transition elements (including transition vesicles); GA cisternae; immature and mature granules (depending on state of development and leukocytic series)	Ackerman (1968); Miller and Herzog (1969); Bainton and Farquhar (1970); Ackerman and Clark (1971)
Female genital tract, and mammary tumor cells, rat	NE; ER; GA cisternae; immature and mature secretory granules (dependency on estrogen)	Churg and Anderson (1974); Anderson et al. (1975)
Root, cucumber	NE; ER; GA cisternae and GA-derived vesicles; vacuoles; PM	Poux (1969)
Catalase		
Lacrymal gland, rat	Absent from ER and GA; present in peroxisomes	Herzog and Fahimi (1976)

Source: From Morré and Ovtracht (1977)

[a]NE, Nuclear envelope; ER, endoplasmic reticulum; GA, Golgi apparatus; PM, plasma membrane

[b]For brevity numerous studies that focus primarily on the plasma membrane localization of these enzymes have been omitted

References for Appendix Table 3

Ackerman, G. A. (1968). Lab. Invest. 19, 220.

Ackerman, G. A., and Clark M. A. (1971). Z. Zellforsch. Mikrosk. Anat. 117, 463.

Anderson, W. A., Kang, Y. H., and DeSombre, E. R. (1975). J. Cell Biol. 64, 668.

Bainton, D. F., and Farquhar, M. G. (1966). J. Cell Biol. 28, 277.

Bainton, D. F., and Farquhar, M. G. (1968). J. Cell Biol. 39, 299.

Bainton, D. F., and Farquhar, M. G. (1970). J. Cell Biol. 45, 54.

Benedetti, E. L., and Delbauffe, D. (1971). In Cell Membranes (G. W. Richter and D. G. Scarpelli, eds.), p. 54. Williams & Wilkins, Baltimore, Maryland.

Benes, F., Higgins, J. A., and Barrnett, R. J. (1973). J. Cell Biol. 57, 613.

Bennett, G., and Leblond, C. P. (1971). J. Cell Biol. 51, 875.

Berjak, P. (1972). Ann. Bot. (London) [N.S.] 36, 73.

Boutry, J.-M., and Novikoff, A. B. (1975). Proc. Natl. Acad. Sci. U.S.A. 72, 508.

Calvert, R. (1973). Anat. Rec. 177, 359.

Calvert, R. (1974). Anat. Rec. 180, 663.

Cheetham, R. D., Morré, D. J., Pannek, C., and Friend, D. S. (1971). J. Cell Biol. 49, 899.

Churg, A., and Anderson, W. A. (1974). J. Cell Biol. 62, 449.

Coulomb, P., Coulomb, C., and Coulon, J. (1972). J. Microsc. (Paris) 13, 263.

Dauwalder, M., Whaley, W. G., and Kephart, J. E. (1969). J. Cell Sci. 4, 455.

Ericsson, J. L. E. (1996). J. Histochem. Cytochem. 14, 361.

Essner, E. (1971). J. Histochem. Cytochem. 19, 216.

Essner, E., and Novikoff, A. B. (1962). J. Cell Biol. 15, 289.

Fahimi, H. D. (1970). J. Cell Biol. 43, 167.

Farquhar, M. G., Bergeron, J. J. M., and Palade, G. E. (1974). J. Cell Biol. 60, 8.

Friend, D. S., and Farquhar, M. G. (1967). J. Cell. Biol. 35, 357.

Goldfischer, S., Essner, E., and Schiller, B. (1971). J. Histochem. Cytochem. 19, 349.

Herzog, V., and Fahimi, H. D. (1976). Histochemistry 46, 273.

Herzog, V., and Miller, F. (1970). Z. Zellforsch. Mikrosk. Anat. 107, 403.

Herzog, V., and Miller, F. (1972). J. Cell Biol. 53, 662.

Herzog, V., Sies, H., and Miller, F. (1976). J. Cell Biol. 70, 692.

Higgins, J. A., and Barrnett, R. J. (1971). J. Cell Biol. 50, 102.

Higgins, J. A., and Barrnett, R. J. (1972). J. Cell Biol. 55, 282.

Holtzman, E., Novikoff, A. B., and Villaverde, H. (1967). J. Cell Biol. 33, 419.

Hopkins, C. R. (1969). Tissue & Cell 1, 653.

Howell, S. L., and Whitfield, M. (1972). J. Histochem. Cytochem. 20, 873.

Hugon, J. S. (1970). J. Histochem. Cytochem. 18, 361.

Hugon, J. S., Maestracci, D., and Ménard, D. (1971). J. Histochem. Cytochem. 19, 515.

Hugon, J. S., Maestracci, D., and Ménard, D. (1972). Histochemie 29, 189.

Hugon, J. S., Maestracci, D., and Ménard, D. (1973). J. Histochem. Cytochem. 21, 426.

Kessel, R.G., Panje, W. R., and Decker, M. L. (1969). J. Ultrastruct. Res. 27, 319.

Lemay, A., and Jarett, L. (1975). J. Cell Biol. 65, 39–50.

Leskes, A., Siekevitz, P., and Palade, G. E. (1971a). J. Cell Biol. 49, 264.

Leskes, A., Siekevitz, P., and Palade, G. E. (1971b). J. Cell Biol. 49, 288.

Levine, A. M., Higgins, J. A., and Barrnett, r. J. (1972). J. Cell Sci. 11, 855.

Marchesi, V. T., and Palade, G. E. (1967). J. Cell Biol. 35, 385.

Miller, F., and Herzog, V. (1969). Z. Zellforsch. Mikrosk. Anat. 97, 84.

Morré, D. J., Keenan, T. W., and Huang, C. M. (1974a). Adv. Cytopharmacol. 2, 107.

Morré, D. J., Yunghans, W. N., Keenan, T. W., and Vigil, E. L. (1974b). In Methodological Developments in Biochemistry. Vol. 4. Subcellular Studies (E. Reid, ed.), p. 195. Longmans, Green, New York.

Novikoff, P. M., Novikoff, A. B., Quintana, N., and Hauw, J. (1971). J. Cell Biol. 50, 859.

Pelletier, G., and Novikoff, A. B. (1972). J. Histochem. Cytochem. 20, 1.

Poux, N. (1966). J. Histochem. Cytochem. 14, 932.

Poux, N. (1969). J. Microsc. (Paris) 8, 855.

Poux, N. (1970). J. Microsc. (Paris) 9, 407.

Sanders, E. J., and Singal, P. K. (1975). Exp. Cell Res. 93, 219.

Schäfer, D. and Hündgen, M. (1971). Histochemie 26, 362.

Smith, R. E., and Fishman, W. H. (1969). J. Histochem. Cytochem. 17, 1.

Strum, J. M., and Karnovsky, M. J. (1970). J. Cell Biol. 44, 655.

Thiéry, G., and Rambourg, A. (1976). J. Microsc. Biol. Cell (Paris) 26, 103.

Wagner, R. C., Kreiner, P., Barrnett, R. J., and Bitensky, M. W. (1972). Proc. Natl. Acad. Sci. U.S.A. 69, 3175.

Wanson, J.-C., Drachmans, P., May, C., Penasse, W., and Popowoki, A. (1975). J. Cell Biol. 66, 23.

Wetzel, B. K., Spicer, S. S., and Horn, R. G. (1967). J. Histochem. Cytochem. 15, 311.

Wilkinson, F. W., Nyquist, S. E., Merritt, W. D., and Morré, D. J. (1972). Proc. Indiana Acad. Sci. 81, 121.

Wise, G. E., and Flickinger, C. J. (1971). Exp. Cell Res. 67, 323.

Zaar, K., and Schnepf, E. (1969). Planta 88, 224.

Zerban, H., and Werz, G. (1975). Cytobiologie 12, 13.

Appendix Table 4. Membrane differentiations in virus-infected cells.

Virus group/ specific agent	Host	Membrane affected[a]	Membrane alteration	References
RNA viruses Orthomyxovirus				
Influenza (A0)	Ehrlich ascites tumor cells	PM	Fuzzy coat containing radial spikes on the outside and a dark layer on the inside at budding sites	Bächi et al. (1969)
	Hamster and bovine kidney cells	SM, ER	Association of viral glycoproteins with isolated smooth and rough cytoplasmic membranes	Compans (1973)
		PM	Seven viral-specific structural proteins found in associations with PM	Lazarowitz et al. (1971)
Influenza (avian)	Cultured chick embryo fibroblasts	PM	Hemagglutinin antigen appears at the cell surface as visualized by fluorescent antibodies	Breitenfeld and Schäfer (1957)
		SM, ER	Association of viral glycoproteins with isolated smooth and rough cytoplasmic membranes	Klenk et al. (1974)
Influenza (WSN strain)	Cultured chick embryo fibroblasts	PM	Surface spikes at points of budding virions; adsorption of erythrocytes to cell surfaces	Compans and Dimmock (1969), and references cited
	Choriallantoic membranes of chick embryos	PM	Electron microscope localizations of viral antigens by ferritin-labeled antibody	Duc-Nguyen et al. (1966); Morgan et al. (1961a, 1962a, b)

(continued)

Appendix Table 4. (continued)

Virus group/ specific agent	Host	Membrane affected[a]	Membrane alteration	References
		NE, ER	Electron microscope localizations of viral ntigens by ferritin-| labeled antibody in frozen cells	Morgan et al. (1961b)
Paramyxovirus				
Newcastle disease	Chick embryo fibroblasts	PM	Increased permeability to sucrose-[^{14}C] and dextran-[^{14}C]	Katzman and Wilson (1974)
	HeLa cells	PM	Bipolar hemadsorption	Marcus (1962)
Parainfluenza	Bovine kidney cells	PM	Increased phosphatidylethanolamine/phosphatidylcholine ratio	Klenk and Choppin (1970)
Rubella	Hamster kidney cells	GA, PM	Virions budding into the GA; binding of ferritin-conjugated antibodies to the PM	Higashi (1973)
Rhabdovirus				
Vesticular stomatitis	L cells	ER, PM	Viral spike protein associated with rough ER- and PM-rich fractions. Nonglycosylated envelope protein most abundant in ER-derived fractions	Wagner et al. (1970, 1972a, b)
		PM	Electron microscope localization of viral antigens by ferritin-labeled antibody and by fluorescent light microscopy using fluorescein-conjugated apoferritin	Wagner et al. (1971)
	HeLa cells	PM	Virus structural proteins associated with isolated plasma membranes	Cohen et al. (1971)
Leukovirus				
Rous sarcoma	Hamster kidney cells	ER, PM	Association of virus-specific antigens	Fleissner (1970)
	Chick embryo cells	PM	Ferritin-labeled antibody to viral envelope antigens located in discrete regions	Gelderblom et al. (1972)
		PM	Increased cholesterol/phospholipid ratio and changes in phospholipid amounts	Quigley et al. (1971, 1972)

(continued)

Appendix Table 4. (continued)

Virus group/ specific agent	Host	Membrane affected[a]	Membrane alteration	References
Mouse mammary tumor	Mouse mammary tumors	PM	Protruding 10-nm spikes characteristic of B particles	Hairstone et al. (1964)
			Surface spikes and binding of mammary tumor virus antibody	Kramarsky et al. (1970)
			Localization of viral antigens by ferritin-labeled antibody	Tanaka and Moore (1967)
			Loss of intramembrane particles	Sheffield (1974)
Arbovirus group A				
Sindbis	Chick embryo fibroblasts	PM	Presence of viral proteins shown by polyacrylamide gel electrophoresis	Bose and Brundige (1972)
			Incorporation of viral proteins	Pfefferkorn and Clifford (1964)
			Increased cholesterol/phospholipid ratio	Hirschberg and Robbins (1974)
			Formation of virus-specific subunit structure by freeze-fracture and surface replica	Birdwell et al. (1973); Brown et al. (1972)
	Chick embryo and hamster kidney cells	PM	Exclusion of phosphatidylinositol from viral envelope	David (1971)
Semliki forest	Chick embryo cells	PM	Coating of projections to outer surface of the PM during virus budding	Acheson and Tamm (1967)
DNA viruses				
Pox virus				
Vaccinia	HeLa cells	PM	Electron microscope localization of viral antigens by ferritin-labeled antibody	Morgan et al. (1961 a–c)
			Presence of virus-induced, nonvirion proteins and glycoproteins	Weintraub and Dales (1974)
			Virally induced glycoproteins associated with crude subcellular membrane fraction	Garon and Moss (1971); Moss et al. (1971)

(continued)

Appendix Table 4. (continued)

Virus group/ specific agent	Host	Membrane affected[a]	Membrane alteration	References
			Hemagglutinin protein found evenly distributed over the PM except in regions where virus is being released	Ichihashi and Dales (1971)
Herpes virus Herpes simplex	Human epidermoid carcinoma	NE	Thickening of inner membrane of NE at sites of virus envelopment	Shipkey et al. (1967)
		ER	Thickening of membranes in apposition to nucleocapsids	Schwartz and Roizman (1969)
		ER, PM	Glycoproteins specified by virus synthesized and associated with membranes including ER and the PM	Spear and Roizman (1970); Spear et al. (1970)
		PM	Change in "social behavior" of infected cells correlated with appearance of new cell surface antigens	Keller et al. (1970)
			Altered antigenic reactivity of cells	Roane and Roizman (1964)
			Twelve virus-specific proteins, nine of which are glycosylated	Heine et al. (1972)
			Presence of viral proteins in host PM	Heine and Roizman (1973)
	HeLa cells	NE, ER, PM	Electron microscope localization of viral antibodies using ferritin-labeled antibody and frozen cells	Nii et al. (1968)
		PM	Presence of ATPase in the PM and mature virion by electron microscope cytochemistry	Epstein and Holt (1963)
	Hamster kidney cells	NE	Thickening of outer membrane of nuclear envelope at sites of virus envelopment	Darlington and Moss (1968)

(continued)

Appendix Table 4.   (continued)

Virus group/ specific agent	Host	Membrane affected[a]	Membrane alteration	References
		PM	Virion-associated antigen CP-1 detected	Reed et al. (1975)
Marek's disease	Chick embryo fibroblasts	PM	New virus-induced pro- teins by polyacryla- mide electrophoresis and immunodiffusion	Kaaden and Dietzschold (1974)
Cytomegalovirus	MAF cells	NE	Thickening of inner membrane of nuclear envelope at sites of virus envelopment	McGavran and Smith (1965)
Pseudorabies	Rabbit kidney cells	NE	Increased labeling of inner membrane following choline-[^{3}H] administration as a result of virus infection	Ben-Porat and Kaplan (1972)

Source: From Morré and Ovtracht (1977)

[a]NE, Nuclear envelope; ER, endoplasmic reticulum; GA, Golgi apparatus; SM, smooth membranes; PM, plasma membrane

References for Appendix Table 4

Acheson, N. H., and Tamm, I. (1967). Virology 32, 128.
Bächi, T., Berhard, W., Lindeman, J., and Mühlethaler, K. (1969). J. Virol. 4, 769.
Ben-Porat, T., and Kaplan, A. S. (1972). Nature (London) 235, 165.
Birdwell, C. R., Strauss, E. G., and Strauss, J. H. (1973). Virology 56, 429.
Bose, H. R., and Brundige, M. A. (1972). J. Virol. 9, 785.
Breitenfeld, P. M., and Schäfer, W. (1957). Virology 4, 328.
Cohen, G. H., Atkinson, P. H., and Summers, D. F. (1971). Nature (London) 231, 121.
Compans, R. W. (1973). Virology 51, 56.
Compans, R. W., and Dimmock, N. J. (1969). Virology 39, 499.
Darlington, R. W., and Moss, L. H. (1968). J. Virol. 2, 48.
David, A. E. (1971). Virology 46, 711.
Duc-Nguyen, H., Rose, H. M., and Morgan, C. (1966). Virology 28, 404.
Epstein, M. A., and Holt, S. J. (1963). J. Cell Biol. 19, 337.
Fleissner, E. (1970). J. Virol. 5, 14.
Garon, C., and Moss, B. (1971). Virology 46, 233.
Gelderblom, H., Bauer, H., and Graf, T. (1972). Virology 47, 416.
Hairstone, M. A., Sheffield, J. B., and Moore, D. H. (1964). J. Natl. Cancer Inst. 33, 825.
Heine, J. W., and Roizman, B. (1973). J. Virol. 9, 431.
Heine, J. W., Spear, P. G., and Roizman, B. (1972). J. Virol. 9, 431.
Higashi, N. (1973). Prog. Med. Virol. 15, 331.
Hirschberg, C. B., and Robbins, P. W. (1974). Virology 61, 602.
Ichihashi, Y., and Dales, S. (1971). Virology 46, 533.
Kaaden, O. R., and Dietzschold, B. (1974). J. Gen. Virol. 25, 1.
Katzman, J., and Wilson, D. E. (1974). J. Gen. Virol. 24, 101.
Keller, J. M., Spear, P. G., and Roizman, B. (1970). Proc. Natl. Acad. Sci. U.S.A. 65, 865.
Klenk, H. D., and Choppin, P. W. (1970). Virology 40, 939.

Klenk, H. D., Wöllert, W., Rott, R., and Scholtissek, C. (1974). Virology 57, 28.

Kramarsky, B., Lasfargues, E. Y., and Moore, D. H. (1970). Cancer Res. 30, 1102.

Lazarowitz, S. G., Compans, R. W., and Choppin, P. W. (1971). Virology 46, 830.

Marcus, P. I. (1962). Cold Spring Harbor Symp. Quant. Biol. 27, 351.

McGavran, M. H., and Smith, M. G. (1965). Exp. Mol. Pathol. 4, 1.

Morgan, C., Hsu, K. C., Rifkind, R. A., Knox, A. W., and Rose, H. M. (1961a). J. Exp. Med. 144, 825.

Morgan, C., Hsu, K. C., Rifkind, R. A., Knox, A. W., and Rose, H. M. (1961b). J. Exp. Med. 144, 833.

Morgan, C., Rifkind, R. A., Hsu, K. C., Holden, M., Seegal, B. C., and Rose, H. M. (1961c). Virology 14, 292.

Moss, B., Rosenblum, E. N., and Garon C. F. (1971). Virology 46, 221.

Nii, S., Morgan, C., Rose, H. M., and Hsu, K. C. (1968). J. Virol. 2, 1172.

Pfefferkorn, E. R., and Clifford, R. L. (1964). Virology 23, 217.

Quigley, J. P., Rifkin, D. B., and Reich, E. (1971). Virology 46, 106.

Quigley, J. P., Rifkin, D. B., and Reich, E. (1972). Virology 50, 550.

Reed, C. L., Cohen, G. H., and Rapp, F. (1975). J. Virol. 15, 668.

Roane, P. R., and Roizman, B. (1964). Virology 22, 1.

Schwartz, J., and Roizman, B. (1969). J. Virol. 4, 879.

Sheffield, J. B. (1974). Virology 57, 287.

Shipkey, F. H., Erlandson, R. A., Bailey, R. B., Babcock, V. I., and Southam, C. M. (1967). Exp. Mol. Pathol. 6, 39.

Spear, P. G., and Roizman, B. (1970). Proc. Natl. Acad. Sci. U.S.A. 66, 730.

Spear, P. G., Keller, J. M., and Roizman, B. (1970). J. Virol. 5, 123.

Tanaka, H., and Moore, D. H. (1967). Virology 33, 197.

Wagner, R. C., Kreiner, P., Barrnett, R. J., and Bitensky, M. W. (1972a). Proc. Natl. Acad. Sci. U.S.A. 69, 3175.

Wagner, R. R., Kiley, M. P., Snyder, R. M., and Schnaitman, C. A. (1972b). J. Virol. 9, 672.

Wagner, R. R., Snyder, R. M., and Yanazaki, S. (1970). J. Virol. 5, 548.

Weintraub, S., and Dales, S. (1974). Virology 60, 96.

References

Adams, E. C. and A. T. Hertig. 1964. Studies on guinea pig oocytes. I. Electron microscopic observations on the development of cytoplasmic organelles in oocytes of primordial and primary follicles. J. Cell Biol. 21: 397–427.

Albert, W. H. W. and D. A. L. Davis. 1973. H-2 antigens on nuclear membranes. Immunology 24: 841–850.

Albertson, R., B. Riggs and W. Sullivan. 2005. Membrane traffic: a driving force in cytokinesis. Trends Cell Biol. 15: 92–101.

Allan, D. and M. J. Obradors. 1999. Enzyme distributions in subcellular fractions of BHK cells infected with Semliki forest virus: evidence for a major fraction of sphingomyelin synthetase in the trans-Golgi network. Biochim. Biophys. Acta 1450: 277–287.

Allan, V. J. and T. E. Kries. 1986 A microtubule-binding protein associated with membranes of the Golgi apparatus. J. Cell Biol. 103: 2229–2239.

Allen, J. M. and J. J. Slater. 1961. A cytochemical study of Golgi-associated thiamine pyrophosphatase in the epididymis of the mouse. J. Histochem. Cytochem. 9: 418–437.

Alroy, J., F. B. Merk, D. J. Morré and R. S. Weinstein. 1982. Membrane differentiation in the Golgi apparatus of the mammalian urinary bladder. Anat. Rec. 203: 429–440.

Antoine, J. C. and C. Jouanne. 1986. Multiple effects of the phenylhydrazone derivative FCCP on the secretory pathway in rat plasma cells. Eur. J. Cell Biol. 42: 68–73.

Aridor, M., S. I. Bannykh, T. Rowe and W. E. Balch. 1995. Sequential coupling between COPII and COPI vesicle coats in endoplasmic reticulum to Golgi transport. J. Cell Biol. 131: 875–893.

Baker, D. L., L. Wuestehube, R. Schekman, D. Botstein and N. Segev. 1990. GTP-binding Ypt1 protein and Ca^{2+} function independently in a cell-free protein transport reaction. Proc. Natl. Acad. Sci. U.S.A. 87: 355–359.

Baker, J. R. 1944. The structure and chemical composition of the Golgi element. Quart. J. Micro. Sci. 85: 1–71.

Baker, J. R. 1955. What is the "Golgi Controversy?" J. R. Microsc. Soc. 74–217–221.

Baker, J. R. 1957. The Golgi Controversy. Symp. Soc. Exp. Biol. 10:1–10.

Baker, J. R. 1963. New developments in the Golgi controversy. J. R. Microsc. Soc. 82: 145–157.

Balch, W. E. 1989. Biochemistry of interorganelle transport. J. Biol. Chem. 264: 16965–16968.

Balch, W. E. 1990. Small GTP-binding proteins in vesicular transport. Trends Biochem. Sci. 15: 473–477.

Balch, W. E. and D. S. Keller. 1986. ATP-coupled transport of vesicular stomatitis virus G protein. Functional boundaries of secretory compartments. J. Biol. Chem. 261(34): 14690–14696.

Balch, W. E. and J. E. Rothman. 1985. Characterization of protein transport between successive compartments of the Golgi apparatus: asymmetric properties of donor and acceptor activities in a cell-free system. Biochim. Biophys. Acta 240: 413–425.

Balch, W. E., E. Fries, W. Dunphy, L. J. Urbani and J. E. Rothman. 1983. Transport-coupled oligosaccharide processing in a cell-free system. Methods Enzymol. 98: 37–47.

Balch, W. E., W. G. Dunphy, W. A. Braell and J. E. Rothman. 1984a. Reconstitution of the transport of protein between successive compartments of the Golgi measured by the coupled incorporation of N-acetylglucosamine. Cell 39: 405–416.

Balch, W. E., B. S. Glick and J. E. Rothman. 1984b. Sequential intermediates in the pathway of inter-compartmental transport in a cellfree system. Cell 39: 525–536.

Balch, W. F., K. R. Wagner and D. S. Keller. 1987. Reconstitution of transport of vesicular stomatitis virus G protein from endoplasmic reticulum to the Golgi complex using a cell-free system. J. Cell Biol. 104: 749–760.

Balch, W. E., J. M. McCaffery, H. Plutner and M. G. Farquhar. 1994. Vesicular stomatitis virus glycoprotein is sorted and concentrated during export from the endoplasmic reticulum. Cell 76: 841–852.

Bannykh, S. I. and W. E. Balch. 1997. Membrane dynamics at the endoplasmic reticulum-Golgi interface. J. Cell Biol. 138: 1–4.

Bannykh, S. I., T. Rowe and W. E. Balch. 1996. The organization of endoplasmic reticulum export complexes. J. Cell Biol. 135: 19–35.

Bannykh, S. I., H. Plutner, J. Matteson and W. E. Balch. 2005. The role of ARF1 and Rab GTPases in polarization of the Golgi stack. Traffic 6: 803–819.

Barlowe, C. 1995. COPII: a membrane coat that forms endoplasmic reticulum- derived vesicles. FEBS Lett. 369: 93–96.

Barlowe, C. and R. Schekman. 1993. SEC12 encodes a guanine-nucleotide exchange factor essential for transport vesicle budding from the ER. Nature 365: 347–349.

Barlowe, C., L. Orci, T. Yeung, M. Hosobuchi, S. Hamamoto, N. Salama, M. F. Rexach, M. Ravazzola, M. Amherdt and R. Schekman. 1994. COPII: a membrane coat formed by sec proteins that drive vesicle budding from the endoplasmic reticulum. Cell 77: 895–907.

Barr, F. A. and B. Short. 2003. Golgins in the structure and dynamics of the Golgi apparatus. Curr. Opin. Cell Biol. 15: 405–413.

Barr, R., K. Safranski, I. L. Sun, F. L. Crane and D. J. Morré. 1984. An electrogenic proton pump associated with the Golgi apparatus of mouse liver driven by NADH and ATP. J. Biol. Chem. 259: 14064–14067.

Baudhuin, P., H. Beaufay, Y. Rahman-Li, O. Z. Sellinger, R. Wattiaux, P. Jacques and C. De Duve. 1964. Tissue fractionation studies. 17. Intracellular distribution of monoamine oxidase, aspartate aminotransferase, alanine aminotransferase, D-amino acid oxidase and catalase in rat liver tissue. Biochem. J. 92: 179–184.

Baumrucker, C. R. and T. W. Keenan. 1975. Membranes of mammary gland. X. Adenosine triphosphate dependent calcium accumulation by Golgi apparatus rich fractions from bovine mammary gland. Exp. Cell Res. 90: 253–260.

Beams, H. W. and R. L. King. 1935. The effect of ultracentrifuging on the cells of the root tip of the bean (*Phaseolus vulgaris*). Proc. R. Soc. Lond B 118: 264–276.

Beams, H. W., T. N. Tahmisian, R. D. Devine and E. Anderson. 1956. Electron microscope studies on the dictyosomes and acroblasts in the male germ cells of the cricket. J. Biophys. Biochem. Cytol. 2: 123–128.

Becker, F. F. 1970. The normal hepatocyte in division: regeneration of the mammalian liver. In: Progress in Liver Diseases, Vol. 3. H. Popper and F. Schaffner (eds.). Grune and Stratton, New York. pp. 60–78.

Beckers, C. J. M. and W. E. Balch. 1989. Calcium and GTP: essential components in vesicular trafficking between the endoplasmic reticulum and the Golgi apparatus. J. Cell Biol. 108: 1245–1256.

Beckers, C. J. M., D. S. Keller and W. E. Balch. 1987. Semi-intact cells permeable to macromolecules: use in reconstitution of protein transport from the endoplasmic reticulum to the Golgi complex. Cell 50: 523–534.

Beckers, C. J. M., M. R. Block, B. S. Glick, J. E. Rothman and W. E. Balch. 1989. Vesicular transport between the endoplasmic reticulum and the Golgi stack requires the NEM-sensitive fusion protein. Nature 339: 397–398.

Beckers, C. J. M., H. Plutner, H. W. Davidson and W. Balch. 1990. Sequential intermediates in the transport of protein between the endoplasmic reticulum and the Golgi. J. Biol. Chem. 263: 18298–18310.

Behnia, R. and S. Munro. 2005. Organelle identity and the signposts for membrane traffic. Nature 438: 597–604.

Benedetti, E. L. and P. Emmelot. 1967. Studies on plasma membranes. IV. The ultrastructural localization and content of sialic acid in plasma membranes isolated from rat liver and hepatomas. J. Cell Sci. 2: 499–512.

Bergen, W. G. and D. B. Bates. 1984. Ionophores: their effect on production efficiency and mode of action. J. Anim. Sci. 58: 1465–1483.

Berger, E. G. 1987. The Golgi apparatus: From discovery to contemporary studies. In: The Golgi Apparatus. E. G. Berger and J. Roth (eds.) Birkhaüser Basel. pp. 1–35.

Bergeron, J. J. M. 1979. Golgi fractions from livers of control and ethanol-intoxicated rats. Enzymic and morphological properties following rapid isolation. Biochim. Biophys. Acta 555: 493–503.

Bergeron, J. J. M., J. H. Ehrenreich, P. Siekevitz and G. E. Palade. 1973. Golgi fractions prepared from rat liver homogenates. II. Biochemical characterization. J. Cell Biol. 59: 73–88.

Bergmann, J. E., K. T. Tokuyasu and S. J. Singer. 1981. Passage of an integral membrane protein, the vesicular stomatitis virus glycoprotein, through the Golgi apparatus en route to the plasma membrane. Proc. Natl. Acad. Sci. U.S.A. 78: 1746–1750.

Bingham, E. W., H. M. Farrel and J. J. Basch. 1972. Phosphorylation of casein. Role of the Golgi apparatus. J. Biol. Chem. 247: 8193–8194.

Bloom, G. S. and T. A. Brashear. 1989. A novel 58-kDa protein associates with the Golgi apparatus and microtubules. J. Biol. Chem. 264: 16083–16092.

Blume-Jensen, P. and T. Hunger. 2001. Oncogenic kinase signaling. Nature 411: 355–365.

Blobel, G. and B. Dobberstein. 1975. Transfer of proteins across membranes. I. Presence of proteolytically processed and unprocessed nascent immunoglobulin light chains on membrane bound ribosomes of murine myeloma. J. Cell Biol. 67: 835–851.

Block, M. R., B. S. Glick, C. A. Wilcox, F. T. Wieland and J. E. Rothman. 1988. Purification of an N-ethylmaleimide-sensitive protein catalyzing vesicular transport. Proc. Natl. Acad. Sci. U.S.A. 33: 7852–7856.

Borgese, N. and J. Meldolesi. 1980. Localization and biosynthesis of NADH-cytochrome b_5 reductase, an integral membrane protein, in rat liver cells. I. Distribution of the enzyme activity in microsomes, mitochondria, and Golgi apparatus. J. Cell Biol. 85: 501–515.

Bourne, G. H. and H. B. Tewari. 1964. Mitochondria and the Golgi complex. In: Cytology and Cell Physiology. G. H. Bourne (ed.). Academic Press, New York-London. pp. 377–421.

Bourne, H. R. 1988. Do GTPases direct membrane traffic in secretion? Cell 53: 669–671.

Bourne, H. R., D. A. Sanders and F. McCormick. 1991. The GTPase superfamily: conserved structure and molecular mechanism. Nature 349: 117–127.

Bowen, R. H. 1928. Studies on the structure of plant protoplasm. I. The osmiophilic platelets. Z. Zellforsch. Mikroskop. Anat. 6: 689–725.

Bowen, R. H. 1929. The cytology of glandular secretion. Q. Rev. Biol. 4:299–324; 484–519.

Bracker, C. E. and S. N. Grove. 1971. Continuity between cytoplasmic endomembranes and outer mitochondrial membranes in fungi. Protoplasma 73: 15–34.

Bracker, C. E., D. J. Morré and S. N. Grove. 1996. Structure, differentiation, and multiplication of Golgi apparaus in fungal hyphae. Protoplasma 194: 250–274.

Brandan, E. and B. Fleischer. 1982. Orientation and role of nucleoside-diphosphatase and 5 -nucleotidase in Golgi vesicles from rat liver. Biochemistry 21: 4640–4645.

Bretscher, M. S. and S. Munro. 1993. Cholesterol and the Golgi apparatus. Science 261: 1280–1281.

Bretz, R. H., H. Bretz and G. E. Palade. 1980. Distribution of terminal glycosyltransferases in hepatic Golgi fractions. J. Cell Biol. 84:87–101.

Brightman, A. O., P. Navas, N. M. Minnifield and D. J. Morré. 1992. Pyrophosphate-induced acidification of trans cisternal elements of rat liver Golgi apparatus. Biochim. Biophys. Acta 1104: 188–194.

Brown, R. M. 1969. Observations on the relationship of the Golgi apparatus to wall formation in the marine chrysophycean alga, *Pleurochrysis scherffelii* Pringsheim. J. Cell Biol. 41: 109–123.

Brown, R. M. and H. J. Arnott. 1971. A photographic method for producing true three-dimensional electron micrographs. Protoplasma 72: 105–107.

Brown, W. J. and M. G. Farquhar. 1984. The mannose-6-phosphate receptor for lysosomal enzymes is concentrated in cis Golgi cisternae. Cell 36: 295–307.

Brown, R., R. H. Gray and I. A. Bernstein. 1985. Retinoids alter the direction of differentiation in primary cultures of cutaneous keratinocytes. Differentiation 28: 268–278.

Brunet, A., A. Bonni, M. J. Zigmond, M. Z. Lin, P. Juo, L. S. Hu, M. J. Anderson, K. C. Arden, J. Blenis and M. E. Greenberg. 1999. Akt promotes cell survival by phosphorylating and inhibiting a Forkhead transcription factor. Cell 96: 857–868.

Burkhardt, J. N. K. and Y. Argon. 1989. Intracellular transport of the glycoprotein of VSV is inhibited by CCCP at a late stage of post-translational processing. J. Cell Sci. 92: 633–642.

Buvat, R. 1957a. Formations de Golgi dans les cellules radiculares d'*Allium cepa*. L. C. R. Acad. Sci. (Paris) 244: 1401–1403.

Buvat, R. 1957b. Relations entre l'ergastroplasme et l'appareil vacuolaire. C. R. Acad. Sci. (Paris) 245: 350–352.

Cajal, S. R. 1908. Les conduits de Golgi-Holmgren du protoplasma nerveux et le réseau pericellulaire de la membrane. Trav. Lab. Rech. Biol. Madr. 6: 123–135.

Cajal, S. R. 1914. Algunas variaciones fisiologicas y patalogicas del parato reticular de Golgi. Trab. Lab. Inv. Biol. Madr. 12:127–227.

Cajal, S. R. 1923. Recuerdos de mi vida, 3rd edition. Juan Pueyo: Madrid. (Translation: Recollections of My Life. E. Horne Craigie and Juano Cano, Memoirs of the American Philosophical Society, Vol. VIII 1937. The American Philosophical Society: Philadelphia).

Capasso, J. M., C. Abeijon and C. B. Hirschberg. 1988. An intrinsic membrane glycoprotein of the Golgi apparatus with *0*-linked *N*-acetylglucosamine facing the cytosol. J. Biol. Chem. 263: 19778–19782.

Capasso, J. M., T. W. Keenan, C. Abeijon and C. B. Hirschberg. 1989. Mechanism of phosphorylation in the lumen of the Golgi apparatus. Translocation of adenosine 5′-triphosphate into Golgi vesicles from rat liver and mammary gland. J. Biol. Chem. 264: 5233–5240.

Carlton, J. G. and P. J. Cullen. (2005). Coincidence detection in phosphoinositide signaling. Trends Cell Biol. 15, 540–547.

Caro, L. and G. E. Palade. 1964. Protein synthesis, storage, and discharge in the pancreatic exocrine cell. An autoradiographic study. J. Cell Biol. 20: 473–495.

Castle, J. D., J. D. Jamieson and G. E. Palade. 1972. Radioautographic analysis of the secretory process in the rabbit parotid gland ascinar cell. J. Cell Biol. 53: 290–311.

Chalpowski, F. J. and R. N. Band. 1971. Assembly of lipids into membranes in *Acanthamoeba palestinensis*. II. The origin and fate of glycerol-^{3}H-labeled phospholipids of cellular membranes. J. Cell Biol. 50: 634–651.

Charder, R. and C. Rouiller. 1957. L'ultrastructure de trois algues desmidiées. Études au microscope électronique. Rev. Cytol. Biol. Veg. 18: 153–178.

Charest, A., V. Kheifets, J. Park, K. Lane, K. McMahon, C. L. Nutt and D. Housman. 2003. Oncogenic targeting of an activated tyrosine kinase to the Golgi apparatus in a glioblastoma. Proc. Natl. Acad. Sci. U.S.A. 100: 916–921.

Chavrier, P., R. G. Parton, H. P. Hauri, K. Simons and M. Zerial. 1990. Localization of low molecular weight GTP binding proteins to exocytic and endocytic compartments. Cell 62: 317–329.

Cheetham, R. D. and D. J. Morré. 1970. Di- and tri-nucleotidase activities of rat liver cytomembranes. Proc. Indiana Acad. Sci. for 1969. 79: 107–109.

Cheetham, R. D., D. J. Morré and W. N. Yunghans. 1970. Isolation of a Golgi apparatus-rich fraction from rat liver. II. Enzymatic characterization and comparison with other cell fractions. J. Cell Biol. 44: 492–499.

Cheetham, R. D., D. J. Morré, C. Pannek and D. S. Fried. 1971. Isolation of a Golgi apparatus-rich fraction from rat liver. IV. Thiamine pyrophosphatase. J. Cell Biol. 49: 899–905.

Chiu, V. K., T. Bivona, A. Hach, J. B. Sajous, J. Silletti, H. Wiener, R. L. Johnson, 2nd, A. D. Cox and M. R. Philips. 2002. Ras signaling on the endoplasmic reticulum and the Golgi. Nat. Cell Biol. 4: 343–350.

Clary, D. O. and J. E. Rothman. 1990. Purification of three related peripheral membrane proteins needed for vesicular transport. J. Biol. Chem. 265: 10109–10117.

Clary, D. O., I. C. Griff and J. E. Rothman. 1990. SNAPs, a family of NSF attachment proteins involved in intracellular membrane fusion in animals and yeast. Cell 61: 709–721.

Claude, A. 1970. Growth and differentiation of cytoplasmic membranes in the course of lipoprotein granule synthesis in the hepatic cell. I. Elaboration of elements of the Golgi complex. J. Cell Biol. 47: 745–766.

Clermont, Y., A. Rambourg and L. Hermo. 1994. Connections between the various elements of the *cis*- and mid-compartments of the Golgi apparatus of early rat spermatids. Anat. Rec. 240: 469–480.

Cope, G. H. and M. A. Williams. 1973. Quantitative analyses of the constituent membranes of parotid acinar cells and of the changes evident after induced exocytosis. Z. Zellforsch. Mikrosk. Anat.145: 311–330.

Copeland, C. S., K. P. Zimmer, K. R. Wagner, G. A. Healey, I. Mellman and A. Helenius. 1988. Folding, trimerization and transport are sequential events in the biogenesis of influenza virus hemagglutinin. Cell 53: 197–209.

Cory, S. and J. M. Adams. 2002. The Bcl-2 family: regulators of the cellular life-or-death switch. Nat. Rev. Cancer. 2: 647–656.

Cox, R., R. J. Mason-Gamer, C. L. Jackson and N. Segev. 2004. Phylogenetic analysis of sec7-domain-containing Arf nucleotide exchangers. Mol. Biol. Cell. 15: 1487–1505.

Crane, F. L. and D. J. Morré. 1977. Evidence for coenzyme Q function in Golgi membranes. In: Biomedical and Clinical Aspects of Coenzyme Q. K. Folkers and Y. Yamamura (eds.). Elsevier Scientific, Amsterdam-Oxford-New York. pp. 3–14.

Creek, K. E., V. P. Walter, D. Evers, E. Yeo, W. L. Elliott, P. F. Heinstein, D. M. Morré and D. J. Morré. 1984. Sialoglycoconjugate changes during 2-acerylaminofluorene-induced hepatocarcinogenesis in the rat. Biochim. Biophys. Acta 793: 133–144.

Croze, E. M. and D. J. Morré. 1981. Flow kinetics of mouse histocompatibility antigens. Proc. Natl. Acad. Sci. U.S.A. 78: 1547–1551.

Croze, E. M. and D. J. Morré. 1984. Isolation of plasma membrane Golgi apparatus and endoplasmic reticulum fractions from single homogenates of mouse liver. J. Cell Physiol. 119: 46–57.

Croze, E. M., D. J. Morré, D. M. Morré, J. Kartenbeck and W. W. Franke. 1982. Distribution of clathrin and spiny-coated vesicles on membranes within mature Golgi apparatus elements of mouse liver. European J. Cell Biol. 28: 130–138.

Cunningham, W. P., D. J. Morré and H. H. Mollenhauer. 1966. Structure of isolated plant Golgi apparatus revealed by negative staining. J. Cell Biol. 28: 169–179.

Cunningham, W. P., L. A. Saehelin, R. W. Rubin, R. Wilkins and M. Bonneville. 1974. Effects of phosphotungstate negative staining on the morphology of the isolated Golgi apparatus. J. Cell Biol. 62: 491–504.

Dai, S., D. J. Morré, C. C. Geilen, B. Almond-Roesler, C. E. Orfanos and D. M. Morré. 1997. Inhibition of plasma membrane NADH oxidase activity and growth of HeLa cells by natural and synthetic retinoids. Mol. Cell. Biochem. 166: 101–109.

Dallner, G. 1978. Isolation of microsomal subfractions by use of density gradients. Methods Enzymol. 52: 71–83.

Dalton, A. J. 1961. Golgi apparatus and secretion granules. In: The Cell. J. Brachet and A. E. Mirsky (eds.). Academic Press, New York. Vol. 2, Chapter 8, pp. 603–617.

Dalton, A. J. 1964. An electron microscopic study of a series of chemically induced hepatomas, In: Cellular Control Mechanisms and Cancer. P. Emmelot and O. Muhlbock (eds.). Elsevier, Amsterdam. pp. 211–225.

Dalton, A. J. and M. D. Felix. 1953. Studies on the Golgi substance of the epithelial cells of the epididymis and duodenum of the mouse. Am. J. Anat. 92:277–293.

Dalton, A. J. and M. D. Felix. 1954. Cytologic and cytochemical characteristics of the Golgi substance of epithelial cells of the epididymis—in situ, in homogenates, and after isolation. Amer. J. Anat. 94: 171–208.

Dalton, A. J. and M. D. Felix. 1956. A comparative study of the Golgi complex. J. Biophys. Biochem. Cytol. Suppl. 2: 79–84.

Dalton, A. J. and M. D. Felix. 1957. Electron microscopy of mitochondria and the Golgi complex. Symp. Soc. Exp. Biol. 10: 148–159.

Davidson, H. W. and W. E. Balch. 1992. Use of two-stage incubations to define sequential intermediates in endoplasmic reticulum to Golgi transport. Methods Enzymol. 219: 261–267.

Davidson, H. W. and W. E. Balch. 1993. Differential inhibition of multiple vesicular transport steps between the endoplasmic reticulum and trans Golgi network. J. Biol. Chem. 268: 4216–4226.

de Duve, C. 1964. Principles of tissue fractionation. J. Theoret. Biol. 6: 33–59.

De Lemos-Chiarandini C., N. E. Ivessa, V. H. Black, Y. S. Tsao, I. Gumper and G. Kreibich. 1992. Golgi-related structure remains after the brefeldin A-induced formation of an ER-Golgi hybrid compartment. Eur. J. Cell Biol. 58: 187–201.

d'Enfert, C. L., J. Wuestehube, T. Lia and K. Schekman. 1991. Secl2p-dependent membrane binding of the small GTP-binding protein Sarlp promotes formation of transport vesicles for the ER. J. Cell Biol. 114: 663–670.

Deng, Y. and K. DeCourcy. 1992. Intermixing of resident Golgi membrane proteins in rat-hamster polykaryons appears to depend on organelle coalescence. Eur. J. Cell. Biol. 57: 1–11.

Deutscher, S. L., K. E. Creek, M. Merion and C. B. Hirschberg. 1983. Subfractionation of rat liver Golgi apparatus: separation of enzyme activities involved in biosynthesis of phosphomannosyl recognition marker in lysosomal enzymes. Proc. Natl. Acad. Sci. U.S.A. 80: 3938–3942.

Dinter, A. and E. C. Berger. 1998. Golgi-disturbing agents. Histochem. Cell Biol. 109: 571–590.

Dobberstein, B., H. Garoff and G. Warren. 1979. Cell-free synthesis and membrane insertion of mouse H-2Dd histocompatibility antigen and beta 2-microglobulin. Cell 17: 759–769.

Doble, B. W. and J. R. Woodgett. 2003. GSK-3: tricks of the trade for a multi-tasking kinase. J. Cell Sci. 116: 1175–1186.

Dod, B. J. and Gray, G. M. 1968. The lipid composition of rat-liver plasma membranes. Biochim. Biophys. Acta 150: 397–404.

Doms, R. W., G. Russ and J. W. Yewdell. 1989. Brefeldin A redistributes resident and itinerant Golgi proteins to the endoplasmic reticulum. J. Cell. Biol. 109: 61–72.

Doms, R. W., D. S. Keller, A. Helenius and W. E. Balch. 1987. Role for adenosine triphosphate in regulating the assembly and transport of vesicular stomatitis virus G protein trimers. J. Cell Biol. 105: 1957–1969.

Donaldson, J. G. and R. D. Klausner. 1994. ARF: a key regulatory switch in membrane traffic and organelles structure. Curr. Opin. Cell Biol. 6: 527–532.

Donaldson, J. G., J. Lippincott-Schwartz, G. S. Bloom, T. E. Kreis and R. D. Klausner. 1990. Dissociation of a110-kD peripheral membrane protein from the Golgi apparatus is an early event in Brefeldin A action. J. Cell. Biol. 111:2295–2306.

Donaldson, J. G., D. Finazzi and R. D. Klausner. 1992. Brefeldin A inhibits Golgi membrane-catalysed exchange of guanine nucleotide onto ARF protein. Nature 360: 350–352.

Driouich, A. and L. A. Staehelin. 1997. The plant Golgi apparatus; Structural organization and functional properties. In: The Golgi Apparatus. E. G. Berger and J. Roth (eds). Birkhäuser, Basel. pp. 275–301.

Driouich, A., S. Levy, L. A. Staehelin and L. Faye. 1994. Structural and functional organization of the Golgi apparatus in plant cells. Plant Physiol. Biochem. 32: 731–749.

Du, X., I. Kristiana, J. Wong and A. J. Brown. 2006. Involvement of Akt in ER-to-Golgi transport of SCAP/SREBP: a link between a key cell proliferative pathway and membrane synthesis. Mol. Biol. Cell. 17: 2735–2745.

Dunkle, S., T. Reust, D. D. Nowack, L. Waits, M. Paulik, D. M. Morré and D. J. Morré. 1992. Temperature and acceptor specificity of cell-free vesicular transfer from transitional endoplasmic reticulum to cis Golgi apparatus. Biochem. J. 288: 969–976.

Dunphy, W. G. and J. E. Rothman. 1985. Compartmental organization of the Golgi stack. Cell 42: 13–21.

Dunphy, W. G., R. Brands and J. E. Rothman. 1985. Attachment of terminal *N*-acetylglucosamine to asparagine-linked oligosaccharides occurs in central cisternae of the Golgi stack. Cell 40: 463–472.

Durieux, I., M. B. Martel and R. Got. 1990. Effect of phospholipids on UDP-glucose: ceramide glucosyltransferase from Golgi membranes. Int. J. Biochem. 22: 709–715.

Ehrenreich, J. H., J. J. M. Bergeron, P. Siekevitz and G. E. Palade. 1973. Golgi fractions prepared from rat liver homogenates. I. Isolation procedure and morphological characterization. J. Cell Biol. 59: 45–72.

Elder, J. H. and D. J. Morré. 1976a. Synthesis *in vitro* of intrinsic membrane proteins by free, membrane-bound, and Golgi apparatus-associated polyribosomes from rat liver. J. Biol. Chem. 251: 5054–5068.

Elder, J. H. and D. J. Morré. 1976b. Distribution of glycoproteins among subcellular fractions from rat liver. Cytobiologie 13: 279–284.

Elhammer, A. and S. Kornfeld. 1984. Two enzymes involved in the synthesis of 0-linked oligosaccharides are localized on membranes of different densities in mouse lymphoma BW5147 cells. J. Cell Biol. 98: 327–331.

Elliott, A. M. and R. G. Zieg. 1968. A Golgi apparatus associated with mating in *Tetrabymena pyriformis*. J. Cell Biol. 36: 391–398.

Elliott, W. L., D. P. Sawick, K. E. Creek, S. L. Deutscher, J. F. Quinn, E. Yeo, W. R. Webb, D. M. Morré, D. D. Harrington, P. F. Heinstein, J. M. Cassday and D. J. Morré. 1984. Early biochemical alterations induced by 2-acetylaminofluorene in rat liver. Int. J. Biochem. 16: 947–956.

Elrod-Erickson, J. J. and C. A. Kaiser. 1996. Genes that control the fidelity of endoplasmic reticulum to Golgi transport identified as suppressors of vesicle budding mutations. Mol. Biol. Cell 7: 1043–1058.

Emmelot, P., C. J. Bos, R. P. Van Hoeven and W. J. van Blitterswizk. 1974. Isolation of plasma membranes of rat and mouse livers and hepatomas. Methods Enzymol. 31A: 75–90.

Eppler, C. M. and D. J. Morré. 1982. Flow kinetics of a nucleoside phosphatase common to endoplasmic reticulum, Golgi apparatus, and plasma membrane of rat liver. Euro. J. Cell Biol. 29: 13–23.

Erdmann, R., F. F. Wiebel, A. Flessau, J. Rytka, A. Beyer, K. U. Fröhlick and W. H. Kunau. 1991. *PAS I*, a yeast gene required for preoxisome biogenesis, encodes a member of a novel family of putative ATPases. Cell 64: 499–510.

Espenshade, P., R. E. Gimeno, E. Holzmacher, P. Teung and C. A. Kaiser. 1995. Yeast SEC16 gene encodes a multidomain vesicle coat protein that interacts with sec 23p. J. Cell Biol. 131: 311–324.

Evans, W. H. 1970. Glycoproteins of mosue liver smooth microsomal and plasma membrane fractions. Biochim. Biophys. Acta 211: 578–581.

Falk, H. 1969. Fusiform vesicles in plant cells. J. Cell Biol. 43:164–174.

Fambrough, D. M. and P. N. Devreotes. 1978. Newly synthesized acetylcholine receptors are located in the Golgi apparatus. J. Cell Biol. 76: 237–244.

Farquhar, M. G. 1985. Progress in unraveling pathways of Golgi traffic. Annu. Rev. Cell. Biol. 1: 447–488.

Farquhar, M. G. and G. E. Palade. 1981. The Golgi apparatus (complex)-(1954–1981)-from artifact to center stage. J. Cell Biol. 91: 77s–103s.

Farquhar, M. G., J. J. M. Bergeron and G. E. Palade. 1974. Cytochemistry of Golgi fractions prepared from rat liver. J. Cell Biol. 60: 8–25.

Fasshauer, D. W. Antonin, V. Subramaniam and R. Jahn. 2002. SNARE assembly and disassembly exhibit a pronounced hysteresis. Nat. Struct. Biol. 9: 144–151.

Favard, P., L. Ovtracht and N. Carasso. 1971. Observations de spécimens biologiques en microscopie électronique a hauté tension. I. Coupes épaisses. J. Microsc. (Paris) 12: 301–316.

Fiedler, K., M. Veit, M. A. Stamnes and J. E. Rothman. 1996. Bimodal interaction of coatomer with a family of putative cargo and vesicle coat protein receptions, the p24 proteins. Science 273: 1396–1398.

Fleischer, B. 1977. Localization of some glycolipid glycosylating enzymes in the Golgi apparatus of rat kidney. J. Supramol. Struct. 7: 79–89.

Fleischer, B. and S. Fleischer. 1970. Preparation and characterization of Golgi membranes from rat liver. Biochim. Biophys. Acta 219: 301–319.

Fleischer, B. and F. Zambrano. 1973. Localization of cerebroside-sulfotransferase activity in the Golgi apparatus of rat kidney. Biochem. Biophys. Res. Commun. 52: 951–958.

Fleischer, B. and F. Zambrano. 1974. Golgi apparatus of rat kidney. Preparation and role in sulfatide formation. J. Biol. Chem. 249: 5995–6003.

Fleischer, B., S. Fleischer and H. Ozawa. 1969. Isolation and characterization of Golgi membranes from bovine liver. J. Cell Biol. 43: 59–79.

Fleischer, S., B. Fleischer, A. Assi and B. Chance. 1971. Cytochrome b_5 and P_{-450} in liver cell fractions. Biochim. Biophys. Acta 222: 194–200.

Fleischer, B., F. Zambrano and S. Fleischer. 1974. Biochemical characterization of the Golgi complex of mammalian cells. J. Supramol. Struct. 2: 737–750.

Flickinger, C. J. 1969a. The development of Golgi complexes and their dependence upon the nucleus in Amebae. J. Cell Biol. 43: 250–262.

Flickinger, C. J. 1969b. Fenestrated cisternae in the Golgi apparatus of the epididymus. Anat. Rec. 163: 39–54.

Flickinger, C. J. 1969c. Fenestrated cisternae in the Golgi apparatus of the epididymus. Anat. Rec. 163: 39–54.

Flickinger, C. J. 1970. The fine structure of the nuclear envelope in amebae: alterations following nuclear transplantation. Expt. Cell Res. 60: 225–236.

Flickinger, C. J. 1974. The role of endoplasmic reticulum in the repair of amebae nuclear envelope damaged microsurgically. J. Cell Sci. 14: 421–437.

Franke, W. W. 1974. Structure, biochemistry and functions of the nuclear envelope. Int. Rev. Cytol. 4 (Suppl): 71–236.

Franke, W. W. and J. Kartenbeck. 1971. Outer mitochondrial membrane continuous with endoplasmic reticulum. Protoplasma 73: 35–41.

Franke, W. W. and J. Kartenbeck. 1976. Some principles of membrane differentiation. In: Progress in Differentiation Research. Müller-Bérat (ed.). Elsevier/North-Holland, New York. pp. 213–243.

Franke, W. W. and U. Scheer. 1972. Structural details of dictyosomal pores. J. Ultrastruct. Res. 40: 132–144.

Franke, W. W., W. A. Eckert and S. Krien. 1971a. Cytomembrane differentiation in a ciliate, Tetrahymena pyriformis. I. Endoplasmic reticulum and dictyosomal equivalents. Z. Zellforsch. 119: 577–604.

Franke, W. W., D. J. Morré, B. Deumling, R. Cheetham, J. Kartenbeck, E. D. Jarasch and H. W. Zentgraf. 1971b. Synthesis and turnover of membrane proteins in rat liver: an examination of the membrane flow hypothesis. Z. Naturforsch. 26b: 1031–1039.

Franke, W. W., J. Kartenbeck, S. Krien, W. J. VanDerWoude, U. Scheer and D. J. Morré. 1972. Inter- and intracisternal elements of the Golgi apparatus: a system of membrane-to-membrane cross-links. Z. Zellforsch. 132: 365–380.

Frantz, C., J. C. Roland, F. A. Williamson and D. J. Morré. 1973. Differentiation in vitro des membranes des dictyosomes. C. R. Acad. Sci. Paris. 277: 1471–1474.

Franz, C. P., E. M. Croze, D. J. Morré and G. Schreiber. 1981. Albumin secreted by rat liver bypasses Golgi apparatus cisternae. Biochim. Biophys. Acta 678: 395–402.

Freedman, R. A., M. M. Weisner and K. J. Isselbacher. 1977. Calcium translocation by Golgi and lateral-basal membrane vesicles from rat intestine: decrease in vitamin D-deficient rats. Proc. Natl. Acad. Sci. U.S.A. 74: 3612–3616.

Freeman, M. R. and K. R. Solomon. 2004. Cholesterol and prostate cancer. J. Cell. Biochem. 91: 54–69.

Friend, D. S. 1965. The fine structure of Brunner's gland in the mouse. J. Cell Biol. 25: 563–576.

Friend, D. S. and M. J. Murray. 1965. Osmium impregnation of the Golgi apparatus. Am. J. Anat. 117: 135–149.

Friend, D. S. and M. G. Farquhar. 1967. Functions of coated vesicles during protein absorption in the rat vas deferens. J. Cell Biol. 35: 357–376.

Fries, E. and I. Lindstrom. 1986. The effects of low temperatures on intracellular transport of newly synthesized albumin and haptoglobin in rat hepatocytes. Biochem. J. 237: 33–39.

Fries, E. and J. E. Rothman. 1980. Transport of vesicular stomatitis virus glycoprotein in a cell-free extract. Proc. Natl. Acad. Sci. U.S.A. 77: 3870–3874.

Fuchs, H. 1902. Über das Epithel im Nebenhoden der Maus. Anat. Hefte 19:313–347.

Fulton, D., J. Fontana, G. Sowa, J. P. Gratton, M. Lin, K. X. Li, B. Michell, B. E. Kemp, D. Rodman and W. C. Sessa. 2002. Localization of endothelial nitric-oxide synthetase phosphorylated on serine 1173 and nitric oxide in Golgi and plasma membrane defines the existence of two pools of active enzyme. J. Biol. Chem. 277: 4277–4284.

Futerman, A. H. and R. E. Pagano. 1991. Determination of the intracellular sites and topology of glucosylceramide synthesis of rat liver. Biochem. J. 280: 295–302.

Gatenby, J. B. 1931. Cytological studies of the acinar cells of the pancreas of the mouse. Am. J. Anat. 48: 421.

Gatenby, J. B. 1955. The Golgi apparatus. J. R. Microsc. Soc. 74: 134–161.

Geuze, J. H. and D. J. Morré. 1991. Trans Golgi reticulum. J. Electron. Microsc. Tech. 17: 24–34.

Gibeaut, D. M. and N. Carpita. (1994). Biosynthesis of plant cell wall polysaccharides. FASEB J. 8: 904–915.

Gilchrist, A., C. E. Au, J. Hiding, A. W. Bell, J. Fernadez-Rodrigues, S. Lesimple, H. Nagaya, L. Roy, S. J. C. Gosline, M. Hallett, J. Paiement, R. K. Kearney, T. Nilsson and J. J. M. Bergeron. 2006. Quantitative proteomic analysis of the secretory pathway. Cell 127: 1265–1281.

Gimeno, R. E., P. Espenshade and C. A. Kaiser. 1996. COPII coat subunit interactions: Sec24p and Sec23p bind to adjacent regions of Sec16p. Mol. Biol. Cell 7: 1815–1823.

Glaumann, H., A. Bergstrand and J. L. E. Ericsson. 1975. Studies on the synthesis and intracellular transport of lipoprotein particles in rat liver. J. Cell Biol. 64: 356–377.

Gleeson, P. A. 1998. Targeting of proteins to the Golgi apparatus. Histochem. Cell Biol. 109: 517–532.

Glick, B. S. 2002. Can the Golgi form de novo? Nat Rev. 3: 615–619.

Goldenberg, H., F. L. Crane and D. J. Morré. 1979. NADH-oxidoreductase of mouse liver plasma membranes. J. Biol. Chem 254: 2491–2498.

Goldfischer, S. 1982. The internal reticular apparatus of Camillo Golgi. J. Histochem. Cytochem. 30: 717–733.

Golgi, C. 1898. Sur la structure de la cellules nerveuses des ganglions spinaux. Arch. Ital. Biol. 30: 60–71. (Originally published in Boll. Soc. Med.-Chir. di Pavia, 1898).

Goud, B. 2002. How Rab proteins link motors to membranes. Nat. Cell Biol. 4: E77–E78.

Graham, J. M., M. C. B. Sumner, D. H. Curtis and C. A. Pasternak. 1973. Sequence of events in plasma membrane assembly during the cell cycle. Nature 246: 291–295.

Green, J., G. Griffiths, D. Louvard, P. Quinn and G. Warren. 1981. Passage of viral membrane proteins through the Golgi complex. J. Mol. Biol. 152: 663–698.

Griff, I. C., R. Schekman, J. E. Rothman and C. A. Kaiser. 1992. The yeast SEC17 gene product is functionally equivalent to mammalian alpha-SNAP protein. J. Biol. Chem. 267: 12106–12115.

Griffiths, G. and K. Simons. 1986. The trans Golgi network: sorting at the exit site of the Golgi complex. Science 234: 438–443.

Griffiths, G., S. Pfeiffer, K. Simons and K. Matlin. 1985. Exit of newly synthesized membrane proteins from the trans cisterna of the Golgi complex to the plasma membrane. J. Cell Biol. 101: 949–964.

Grimstone, A. V. 1959. Cytoplasmic membranes and the nuclear membrane in the flagellate *Trichonympha*. J. Biophys. Biochem. Cytol. 6: 369–378.

Grove, S. N., C. E. Bracker and D. J. Morré. 1968. Cytomembrane differentiation in the endoplasmic reticulum-Golgi apparatus-vesicle complex. Science 161: 171–173.

Grove, S. N., C. E. Bracker and D. J. Morré. 1970. An ultrastructural basis for hyphal tip growth in *Pythium ultimum*. Am. J. Bot. 57: 245–266.

Gruenberg, J. and K. E. Howell. 1989. Membrane traffic in endocytosis: insights from cell-free assays. Annu. Rev. Cell Biol. 5: 453–481.

Hakomori, S.-I. 1981. Glycosphingolipids in cellular interaction, differentiation, and oncogenesis. Annu. Rev. Biochem. 50: 733–736.

Hamilton, R. L., D. J. Morré, R. Mahley and V. S. LeQuire. 1967. Morphological studies of a Golgi apparatus-rich cell fraction isolated from rat liver. J. Cell Biol. 35: 53A.

Hammond, R. G., R. R. Majewski, K. E. Muse, T. D. Oberley, L. W. Morrisey and A. M. Amendt-Raduege. 1994. Energy transfer assays of rat renal cortical endosome fusion: evidence for superfusion. Am. J. Physiol. 267: F1021–F1033.

Hanada, K., K. Kumagai, S. Yasuda, Y. Miura, M. Kawano, M. Fukasawa and M. Nishijima. 2003. Molecular machinery for non-vesicular trafficking of ceramide, Nature 426: 803–809.

Hannig, K. and H. G. Heidrich. 1977. Continuous free-flow electrophoresis and its application to biology. Cell Separation Methods. IV. Electrophoretic Methods. H. Bloemendal (ed.). Elsevier/North Holland Biomedical Press, Amsterdam. pp. 93–116.

Hannig, K. and H. G. Heidrich. 1990. Free-Flow Electrophoresis. GIT Verlag, Darmstadt. P. 119.

Hansen G. H., L. L. Niels-Christiansen, E. Thorsen, L. Immerdal and E. M. Danielsen. 2000. Cholesterol depletion of enterocytes: effect on the Golgi complex and apical membrane trafficking. J. Biol. Chem. 275: 5136–5142.

Harryson, P., D. J. Morré and A. S. Sandelius. 1996. Cell-free transfer of phosphatidylinositol between membrane fractions isolated from soybean. Plant Physiol. 110: 631–637.

Hartel-Schenk, S., N. Minnifield, W. Reutter, C. Hanski, C. Bauer and D. J. Morré. 1991. Distribution of glycosyltransferases among Golgi apparatus subfractions from liver and hepatomas of the rat. Biochim. Biophys. Acta 1115: 108–122.

Harter, C. 1995. COP-coated vesicles in intracellular protein transport. FEBS Lett. 369: 89–92.

Hay, J. C., D. S. Chao, C. S. Kuo and R. H. Scheller. 1997. Protein interactions regulating vesicle transport between the endoplasmic reticulum and Golgi apparatus in mammalian cells. Cell 89: 149–158.

Hayman, M. J. and H. Beug. 1984. Identification of a form of the avian erythroblastosis virus erb-B gene product at the cell surface. Nature 309: 460–462.

Heidrich, H. G. 1981. Free-flow electrophoresis in malaria research. In: Electrophoresis '81. R. C. Allen and P. Arnaud (eds.). Walter de Gruyter, Berlin/New York. pp. 859–870.

Heitz, E. 1957a. Die Struktur der Chondriosomen und Plastiden im Wurzelmeristem von Zea mais und Vicia faba. Z. Naturf. 12b: 283–286.

Heitz, E. 1957b. Die strukturellen Beziehunger zwischen pflanzlichen und tierischen Chondriosomen. Z. Naturf. 12b: 576–578.

Heitz, E. 1957c. Über Plasmastrukturen bei Antirrhinum majus und Zea mais. Z. Naturforsch. 12b:579–583.

Hellgren, L. and D. J. Morré. 1992. ATP-induced budding of nuclear envelope in vitro. Protoplasma 167: 238–241.

Helms, J. B. and J. E. Rothman. 1992. Inhibition by brefeldin A of a Golgi membrane enzyme that catalyses exchange of guanine nucleotide bound to ARF. Nature 360: 352–354.

Helvoort, A. van, W. Stoorvogel, G. van Meer and K. N. J. Burger. 1997. Sphingomyelin synthase is absent from endosomes. J. Cell Sci. 110: 781–788.

Hendricks, L. C., S. L. McClanahans, M. McCaffery, G. E. Palade and M. G. Farquhar. 1992. Golgi proteins persist in the tubulovesicular remnants found in brefeldin A-treated pancreatic acinar cells. Eur. J. Cell Biol. 58: 202–213.

Herberman, R. and C. A. Stetson. 1965. The expression of histocompatibility antigens on cellular and subcellular membranes. J. Exp. Med. 121: 533–549.

Hess, K. A., D. J. Morré and W. D. Merritt. 1979. Lipoprotein secretion by rat liver Golgi apparatus. Lipoprotein particles and lipase activity. Cytobiologie 18: 431–449.

Heuser, J. E. and T. E. Reese. 1973. Evidence for recycling of synaptic vesicle membranes during transmitter release at the frog neuromuscular junction. J. Cell Biol. 57: 315–344.

Heyningen, H. van. 1965. Correlated light and electron microscope observations on glycoprotein-containing globules in the follicular cells of the thyroid gland of the rat. J. Histochem. Cytochem. 13: 286–296.

Hicke, L. and R. Schekman. 1990. Molecular machinery required for protein transport from the endoplasmic reticulum to the Golgi complex. Bioessays 12(6): 253–258.

Hicks, R. M. 1966. The function of the Golgi complex in transitional epithelium. Synthesis of the thick cell membrane. J. Cell Biol. 30: 623–643.

Higgens, J. A. and J. K. Fieldsend. 1987. Phosphatidylcholine synthesis for incorporation into membranes or for secretion as plasma lipoproteins by Golgi membranes of rat liver. J. Lipid Res. 28: 268–278.

Higgens, J. A. and J. L. Hutson. 1984. The roles of Golgi and endoplasmic reticulum in the synthesis and assembly of lipoprotein lipids in rat hepatocytes. J. Lipid Res. 25: 1295–1305.

Hill, F. G. and D. E. Outka. 1974. The structure and origins of mastigonemes in Ochromonas minute and Monas sp. J. Protozool. 21: 299–312.

Hino, Y., A. Asano and R. Sato. 1978a. Biochemical studies on rat liver Golgi apparatus. III. Subfractionation of fragmented Golgi apparatus by counter-current distribution. J. Biochem. 83: 935–942.

Hino, Y., A. Asano, R. Sato and S. Shimizu. 1978b. Biochemical studies of rat liver Golgi apparatus. I. Isolation and preliminary characterization. J. Biochem. (Tokyo) 83: 909–923.

Hirsch, G. C. 1939. Form und Stoffwechsel der Golgi Körper. Protoplasma Monographieren, Vol. 18. Gerbruder Borntraeger, Berlin.

Hirschberg, C. B. 1997a. Transporters of nucleotide sugars, nucleotide sulfate and ATP in the Golgi apparatus membrane: where next? Glycobiology 7: 169–171.

Hirschberg, C. B. 1997b. Transport of nucleotide sugars, nucleotide sulfate and ATP into the lumen of the Golgi apparatus. In: The Golgi Apparatus. E. G. Berger and J. Roth (eds.). Birkhäuser, Basel. pp. 163–178.

Hirschberg, C. B. and M. D. Snider. 1987. Topography of glycosylation in the rough endoplasmic reticulum and Golgi apparatus. Annu. Rev. Biochem. 56: 63–87.

Hirschberg, C. B., P. W. Robbins and C. Abeijon. 1998. Transporters of nucleotide sugars, ATP and nucleotide sulfate in endoplasmic reticulum and Golgi apparatus. Annu. Rev. Biochem. 67: 49–69.

Hodge, A. J., J. D. McLean and F. V. Mercer. 1956. A possible mechanism for the morphogenesis of lamellar systems in plant cells. J. Biophys. Biochem. Cytol. 2: 597–607.

Hodson, S. 1978. The ATP-dependent concentration of calcium by a Golgi apparatus-rich fraction isolated from rat liver. J. Cell Sci. 30: 117–128.

Holmes, K. V., E. W. Doller and L. S. Sturman. 1981. Tunicamycin resistant glycosylation of a coronavirus glycoprotein: determination of a novel type of viral glycoprotein. Virology 115: 334–344.

Holmgren, E. 1902. Einige Worte über das "Trophospongium" verscheidener Zellarten. Anat. Anz. 20: 433–440.

Howell, K. E. and G. E. Palade. 1982. Hepatic Golgi fractions resolved into membrane and content subfractions. J. Cell Biol. 92: 822–832.

Hruban, Z. 1979. Ultrastructure of hepatocellular tumors, In: Liver Carcinogenesis. K. Lapis and J. V. Johannessen (eds.). McGraw-Hill, New York. pp. 403–431.

Hruban, Z., H. Swift and M. Rechcigl, Jr. 1965. Fine structure of transplantable hepatomas of the rat. J. Natl. Cancer Inst. 35: 459–495.

Hruban, Z., Y. Mochizuki, A. Slesers and H. P. Morris. 1972. A comparative study of cellular organelles of Morris hepatomas. Cancer Res. 32: 853–867.

Huang, C. M., H. Goldenberg, C. Frantz, D. J. Morré, T. W. Keenan and F. L. Crane. 1979. Comparison of NADH-linked cytochrome c reducatses of endoplasmic reticulum, Golgi apparatus and plasma membrane. Int. J. Biochem. 10: 723–731.

Hudgin, R. L., P. K. Murray, L. Pinteric, H. P. Morris and H. Schachter. 1971. The use of nucleotide-sugar: glycoprotein glycosyl-transferases to assess Golgi apparatus function in Morris hepatomas. Can. J. Biochem. 49: 61–70.

Hunt, L., D. L. Hacker, F. Grosjean, M. De Jesus, L. Uebersax, M. Jordan and F. M. Wurm. 2004. Low-temperature pausing of cultivated mammalian cells. Biotechnol. Bioengin. 89: 157–163.

Hunter, T. 1985. Oncogene and growth control. Trends Biochem. Sci. 10: 275–280.

Huttner, W. B. 1987. Protein tyrosine sulfation. Trends Biochem. Sci. 12: 361–363.

Hynes, R. O. 1979. Surfaces of normal and malignant cells. Wiley, New York.

Jacobs, E., D. J. Morré, R. de Cabo, M. Sweeting and D. M. Morré. 1996. Response of a protein disulfide isomerase life activity of transitional endoplasmic reticulum to all trans-retinol. Life Sci. 59: 273–284.

Jamieson, J. D. and G. E. Palade. 1967a. Intracellular transport of secretory proteins in the pancreatic exocrine cell. I. Role of the peripheral elements of the Golgi complex. J. Cell Biol. 34: 577–596.

Jamieson, J. D. and G. E. Palade. 1967b. Intracellular transport of secretory proteins in the pancreatic exocrine cell. II. Transport to condensing vacuoles and zymogen granules. J. Cell Biol. 34: 597–615.

Jamieson, J. D. and G. E. Palade. 1968a. Intracellular transport of secretory proteins in the pancreatic exocrine cell. III. Dissociation of intracellular transport from protein synthesis. J. Cell Biol. 39: 580–588.

Jamieson, J. D. and G. E. Palade. 1968b. Intracellular transport of secretory proteins in the pancreatic exocrine cell. IV. Metabolic requirements. J. Cell Biol. 39: 589–603.

Jamieson, J. D. and G. E. Palade. 1971. Condensing vacuole conversion and zymogen granule discharge in pancreatic exocrine cells: metabolic studies. J. Cell Biol. 48: 503–522.

Jarasch, E. D., J. Kartenbeck, G. Bruder, A. Fink, D. J. Morré and W. W. Franke. 1979. β-type cytochromes in plasma membranes isolated from rat liver, in comparison with those of endomembranes. J. Cell Biol. 80: 37–52.

Jarosch, R. 1962. Golgi apparate lebender Planzenzellen im Lichtmikroskop. Protoplasma 60: 406–410.

Jelsema, C. L. and D. J. Morré. 1978. Distribution of phospholipid biosynthetic enzymes among cell components of rat liver. J. Biol. Chem. 253: 7960–7971.

Jenness, R. 1988. Composition of milk. In Fundamentals of Dairy Chemistry, 3rd edition. N. P. Wong, R. Jenness, M. Keeney and E. H. Marth (eds.). Van Nostrand Reinhold, New York. pp. 1–38.

Johnson, K. D. and Chrispeels, M. J. 1987. Substrate specificities of N-acetylglucosaminyl-, fucosyl- and xylosyl-transferases that modify glycoproteins in the Golgi apparatus of bean cotyledons. Plant Physiol. 84: 1301–1308.

Josefsberg, Z., B. I. Posner, B. Patel and J. J. M. Bergeron. 1979. The uptake of prolactin into female rat liver. Concentration of intact hormone in the Golgi apparatus. J. Biol. Chem. 254: 209–214.

Kalen, A., B. Norling, E. L. Applekvist and G. Dallner. 1987. Ubiquinone biosynthesis by microsomal fraction from rat liver. Biochim. Biophys. Acta. 926: 70–78.

Kanwar, K. C. 1961–1962. The Golgi controversy in the light of electron microscopy. La Cellule 62:377–385.

Kaplan, F. and P. Hechtman. 1983. Purification and properties of two enzymes catalyzing galactose transfer to Gm2 ganglioside from rat liver Golgi. J. Biol. Chem. 258: 770–776.

Kappler, R., U. Kristen and D. J. Morré. 1986. Membrane flow in plants: fractionation of growing pollen tubes of tobacco by preparative free-flow electrophoresis and kinetics of labeling of endoplasmic reticulum and Golgi apparatus with [3H]leucine. Protoplasma 132: 38–50.

Kasap, M., S. Thomas, E. Danaher, V. Holton, S. Jiang and B. Storrie. 2004. Dynamic nucleation of Golgi apparatus assembly from the endoplasmic reticulum in interphase HeLa cells. Traffic 5: 595–605.

Kasper, C. B. 1971. Biochemical distinctions between the nuclear and microsomal membranes from rat hepatocytes. The effect of phenobarbital administration. J. Biol. Chem. 246: 577–581.

Kasper, C. B. 1974. Chemical and biochemical properties of the nuclear envelope. In: The Cell Nucleus. H. Busch, (ed.). Academic Press, New York. Vol. 1, pp. 349–384.

Kawamoto, K., Y. Yoshida, H. Tamaki, S. Toril, C. Shinotsuka, S. Yamashina and K. Nakayama. 2002. GBF1, a guanine nucleotide exchange factor for ADP-ribosylation factors, is localized to the cis-Golgi and involved in membrane association of the COPI coat. Traffic 3: 483–495.

Keenan, T. W. 1998. Biochemistry of the Golgi Apparatus. Histochem. Cell Biol. 109: 505–516.

Keenan, T. W. and D. J. Morré. 1970. Phospholipid class and fatty acid composition of Golgi apparatus isolated from rat liver and comparison with other cell fractions. Biochemistry 9:19–25.

Keenan, T. W., D. J. Morré and R. D. Cheetham. 1970. Lactose synthesis by a Golgi apparatus fraction from rat mammary gland. Nature 228: 1105–1106.

Keenan, T. W., D. J. Morré, D. E. Olson, W. N. Yunghans and S. Patton. 1970. Biochemical and morphological comparison of plasma membrane and milk fat globule membrane from bovine mammary gland. J. Cell Biol. 44: 80–93.

Keenan, T. W., C. M. Huang and D. J. Morré. 1972a. Membranes of mammary gland. III. Lipid composition of Golgi apparatus from rat mammary gland. J. Dairy Sci. 55: 51–57.

Keenan, T. W., D. J. Morré and C. M. Huang. 1972b. Distribution of gangliosides among subcellular fractions from rat liver and bovine mammary gland. FEBS Lett. 24: 204–208.

Keenan, T. W., D. J. Morré and S. Basu. 1974. Ganglioside biosynthesis. Concentration of glycosphingolipid glycosyltransferases in Golgi apparatus from rat liver. J. Biol. Chem. 249: 310–315.

Keenan, T. W., M. Sasaki, W. N. Eigel, D. J. Morré, W. W. Franke, I. M. Zulak and A. A. Bushway. 1979. Characterization of a secretory vesicle-rich fraction from lactating mammary gland. Exp. Cell Res. 124: 47–61.

Klausner, R. O., J. G. Donaldson and J. Lippencott-Schwartz. 1992. Brefeldin A: insights into the control of membrane traffic and organelle structure. J. Cell Biol. 116: 1071–1080.

Kleinig, H. 1970. Nuclear membranes from mammalian liver. II. Lipid composition. J. Cell Biol. 46: 396–402.

Kobayashi, T. and R. E. Pagano. 1988. ATP-dependent fusion of liposomes with the Golgi apparatus of perforated cells. Cell 55: 797–805.

Krangel, M. S., H. T. Orr and J. L. Strominger. 1979. Assembly and maturation of HLA-A and HLA-B antigens in vivo. Cell 18: 979–991.

Kuff, E. L. and A. J. Dalton 1959. Biochemical studies of isolated Golgi membranes. In: Subcellular Particles. T. Hayashi (ed.). Ronald Press, New York. pp. 114–126.

Kuge, O., C. Dascher, L. Orci, T. Rowe, M. Amherdt, H. Plutner, M. Ravazzola, G. Tanigawa, J. E. Rothman and W. E. Balch. 1994. Sar1 promotes vesicle budding from the endoplasmic reticulum but not Golgi compartments. J. Cell Biol. 125: 51–65.

Kuhn, N. J. and A. White. 1976. Evidence for specific transport of uridine diphosphate galactose across the Golgi membrane of rat mammary gland. Biochem. J. 154: 243–244.

Kuhn, N. J. and A. White. 1977. The role of nucleotide diphosphatase in a uridine nucleotide cycle associated with lactose synthesis in rat mammary-gland Golgi apparatus. Biochem. J. 168: 423–433.

Lagunoff, D. and H. Wan. 1974. Temperature dependence of mast cell histamine secretion. J. Cell Biol. 61: 809–811.

Lamorte, L. and M. Park. 2001. The receptor tyrosine kinases: role in cancer progression. Surg. Oncol. Clin. North Am. 10: 271–288.

Lane, N. L., L. Caro, L. R. Otero-Vilardebo and G. C. Godman. 1964. On the site of sulfation in colonic goblet cells. J. Cell Biol. 21: 339–352.

La Valette St. George, A. J. H. 1865. Über die Genese der Samenkorper. Part 1. Arch. Mikrosk. Anat. 1:403–414.

La Valette St. George, A. J. H. 1867. Über die Genese der Samenkorper. Part 2. Arch. Mikrosk. Anat. 2:263–273.

Lavoie, C. and J. Paiement. 1996. Vesiculation of smooth endoplasmic reticulum. Mol. Biol. Cell 7: 71a.

Lavoie, C., J. Lanoix, F. W. K. Kan and J. Paiement. 1996. Cell-free assembly of rough and smooth endoplasmic reticulum. J. Cell Sci. 109: 1415–1425.

Lawrence, J. B., P. Moreau, C. Cassagne and D. J. Morré. 1994. Acyl transfer reactions associated with cis Golgi apparatus of rat liver. Biochim. Biophys. Acta 1210: 146–150.

Lazar, T., M. Götte and D. Gallwitz. 1997. Vesicular transport: how many Ypt/Rab-GTPases make a eukaryotic cell? Trends Biochem. Sci. 22: 468–472.

Leblond, C. P. 1950. Distribution of periodic acid-reactive carbohydrates in the adult rat. Am. J. Anat. 86: 1–49.

Leblond, C. P. and G. Bennett. 1977. Role of the Golgi apparatus in terminal glycosylation. In: International Cell Biology 1976–1977. B. R. Brinkley and K. R. Porter (eds.). Rockefeller University Press, New York. pp. 326–336.

Ledbetter, M. C. 1962. Observations on membranes in plant cells fixed with OsO4. In: Proceedings 5th International Congress of Electron Microscopy. S. S. Breese (ed.). Academic Press, New York. pp. 1–10.

Leelavathi, D. E., L. W. Estes, D. S. Feingold and B. Lombardi. 1970. Isolation of a Golgi-rich fraction from rat liver. Biochim. Biophys. Acta 211: 124–138.

Leskes, A., P. Siekevitz and G. E. Palade. 1971a. Differentiation of endoplasmic reticulum in hepatocytes. I. Glucose-6-phosphatase distribution in situ. J. Cell Biol. 49: 264–287.

Leskes, A. P., P. Siekevitz and G. E. Palade. 1971b. Differentiation of endoplasmic reticulum in hepatocytes. II. Glucose-6-phosphatase in rough microsomes. J. Cell Biol. 49: 288–302.

Lippincott-Schwartz, J., L. C. Yuan, J. S. Bonifacino and R. D. Klausner. 1989. Rapid redistribution of Golgi proteins into the ER in cells treated with brefeldin A: evidence for membrane cycling from Golgi to ER. Cell 56: 801–813.

Lockhart, J. A. 1965. An analysis of irreversible plant cell elongation. J. Theor. Biol. 8: 264–275.

Losev, E., C. A. Reinke, J. Jellen, D. E. Strongin, B. J. Bevis and B. S. Glick. 2006. Golgi maturation visualized in living yeast. Nature 44: 1002–1006.

Loud, A. V. 1962. A method for the quantitative estimation of cytoplasmic structures. J. Cell Biol. 15:481–487.

Ludford, R. J. 1925. Cell organs during secretion in the epididymis. Proc. Roy. Soc. London B 98: 354–372.

MacKinlay, A. G., D. W. West and W. Manson. 1977. Specific casein phosphorylation by a casein kinase from lactating bovine mammary gland. Eur. J. Biochem. 76: 233–243.

Mahley, R. W., B. D. Bennett, D. J. Morré, M. E. Gray, W. Thistlethwaite and V. S. LeQuire. 1971. Lipoproteins associated with the Golgi apparatus isolated from epithelial cells of rat small intestine. Lab. Invest. 25: 435–444.

Malhotra, V., L. Orci, B. S. Glick, M. R. Block and J. E. Rothman. 1988. Role of an N-ethylmaleimide sensitive transport component in promoting fusion of transport vesicles with cisternae of the Golgi stack. Cell 54: 221–227.

Malhotra, V. L., T. Serafini, L. Orci, J. C. Shepherd and J. E. Rothman. 1989. Purification of a novel class of coated vesicles mediating biosynthetic protein transport through the Golgi stack. Cell 58: 329–336.

Marinozzi, V. 1967. Reaction de l'acide phosphotungstique avec la mucine et les glycoprotéines des plasmamembranes. J. Micro. 6: 68a.

Matlin, K. S. and K. Simons. 1983. Reduced temperature prevents transfer of a membrane glycoprotein to the cell surface but does not prevent terminal glycosylation. Cell 34: 233–243.

Matsuoka, K. L. Orci, S. V. Bednarak, S. Hamamoto and R. Schekman. 1997. COPII-coated vesicle
formation reconstituted with purified coat proteins and liposomes. Mol. Biol. Cell 8: 405a.

Matsuura-Tokita, K., M. Takeuchi, A. Ichihara, K. Mikuriya and A. Nakano. 2006. Live imaging of
yeast Golgi cisternal maturation. Nature 441: 939–940.

Matyas, G. R. and D. J. Morré. 1983. Coupling of uridine-5′-diphosphate (UDP) formation and nicotinamide dinucleotide (NAD+) reduction for cytochemical localization of glycosyltransferases.
J. Histochem. Cytochem. 31:1175–1182.

Matyas, G. R. and D. J. Morré. 1987. Subcellular distribution and biosynthesis of rat liver gangliosides.
Biochim. Biophys. Acta 921: 599–614.

Matyas, G. R., D. C. Evers, R. Radinsky and D. J. Morré. 1986. Fibronectin binding to gangliosides
and rat liver plasma membranes. Exp. Cell Res. 162: 296–318.

May, A. P., S. W. Whiteheart and W. I. Weis. 2001. Unraveling the mechanism of the vesicle transport
ATPase NSF, the N-ethylmaleimide-sensitive factor. J. Biol. Chem. 276: 21991–21994.

McCarthy. P., C. L. Richardson, W. D. Merritt, D. J. Morré and H. H. Mollenhauer. 1974. Altered
Golgi apparatus architecture in animal and plant tumors. Proc. Ind. Acad. Sci. 84: 179–185.

Meder, D. and K. Simons. 2005. Ras on the roundabout. Science 307: 1731–1733.

Meer, G. van. 1989. Lipid traffic in animal cells. Annu. Rev. Cell Biol. 5: 247–275.

Melancon, P., B. S. Glick, V. Malhotra, P. J. Weidman, T. Serafini, M. L. Gleason, L. Orci and J. E.
Rothman. 1987. Involvement of GTP-binding "G" proteins in transport through the Golgi stack.
Cell 51: 1053–1062.

Meldolesi, J. 1974. Membranes and membrane surfaces. Dynamics of cytoplasmic membranes in
pancreatic acinar cells. Philos. Trans. R. Soc. Lond. B Biol. Sci. 268: 39–53.

Meldolesi, J., J. D. Jamieson and G. E. Palade. 1971. Composition of cellular membranes in the pancreas of the guinea pig. II. Lipids. J. Cell Biol. 49: 130–149.

Melkerson-Watson, L. J. and C. C. Sweeley. 1991. Purification to apparent homogeneity by immu-
noaffinity chromatography and partial characterization of the Gm3 ganglioside-forming enzyme,
CMP-sialic acid:lactosylceramide α2,3-sialyltransferase (SAT-1) from rat liver Golgi. J. Biol.
Chem. 266: 4448–4457.

Merritt, W. D. and D. J. Morré. 1973. A glycosyl transferase of high specific activity in secretory
vesicles derived from Golgi apparatus of rat liver. Biochim. Biophys. Acta 304: 397–407.

Merritt, W. D., D. J. Morré and T. W. Keenan. 1978a. Gangliosides of liver tumors induced by N-2-flu-
orenylacetamide. II. Alterations in biosynthetic enzymes. J. Natl. Cancer Inst. 60: 1329–1337.

Merritt, W. D., C. L. Richardson, T. W. Keenan and D. J. Morré. 1978b. Gangliosides of liver tumors
induced by N-2-fluorenylacetamide. I. Ganglioside alterations in liver tumorigenesis and normal
development. J. Natl. Cancer Inst. 60: 1313–1328.

Michaels, J. F. and C. P. LeBlond. 1976. Transport of glycoprotein from Golgi apparatus to cell surface
by means of "carrier" vesicles, as shown by radioautography of mouse colonic epithelium after
injection of 3H fucose. J. Microsc. Biol. Cell (Paris) 25: 243–248.

Miles, S., H. McManus, K. E. Forsten and B. Storrie. 2001. Evidence that the entire Golgi apparatus
cycles in interphase Hela cells: sensitivity of Golgi matrix proteins to an ER exit block. J. Cell
Biol. 155: 543–555.

Miller. S., L. Carnell and H. H. Moore. 1992. Post-Golgi membrane traffic: brefeldin A inhibits export
from distal Golgi compartments to the cell surface but not recycling. J. Cell Biol. 118: 267–283.

Mironov, A. A., A. Colanzi, R. S. Polishchuk, G. V. Beznoussenko, A. A. Moronov, Jr., A. Fusella, G.
Di Tullio, M. G. Silletta, D. Corda, M. A. De Matteis and A. Luini. 2004. Dicumarol, an inhibitor
of ADP-ribosylation of CtBP3/BARS, fragments golgi non-compact tubular zones and inhibits
intra-golgi transport. Eur. J. Cell Biol. 83(6): 263–279.

Misumi, Y., K. Miki, A. Takatsuki, G. Tamura and T. Ikehara. 1986. Novel blockade by brefeldin A
of intracellular transport of secretory proteins in cultured rat hepatocytes. J. Biol. Chem. 261:
11398–11403.

Misumi, Y., K. Oda, T. Fujiwara, N. Takami, K. Tashiro and Y. Ikehara. 1991. Functional expression
of furin demonstrating its intracellular localization and endoprotease activity for processing
proalbumin and complement pro-C3. J. Biol. Chem. 266: 16954–16959.

Mitchison, J. M. 1971. The Biology of the Cell Cycle. Cambridge University Press, London, UK.

Mitin, N. Y., M. B. Ramocki, A. J. Zullo, C. J. Der, S. F. Konieczny and E. J. Taparowsky. 2004. Identification and characterization of rain, a novel Ras-interacting protein with a unique subcellular localization. J. Biol. Chem. 279: 22353–22361.

Mollenhauer, H. H. 1965. An intercisternal structure in the Golgi apparatus. J. Cell Biol. 24: 504–511.

Mollenhauer, H. H. 1974. Distribution of microtubules in the Golgi apparatus of Euglena gracilis. J. Cell Sci. 15: 89–97.

Mollenhauer, H. H. and B. A. Mollenhauer. 1978. Changes in the secretory activity of the Golgi apparatus during the cell cycle in root tips of maize (Zea mays L.). Planta 138: 113–118.

Mollenhauer, H. H. and D. J. Morré. 1966a. Golgi apparatus and plant secretion. Annu. Rev. Plant Physiol. 17: 27–46.

Mollenhauer, H. H. and D. J. Morré. 1966b. Tubular connections between dictyosomes and forming secretory vesicles in plant Golgi apparatus. J. Cell Biol. 29: 373–376.

Mollenhauer, H. H. and D. J. Morré. 1974. Polyribosomes associated with the Golgi apparatus. Protoplasma 79: 333–336.

Mollenhauer, H. H. and D. J. Morré. 1975. A possible role for intercisternal elements in the formation of secretory vesicles in plant Golgi apparatus. J. Cell Sci. 19: 231–237.

Mollenhauer, H. H. and D. J. Morré. 1976a. Transition elements between endoplasmic reticulum and Golgi apparatus in plant cells. Cytobiologie 13: 297–306.

Mollenhauer, H. H. and D. J. Morré. 1976b. Cytochalasin B, but not colchicine, inhibits migration of secretory vesicles in root tips of maize. Protoplasma 87: 39–48.

Mollenhauer, H. H. and D. J. Morré. 1977. Dictyosome-like structures with cylindrical intersaccular connections (microtubules?) in guinea pig spermatocytes. Amer. J. Anat. 150: 381–394.

Mollenhauer, H. H. and D. J. Morré. 1978a. Polyribosomes associated with forming acrosome membranes in guinea pig spermatids. Science 200: 85–86.

Mollenhauer, H. H. and D. J. Morré. 1978b. Structural compartmentation of the cytosol: zones of exclusion, zones of adhesion, cytoskeletal and intercisternal elements. Subcell. Biochem. 5: 327–359.

Mollenhauer, H. H. and D. J. Morré. 1978c. Structural differences contrast plant and animal Golgi apparatus. J. Cell Sci. 32: 357–362.

Mollenhauer, H. H. and D. J. Morré. 1991. Perspectives on Golgi apparatus form and function. J. Electron Microsc. Tech. 17: 2–14.

Mollenhauer, H. H. and D. J. Morré. 1994. Structure of plant Golgi apparatus. Protoplasma 180: 14–28.

Mollenhauer, H. H. and D. J. Morré. 1998. The tubular network of the Golgi apparatus. Histochem. Cell. Biol. 109: 533–543.

Mollenhauer, H. H., D. J. Morré and L. Bergmann. 1967. Homology of form in plant and animal Golgi apparatus. Anat. Rec. 158: 313–317.

Mollenhauer, H. H., D. J. Morré and C. Totten. 1973. Intercisternal substances of the Golgi apparatus. Unstacking of plant dictyosomes using chaotropic agents. Protoplasma 78: 443–459.

Mollenhauer, H. H., D. J. Morré and W. J. VanDerWoude. 1975. Endoplasmic reticulum-Golgi apparatus associations in maize root tips. Mikroskopie 31: 257–272.

Mollenhauer, H. H., B. S. Hass and D. J. Morré. 1976. Membrane transformations in Golgi apparatus of rat spermatids. A role for thick cisternae and two classes of coated vesicles in acrosome formation. J. Microscop. Biol. Cell. 27: 33–36.

Mollenhauer, H. H., D. J. Morré and L. D. Rowe. 1990. Alteration of intracellular traffic by monensin. Mechanism, specificity and relationship to toxicity. Biochim. Biophys. Acta-Rev. Biomembranes 1031: 225–246.

Mollenhauer, H. H., D. J. Morré and L. R. Griffing. 1991. Post Golgi apparatus structures and membrane removal in plants. Protoplasma 162: 55–60.

Moreau, P. and D. J. Morré. 1991. Cell-free transfer of membrane lipids. Evidence for lipid processing. J. Biol. Chem. 266: 4329–4333.

Moreau, P., M. Rodriguez, C. Cassagne, D. M. Morré and D. J. Morré. 1991. Trafficking and sorting of lipids from endoplasmic reticulum to the Golgi apparatus in a cell-free system from rat liver. J. Biol. Chem. 266: 4322–4328.

Moreau, P., H. Juguelin, C. Cassagne and D. J. Morré. 1992. Molecular basis for low temperature compartment formation by transitional endoplasmic reticulum of rat liver. FEBS Lett. 310: 223–228.

Moreau, P., C. Cassagne, T. W. Keenan and D. J. Morré. 1993. Ceramide excluded from cell-free vesicular lipid transfer from endoplasmic reticulum to Golgi apparatus. Evidence for lipid sorting. Biochim. Biophys. Acta 1146: 9–16.

Morin-Ganet, M. N., A. Rambourg, S. B. Deitz, A. Franzusoff and F. Kepes. 2000. Morphogenesis and dynamics of the yeast Golgi apparatus. Traffic 1: 56–68.

Morré, D. J. 1970. In vivo incorporation of radioactive metabolites by dictyosomes and other cell fractions of onion stem. Plant Physiol. 45: 791–799.

Morré, D. J. 1971. Isolation of Golgi apparatus. Methods Enzymol. 22:130–148.

Morré, D. J. 1973. Isolation and purification of organelles and endomembrane components from rat liver. In: Molecular Techniques and Approaches in Developmental Biology. M. J. Chrispeels (ed.). John Wiley & Sons, New York. pp. 1–27.

Morré, D. J. 1975. Membrane biogenesis. Ann. Rev. Plant Physiol. 26: 441–481.

Morré, D. J. 1976. Occurrence and isolation of Golgi apparatus. Biological Handbook on Cell Biology, Vol. 1. FASEB, Bethesda, MD. pp. 240–244.

Morré, D. J. 1977a. Membrane differentiation and the control of secretion. A comparison of plant and animal Golgi apparatus. In: International Cell Biology 1976–1977. B. R. Brinkley and K. R. Porter (eds.). Rockefeller University Press, New York. pp. 293–303.

Morré, D. J. 1977b. The Golgi apparatus and membrane biogenesis. In: Cell Surface Reviews. Vol. 4. The Synthesis, Assembly and Turnover of Cell Surface Components. G. Poste and G. L. Nicolson (eds.). North-Holland, Amsterdam, New York.-Oxford. pp. 1–83.

Morré, D. J. 1981. An alternative pathway for secretion of lipoprotein particles in rat liver. Eur. J. Cell Biol. 26: 21–25.

Morré, D. J. 1987. The Golgi apparatus. Int. Rev. Cytol. 17: 211–253.

Morré, D. M. 1991. Role of the Golgi apparatus in cellular pathology. J. Elect. Micros. Techniq. 17: 200–211.

Morré, D. J. 1994a. Evolution of the endomembrane system. In: Isopentenods and Other Natural Products, Evolution and Function. W. D. Nes (ed.). American Chemical Society, Washington D.C. ACS Symposium Series 56, pp. 142–162.

Morré, D. J. 1994b. Physical membrane displacement: reconstitution in a cell-free system and relationship to cell growth. Protoplasma 180: 3–13.

Morré, D. J. 1994c. The hormone- and growth factor-stimulated NADH oxidase. J. Bioenerg. Biomemb. 26: 421–433.

Morré, D. J. 1998. Cell-free analysis of Golgi apparatus membrane traffic in rat liver. Histochem. Cell Biol. 109: 487–504.

Morré, D. J. and T. J. Buckhout. 1979. Isolation of Golgi apparatus. In: Plant Organelles, Methodological Surveys Subseries B (Biochemistry). E. Reid (ed.). Ellis Horwood, Chichester, UK. Vol. 9, pp. 207–224.

Morré, D. J. and W. R. Eisinger. 1968. Cell wall extensibility: its control by auxin and relationship to cell elongation. In: Biochemistry and Physiology of Plant Growth Substances. F. Wightman and G. Setterfied (eds.). Runge Press, Ottawa. pp. 625–645.

Morré, D. J. and T. W. Keenan. 1994. Golgi apparatus buds – vesicles or coated ends of tubules? Protoplasma 179: 1–4.

Morré, D. J. and T. W. Keenan. 1997. Membrane flow revisited. BioScience 47: 489–498.

Morré, D. J. and H. H. Mollenhauer. 1964. Isolation of the Golgi apparatus from plant cells. J. Cell Biol. 23:295–305.

Morré, D. J. and H. H. Mollenhauer. 1974. The endomembrane concept: a functional integration of endoplasmic reticulum and Golgi apparatus. In: Dynamic Aspects of Plant Ultrastructure. A. W. Robards (ed.). McGraw-Hill, New York, London. pp. 84–137.

Morré, D. J. and H. H. Mollenhauer. 1976. Interactions among cytoplasm, endomembranes and the cell surface. In: Encyclopedia of Plant Physiology, New Series. C. R. Stocking and U. Heber (eds.). Springer-Verlag, Berlin. Vol. 3, pp. 288–344.

Morré, D. J. and H. H. Mollenhauer. 1983. Dictyosome polarity and membrane differentiation in outer cap cells of the maize root. Europ. J. Cell Biol. 29: 126–132.

Morré, D. J. and H. H. Mollenhauer. 2007. Microscopic morphology and the origins of the membrane maturation model of Golgi apparatus function. Int. Rev. Cytol. 262: 191–218.

Morré, D. J. and D. M. Morré. 1987. Transition vesicles of the cis Golgi apparatus face of rat liver are increased by retinol. Cell Biol. Int. Rep. 11: 89–93.

Morré, D. J. and D. M. Morré. 1989. Mammalian plasma membranes by aqueous two-phase partition. BioTechniques 7: 946–958.

Morré, D. M. and D. J. Morré. 2000. Aqueous two-phase partition applied to the isolation of plasma membranes and Golgi apparatus from cultured mammalian cells. J. Chromatog. B 743: 377–387.

Morré, D. J. and D. M. Morré. 2003. Cell surface NADH oxidases (ECTO-NOX proteins) with roles in cancer, cellular time-keeping, growth, aging and neurodegenerative disease. Free Radical Res. 37: 795–808.

Morré, D. J. and L. Ovtracht. 1977. Dynamics of the Golgi apparatus: membrane differentiation and membrane flow. Int. Rev. Cytol. Suppl. 5:61–188.

Morré, D. J. and L. Ovtracht. 1981. Structure of rat liver Golgi apparatus: relationship to lipoprotein secretion. J. Ultrastruc. Res. 74: 284–295.

Morré, D. J. and M. Paulik. 1993. Low temperature compartment formation in feline immunodeficiency virus-infected and uninfected feline kidney cells. Protoplasma 177: 15–22.

Morré, D. J. and W. J. VanDerWoude. 1974. Origin and growth of cell surface components. In: Macromolecules Regulating Growth and Development. E. D. Hay, T. J. King and J. Papaconstantinou (eds.). Academic Press, New York. pp. 81–111.

Morré, D. J. and E. L. Vigil. 1979. Membrane differentiation within Golgi apparatus of rat hepatocytes. J. Ultrastruc. Res. 68: 317–324.

Morré, D. J., H. H. Mollenhauer and J. E. Chambers. 1965. Glutaraldehyde stabilization as an aid to Golgi apparatus isolation. Exp. Cell Res. 38: 672–675.

Morré, D. J., D. D. Jones and H. H. Mollenhauer. 1967. Golgi apparatus mediated polysaccharide secretion by outer root cap cells of Zea mays. 1. Kinetics and secretory pathway. Planta 74: 286–301.

Morré, D. J., R. Cheetham and W. Yunghans. 1968a. Biochemical characterization of a Golgi apparatus-rich fraction from rat liver. J. Cell Biol. 39: 96a.

Morré, D. J., H. H. Mollenhauer, R. L. Hamilton, R. W. Manley and W. P. Cunningham. 1968b. Golgi apparatus isolation. J. Cell Biol. 39: 157a.

Morré, D. J., L. M. Merlin and T. W. Keenan. 1969. Localization of glycosyltransferase activities in a Golgi apparatus-rich fraction isolated from rat liver. Biochem. Biophys. Res. Commun. 37: 813–819.

Morré, D. J., R. L. Hamilton, H. H. Mollenhauer, R. W. Mahley, W. P. Cunningham, R. D. Cheetham and V. S. LeQuire. 1970a. Isolation of a Golgi apparatus-rich fraction from rat liver. 1. Methods and morphology. J. Cell Biol. 44: 484–490.

Morré, D. J., S. Nyquist and E. Rivera. 1970b. Lecithin biosynthetic enzymes in onions and the distribution of phosphorylcholine-cytidyl transferase among cell fractions. Plant Physiol. 45: 800–804.

Morré, D. J., T. W. Keenan and H. H. Mollenhauer. 1971a. Golgi apparatus function in membrane transformations and product compartmentalization: studies with cell fractions from rat liver. In: Advances in Cytopharmacology. F. Clementi and B. Ceccarelli, (eds.). Raven Press, New York. pp. 159–182.

Morré, D. J., W. D. Merritt and C. A. Lembi. 1971b. Connections between mitochondria and endoplasmic reticulum in rat liver and onion stem. Protoplasma 73: 43–49.

Morré, D. J., H. H. Mollenhauer and C. E. Bracker. 1971c. Origin and continuity of Golgi apparatus. In: Results and Problems in Cell Differentiation. II. Origin and Continuity of Cell Organelles. T. Reinert and H. Ursprung (eds.). Springer-Verlag, Berlin. pp. 82–126.

Morré, D. J., W. W. Franke, B. Deumling, S. E. Nyquist and L. Ovtracht. 1971d. Golgi apparatus function in membrane flow and differentiation: origin of plasma membrane from endoplasmic reticulum. Biomembranes 2: 95–104.

Morré, D. J., R. D. Cheetham, S. E. Nyquist and L. Ovtracht. 1972. A simplified procedure for isolation of Golgi apparatus from rat liver. Prep. Biochem. 2:61–69.

Morré, D. J., T. W. Keenan and C. M. Huang. 1974a. Membrane flow and differentiation: origin of Golgi apparatus membranes from endoplasmic reticulum. In: Advances in Cytopharmacology 2. B. Ceccarelli, F. Clementi and J. Meldolesi (eds.). Proceedings of Advanced Study Institute of Cytopharmacology, Venice, Italy, 17–26 June 1973. Raven Press, New York. pp. 107–125.

Morré, D. J., W. N. Yunghans, E. L. Vigil and T. W. Keenan. 1974b. Isolation of organelles and endomembrane components from rat liver. Methodological Developments in Biochemistry. Subcellular Studies. Longman, London. Vol. 4. pp. 195–236.

Morré, D. J., T. M. Kloppel, W. D. Merritt and T. W. Keenan. 1978a. Glycolipids as indicators of the tumoringenic transformation. J. Supramol. Struct.9: 157–177.

Morré, D. J., E. L. Vigil, C. Frantz, H. Goldenberg and F. L. Crane. 1978b. Cytochemical demonstration of glutaraldehyde-resistant, NADH-ferricyanide oxido-reductase activities in rat-liver plasma membranes and Golgi apparatus. Eur. J. Cell Biol. 18:213–230.

Morré, D. J., G. B. Cline, R. Coleman, W. H. Evans, H. Glaumann, D. R. Headon, E. Reid, G. Siebert and C. C. Widnell. 1979a. Markers for endomembrane components. Eur. J. Cell Biol. 20:195–199.

Morré, D. J., J. Kartenbeck and W. W. Franke. 1979b. Membrane flow and interconversions among endomembranes. Biochem. Biophys. Acta Rev. Biomembranes. 559:71–152.

Morré, D. J., V. Schirrmacher, P. Robinson, K. Hess and W. W. Franke. 1979c. H-2 histocompatibility antigens of subcellular membranes of mouse liver. Exp. Cell Res. 119: 265–275.

Morré, D. M., D. J. Morré and M. Walter. 1981. Vitamin A effects on hepatic Golgi apparatus architecture. Eur. J. Cell Biol. 25: 28–35.

Morré, D. J., W. F. Boss, H. Grimes and H. H. Mollenhauer. 1983a. Kinetics of Golgi apparatus membrane flux following monensin treatment of embryogenic carrot cells. Europ. J. Cell Biol. 30: 25–32.

Morré, D. J., D. M. Morré and H.-G. Heidrich. 1983b. Subfractionation of rat liver Golgi apparatus by free-flow electrophoresis. Eur. J. Cell Biol. 31:263–274.

Morré, D. J., K. Creek, G. R. Matyas, N. Minnifield, I. Sun, P. Baudoin, D. M. Morré and F. L. Crane. 1984a. Free-flow electrophoresis for subfractionation of rat liver Golgi apparatus. Bio. Techniques 2: 224–233.

Morré, D. J., W. F. Boss and H. H. Mollenhauer. 1984b. Distribution of Golgi apparatus-associated polyribosomes across the polarity axis of dictyosomes of wild carrot (Daucus carota L.). Protoplasma 123: 221–225.

Morré, D. J., M. Paulik and D. Nowack. 1986a. Transition vesicle formation in vitro. Protoplasma 132: 110–113.

Morré, D. J., E. Schnepf and G. Deichgräber. 1986b. Inhibition of elongation growth in *Pellia* setae by the monovalent ionophore monensin. Bot. Gaz 147: 252–257.

Morré, D. J., F. L. Crane, I. L., Sun and P. Navas. 1987a. The role of ascorbate in biomembrane energetics. Ann. NY Acad. Sci. 498: 153–171.

Morré, D. J., D. M. Morré, H. H. Mollenhauer and W. Reutter. 1987b. Golgi apparatus cisternae of monensin-treated cells accumulate in the cytoplasm of liver slices. Europ. J. Cell Biol. 43: 235–242.

Morré, D. M., D. J. Morré, S. Bowen, W. Reutter and K. Windel. 1988. Vitamin A excess alters membrane flow in rat liver. Eur. J. Cell Biol. 46: 307–315.

Morré, D. J., N. Minnifield and M. Paulik. 1989a. Identification of the 16°C compartment of the endoplasmic reticulum in rat liver and cultured hamster kidney cells. Biol. Cell 67: 51–60.

Morré, D. J., D. D. Nowack, M. Paulik, A. O. Brightman, K. Thornborough, J. Yim and G. Auderset. 1989b. Transitional endoplasmic reticulum membranes and vesicles isolated from animals and plants. Homologous and heterologous cell-free membrane transfer to Golgi apparatus. Protoplasma 153: 1–13.

Morré, D. J., J. T. Morré, S. R. Morré, C. Sundqvist and A. S. Sandelius. 1991a. Chloroplast biogenesis. Cell-free transfer of envelope monogalactosylglycerides to thylakoids. Biochim. Biophys. Acta 1070: 437–445.

Morré, D. J., C. Penel, D. M. Morré, A. S. Sandelius, P. Moreau and B. Andersson. 1991b. Cell-free transfer and sorting of membrane lipids in spinach. Donor and acceptor specificity. Protoplasma 160: 49–64.

Morré, D. J., C. Penel, D. M. Morré, L. Hellgren, A. S. Sandelius and H. Greppin. 1992a. ATP-dependent cell-free transfer of membrane lipids from nuclei to Golgi apparatus of germinating axes of garden pea. Protoplasma 170: 1–9.

Morre, D. M., H. Spring, M. M. Trendlenburg, B. A. Mollenhauer, H. H. Mollenhauer and D. J. Morré. 1992b. Retinol stimulates Golgi apparatus activity in cultured bovine mammary gland epithelial cells. J. Nutr. 122: 1248–1253.

Morré, D. J., T. W. Keenan and D. M. Morré. 1993. Golgi apparatus isolation and use in cell-free systems. A perspective. Protoplasma 172: 12–26.

Morré, D. J., J. Lawrence, K. Safranski, T. Hammond and D. M. Morré. 1994a. Experimental basis for separation of membrane vesicles by preparative free-flow electrophoresis. J. Chromatogr. 668: 201–214.

Morré, D. J., M. Paulik, J. L. Lawrence and D. M. Morré. 1994b. Inhibition by brefeldin A of NADH oxidation activity of rat liver Golgi apparatus accelerated by GDP. FEBS Lett. 346: 199–202.

Morré, D. M., S. Wang, P.-J. Chueh, J. Lawler, K. Safranski, E. Jacobs and D. J. Morré. 1998. A biochemical basis for retinol stimulation of vesicle budding in vivo and in vitro. Mol. Cell. Biochem. 187: 73–83.

Morré, D. J., C. Kim and C. Hicks-Berger. 2007. ATP-dependent and drug-inhibited vesicle enlargement reconstituted using synthetic lipids and recombinant proteins. BioFactors 28: 105–117.

Mukherjee, S., R. Chiu, S.-M. Leung and D. Shields. 2007. Fragmentation of the Golgi apparatus: an early apoptotic event independent of the cytoskeleton. Traffic 8: 369–378.

Munro, J. R., S. Narasimhan, S. Wetmore, J. R. Riordan and H. Schachter. 1975. Intracellular localization of GDP-l-fucose: glycoprotein and CMP-sialic acid: apolipoprotein glycosyltransferases in rat and pork livers. Arch. Biochem. Biophys. 169: 269–277.

Nahm, L. J. 1940. The problem of Golgi material in plant cells. Bot. Rev. 6:49–71.

Nakamura, N., M. Lowe, T. P. Levine, C. Rabouille and G. Warren. 1997. The vesicle docking protein p115 binds GM130, a cis-Golgi matrix protein, in a mitotically regulated manner. Cell 89: 445–455.

Nakano, A. and M. Muramatsu. 1989. A novel GTP-binding protein, Sar1p, is involved in transport from the endoplasmic reticulum to the Golgi apparatus. J. Cell Biol. 109: 2688–2691.

Nakano, A., D. Brada and R. Scheckman. 1988. A membrane glycoprotein, Sec12p, required for protein transport from the endoplasmic reticulum to the Golgi apparatus in yeast. J. Cell Biol. 107: 851–863.

Nassonov, D. N. 1923. Das Golgische Binnennetz und seine Beziehungen zu der Sekretion. Untersuchungen über einige Amphibiendrusen. Arch. Mikrosk. Anat. 97: 136–186.

Nassonov, D. N. 1924. Das Golgische Binnennetz und seine Beziehungen zu der Sekretion (Forsetsung). Arch. Mikrosk. Anat. 100: 433–472.

Navas, P., N. Minnifield, I. Sun and D. J. Morré. 1986. NADP phosphatase: a marker in free-flow electrophoretic separations for cisternae of the Golgi apparatus midregion. Biochim. Biophys. Acta 881: 1–9.

Neutra, M. and C. P. Leblond. 1966a. Synthesis of the carbohydrate of mucus in the Golgi complex as shown by electron microscope radioautography of goblet cells from rats injected with glucose-H3. J. Cell Biol. 30: 119–136.

Neutra, M. and C. P. Leblond. 1966b. Radioautographic comparison of the uptake of galactose-H3 and glucose-H3 in the Golgi region of various cells secreting glycoproteins or mucopolysaccharides. J. Cell Biol. 30: 137–150.

Nichols, B. J. and H. R. Pelham. 1998. SNAREs and membrane fusion in the Golgi apparatus. Biochim. Biophys. Acta 1404: 9–31.

Nilsson, T. and G. Warren. 1994. Retention and retrieval in the endoplasmic reticulum and the Golgi apparatus. Curr. Opin. Cell Biol. 6: 517–521.

Noguchi, T. and D. J. Morré. 1991a. Membrane flow in plants: preparation and kinetics of labeling of plasma membranes from growing pollen tubes of tobacco. Protoplasma 163: 34–42.

Noguchi, T. and D. J. Morré. 1991b. Vesicular membrane transfer between endoplasmic reticulum and the Golgi apparatus of a green alga, *Micrasterias americana*. A 16°C block and reconstitution in a cell-free system. Protoplasma 162: 128–139.

Novick, P. and M. Zerial. 1997. The diversity of rab proteins in vesicle transport. Curr. Opin. Cell Biol. 9: 496–504.

Novick, P., C. Field and R. Schekman. 1980. Identification of 23 complementation groups required for post-translational events in the yeast secretory pathway. Cell 21: 205–215.

Novikoff, A. B. 1964. GERL, it's form and function in neurons of rat spinal ganglia. Biol Bull (Woods Hole) 127:358.

Novikoff, A. B. and S. Goldfisher. 1961. Nucleoside diphosphatase activity in the Golgi apparatus and its usefulness for cytological studies. Proc. Natl. Acad. Sci. U.S.A. 47: 802–810.

Novikoff, A. B., E. Essner, S. Goldfischer and M. Heus. 1962. Nucleoside-diphosphatase activities of cytomembranes. In: The Interpretation of Ultrastructure. R. J. C. Harris (ed.). Academic Press, New York. pp. 149–192.

Novikoff, P. M., A. B. Novikoff, N. Quintana and J.-J. Hauw. 1971. Golgi apparatus, GERL, and lysosomes of neurons in rat dorsal root ganglia, studied by thick section and thin section cytochemistry. J. Cell Biol. 50: 859–886.

Nowack, D. D., D. M. Morré, M. Paulik, T. W. Keenan and D. J. Morré. 1987. Intracellular membrane tiow: reconstitution of transition vesicle formation and function in a cell-free system. Proc. Natl. Acad. Sci. U.S.A. 84: 6098–6102.

Nowack, D. D., M. Paulik, D. J. Morré and D. M. Morré. 1990. Retiriol modulation of cell-free membrane transfer between endoplasmic reticulum and Golgi apparatus. Biochim. Biophys. Acta 1051: 250–258.

Nyquist, S. E. and D. J. Morré. 1971. Distribution of UDP-glucuronyl transferase among cell fractions of rat liver. J. Cellular Physiol. 78: 9–12.

Nyquist, S. E., R. Barr and D. J. Morré. 1970. Ubiquinones from rat liver Golgi apparatus fractions. Biochim. Biophys. Acta 208: 532–534.

Nyquist, S. E., F. L. Crane and D. J. Morré. 1971a. Vitamin A: concentration in the rat liver Golgi apparatus. Science 173: 939–941.

Nyquist, S. E., J. T. Matschiner and D. J. Morré. 1971b. Distribution of vitamin K among rat liver cell fractions. Biochim. Biophys. Acta 244: 645–649.

Oda, K., S. Hirose, N. Talcami, Y. Misume, A. Takatsuki and Y. Ikehara. 1987. Brefeldin A arrests the intracellular transport of a precursor of complement C3 before its conversion site in rat hepatocytes. FEBS Lett. 214: 135–138.

Oda, K., T. Fujiwara and Y. Ikehara. 1990. Brefeldin A arrests the intracellular transport of viral envelope proteins in primary cultured rat hepatocytes and HepG2 cells. Biochem. J. 265: 161–167.

Orci, L., B. S. Glick and J. E. Rothman. 1986. A new type of coated vesicular carrier that appears not to contain clathrin: its possible role in protein transport within the Golgi stack. Cell 46: 171–184.

Osaki, M., M. Oshimura and H. Ito. 2004. P13K-Akt pathway: its functions and alterations in human cancer. Apoptosis 9: 667–676.

Osowska-Rogers, S., E. Swiezewska, B. Anderson and G. Callner. 1994. The endoplasmic reticulum-Golgi system is a major site of plastoquinone synthesis in spinach leaves. Biochem. Biophys. Res. Commun. 205: 714–721.

Ostermann, J., L. Orci, K. Tani, M. Amherdt, M. Ravazzola, Z. Elazar and J. E. Rothman. 1993. Stepwise assembly of functionally active transport vesicles. Cell 75: 1015–1025.

Ovtracht, L., D. J. Morré and L. M. Merlin. 1969. lsolement de l'appareil de Golgi d'une glande secretrice de mucopolysaccharides de l'escargot (Helix pomatia). J. Microsc. 8: 989–1002.

Ovtracht, L. D., D. J. Morré, R. D. Cheetham and H. H. Mollenhauer. 1973. Subraction of Golgi apparatus from rat liver: method and morphology. J. Microscopie 18:87–102.

Palade, G. E. 1983. Membrane biogenesis: an overview. Methods Enzymol. 69: xxix–lv.

Palade, G. E. and A. Claude. 1949a. The nature of the Golgi apparatus. I. Parallelism between intracellular myelin figures and Golgi apparatus in somatic cells. J. Morphol. 85: 35–70.

Palade, G. E. and A. Claude. 1949b. The nature of the Golgi apparatus. II. Identification of the Golgi apparatus with a complex of myelin figures. J. Morphol. 85: 71–112.

Paquet, M. R., S. R. Pfeffer, I. D. Burczak, B. S. Glick and J. E. Rothman. 1986. Components responsible for transport between successive Golgi cisternae are highly conserved in evolution. J. Biol. Chem. 261: 4367–4370.

Parat, M. and J. Painlevé. 1924a. Observation vitale d'une cellule glandulaire et activité. Nature et rôle de l'appareil interne de Golgi et de l'appareil de Holmgren. C. R. Acad. Sci. (Paris) 179: 612–614.

Parat, M. and J. Painlevé. 1924b. Appareil réticulaire interne de Golgi, trophosponge de Holmgren, et vacuome. C. R. Acad. Sci. 179: 844–846.

Patton, S. 1970. Correlative relationship of cholesterol and sphingomyelin in cell membranes. J. Theor. Biol. 29: 489–491.

Paulik, M., D. D. Nowack and D. J. Morré. 1988. Isolation of a vesicular intermediate in the cell-free transfer of membrane from transitional elements of the endoplasmic reticulum to Golgi apparatus cisternae of rat liver. J. Biol. Chem. 263: 17738–17748.

Paulik, M. A., C. C. Widnell, P. A. Whitaker-Dowling, N. Minnifield, D. M. Morré and D. J. Morré. 1999. Cell-free transfer of vesicular stomatitis virus protein from an endoplasmic reticulum compartment of BHK cells to a rat liver Golgi apparatus compartment for Man8–9 to Man5 processing. Arch. Biochem. Biophys. 367: 265–273.

Paulson, J. C. and K. J. Colley. 1989. Glycosyltransferases. Structure, localization and control of cell type-specific glycosylation. J. Biol. Chem. 264: 17615–17618.

Pavelka, M. and A. Ellinger. 1991. Cytochemical characteristics of the Golgi apparatus. J. Elect. Micro. Tech. 17: 35–50.

Pearse, B. M. F. 1976. Clathrin: a unique protein associated with intracellular transfer of membrane by coated vesicles. Proc. Natl. Acad. Sci. U S A. 73: 1255–1259.

Pearse, B. M. F. and M. S. Robinson. 1990. Clathrin, adapters and sorting. Annu. Rev. Cell Biol. 6: 151–171.

Pecot, M. Y. and V. Malhotra. 2004. Golgi membranes remain segregated from the endoplasmic reticulum during mitosis in mammalian cells. Cell 116: 99–107.

Pecot, M. Y. and V. Malhotra. 2006. The Golgi apparatus maintains its organization independent of the endoplasmic reticulum. Mol. Biol. Cell 17: 5372–5380.

Pelham, H. R. 1990. The retention signal for soluble proteins of the endoplasmic reticulum. Trends Biochem. Sci. 15: 483–486.

Pelham, H. R. 2001. Traffic through the Golgi apparatus. J. Cell Biol. 155(7): 1099–1101.

Pelletier, L., E. Jokitalo and G. Warren. 2000. The effect of Golgi depletion on exocytic transport. Nature Cell Biol. 2: 840–846.

Pelletier, L., C. A. Stern, M. Pypaert, D. Sheff, H. Ngo, N. Roper, C. Y. He, K. Hu, D. Toomre, I. Coppens, D. S. Roos, K. A. Joiner and G. Warren. 2002. Golgi biogenesis in Toxoplasma gondii. Nature 418: 548–552.

Perez, M. and C. B. Hirschberg. 1986. Transport of sugar nucleotides and adenosine 3 -phosphate 5′-phosphosulfate into vesicles derived from the Golgi apparatus. Biochim. Biophys. Acta 864: 213–222.

Perez de Castrol, I., T. G. Bivona, M. R. Philips and A. Pellicer. 2004. Ras activation in Jurkat T cells following low-grade stimulation of the T-cell receptor is specific to N-Ras and occurs only on the Golgi apparatus. Mol. Cell Biol. 24: 3485–3496.

Perner, E. S. 1957. Zum elektronenmikroskopischen Nachweis des "Golgi-Apparates" in Zellen höherer Pflanzen. Naturwiss 44: 336.

Perner, E. S. 1958. Elektronenmikroskopische Untersuchungen zu Cytomorphologie des soganannten "Golgisystems" in Wurzelzellen verschiedener Angiospermen. Protoplasma 49: 407–446.

Perroncito, A. 1910. Contribution à l'étude de la biologie cellulaire. Mitochondres, chromidies et appareil réticulaire interne dans les cellules spermatiques. Le phénomène de la dictyokinèse. Arch. Ital. Biol. 54: 307–345. (Originally published in Redn. R. Ist. Lomb. Sci. e Let. 16–17: 1908–1909.)

Peter, F., H. Plutner, H. Zhu, T. E. Kreis and W. E. Balch. 1993. β-COP is essential for transport of protein from the endoplasmic reticulum to the Golgi in vitro. J. Cell Biol. 122: 1155–1167.

Peters, J. M., M. J. Walsh and W. W. Franke. 1990. An abundant and ubiquitous homo-oligomeric ring-shaped ATPase particle related to the putative vesicle fusion proteins Sec 18p and NSF. EMBO J. 9: 1757–1767.

Peterson, M. and C. P. LeBlond. 1964. Synthesis of complex carbohydrates in the Golgi region as shown by radioautography after injection of glucose. J. Cell Biol. 21: 143–148.

Peyrière, M. 1975. A propos des relations dictyosomes-mitochondries observées chez les rhodophycées floridées. C. R. Hebd. Séances Acad. Sci. Ser. D. 281: 1579.

Pfenninger, K. H. and R. P. Bunge. 1974. Freeze-fracturing of nerve growth cones and young fibers. A study of developing plasma membrane. J. Cell Biol. 63: 180–196.

Pind, S., C. Nuoffer, J. M. McCaffery, H. Plutner, H. W. Davidson, M. G. Farquhar and W. E. Balch. 1994a. RabI and Ca2- are required for the fusion of carrier vesicles mediating endoplasmic reticulum to Golgi transport. J. Cell Biol. 125: 239–252.

Pind, S., J. R. Riordan and D. B. Williams. 1994b. Participation of the endoplasmic reticulum chaperone calnexin (p88 IP9O) in the biogenesis of the cystic fibrosis transmembrane conductance regulator. J. Biol. Chem. 269: 12786–12788.

Piqueras, A. I., M. Somers, T. G. Hammond, K. Strange, H. W. Harris, Jr., M. Gawryl and M. L. Zeidel. 1994. Permeability properties of rat renal lysosomes. Am. J. Physiol. 266: C121–C133.

Platner, G. 1889. Beiträge zur Kenntnis der Zelle und ihrer Teilungs erscheinungen. Arch. Mikrosk. Anat. 33: 180–216.

Plutner, H., H. W. Davidson, J. Saraste and W. E. Balch. 1992. Morphological analysis of protein transport from the endoplasmic reticulum to Golgi membranes in digitonin-permeabilized cells: role of the p58-containing compartment. J. Cell Biol. 119: 1097–1116.

Pohlmann, R., A. Waheed, A. Hasilik and K. von Figura. 1982. Synthesis of phosphorylated recognition marker in lysoomal enzymes is located in the cis part of Golgi apparatus. J. Biol. Chem. 257: 5323–5325.

Pohlentz, G., D. Klein, G. Schwartzmann, D. Schmitz and K. Sandhoff. 1988. Both GA2, GM2, and GD2 synthases and Gmlb, GDla, and GTlb synthases are single enzymes in Golgi vesicles from rat liver. Proc. Natl. Acad. Sci. USA 85:7044–7048.

Porter, K. R. 1957. The submicroscopic morphology of protoplasm. Harvey Lect. 1955–1956. 51: 175–228.

Porter, K. R., K. Kenyon and S. Badenhausen. 1967. Specialization of the unit membrane. Protoplasma 63: 262–274.

Posner, B. I., Z. Josefsberg and J. J. M. Bergeron. 1978. Intracellular polypeptide hormone receptors. Characterization of insulin binding sites in Golgi fractions from the liver of female rats. J. Biol. Chem. 253: 4067–4073.

Pressman, B. C. 1968. Ionophorous antibiotics as models for biological transport. Fed. Proc. 27: 1283–1288.

Rabouille, C. and G. Warren. 1997. Changes in the architecture of the Golgi apparatus during mitosis. In: The Golgi Apparatus. E. G. Berger and J. Roth (eds.). Birkhäuser, Basel. pp. 195–217.

Rabouille, C., T. Misteli, R. Watson and G. Warren. 1995. Reassembly of Golgi stacks from mitotic Golgi fragments in a cell-free system. J. Cell Biol. 129: 605–618.

Rambourg, A. 1971. Morphological and histochemical aspects of glycoproteins at the surface of animal cells. Int. Rev. Cytol. 31: 57–114.

Rambourg, A., W. Hernandez and C. P. LeBlond. 1969. Detection of complex carbohydrates in Golgi apparatus of rat cells. J. Cell Biol. 40: 395–414.

Rambourg, A., Y. Clermont, L. Hermo and D. Segretain. 1987. Tridimensional architecture of the Golgi apparatus and its components in mucous cells of Brunner's gland of the mouse. Am. J. Anat. 179: 95–107.

Ray, T. K., I. Lieberman and A. I. Lansing. 1968. Synthesis of the plasma membrane of the liver cell. Biochem. Biophys. Res. Commun. 31: 54–58.

Ray, P. M., T. L. Shininger and M. M. Ray. 1969. Isolation of β-glucan synthetase particles from plant cells and identification with Golgi membranes. Proc. Natl. Acad. Sci. U.S.A. 64: 605–612.

Razin, S. 1974. Correlation of cholesterol to phospholipid content in membranes of growing mycoplasmas. FEBS Lett. 47: 81–85.

Redman, C. M., S. Yu, D. Bannerjee and H. P. Morris. 1979. In vitro synthesis and secretion of albumin by Morris hepatomas 5123C and 7800. Cancer Res. 39: 101–111.

Reutter, W. and C. Bauer. 1978. Terminal sugars in glycoconjugates: metabolism of free and protein-bound l-fucose, N-acetylneuraminic acid and d-galactose in liver and Morris hepatomas. In: Morris Hepatomas. Mechanisms of Regulation. H. P. Morris and W. E. Criss (eds.). Plenum, New York. pp. 405–437.

Rexach, M. F. and R. W. Schekman. 1991. Distinct biochemical requirements for the budding, targeting and fusion of ER derived transport vesicles. J. Cell Biol. 114: 219–229.

Richardson, C. D. and D. E. Vance. 1976. Biochemical evidence that Semliki Forest virus obtains its envelope from the plasma membrane of the host cell. J. Biol. Chem. 251: 5544–5550.

Richardson, C. L., T. W. Keenan and D. J. Morré. 1977. Ganglioside biosynthesis. Characterization of CMP-N-acetylneuraminic acid: lactosylceramide sialytransferase in Golgi apparatus from rat liver. Biochim. Biophys. Acta 488: 88–96.

Roberts, R. M. and B. O. Yuan. 1974. Chemical modification of the plasma membrane polypeptides of cultured mammalian cells as an aid to studying protein turnover. Biochemistry 13: 4846–4856.

Robinson, D. G. and P. M. Ray. 1977. The reversible cyanide inhibition of Golgi secretion in pea cells. Eur. J. Cell Biol. 15: 65–77.

Robinson, D. N. and J. A. Spudich. 2000. Towards a molecular understanding of cytokinesis. Trends Cell Biol. 10: 228–237.

Rodriguez, M., P. Moreau, M. Paulik, J. Lawrence, D. J. Morré and D. M. Morré. 1992. NADH-activated cell-free transfer between Golgi apparatus and plasma membranes of rat liver. Biochim. Biophys. Acta 1107: 131–138.

Rocks, O., A. Peyker and P. H. Bastiaens. 2006. Spatio-temporal segregation of Ras signals: one ship, three anchors, many harbors. Curr. Opin. Cell Biol. 18: 351–357.

Roelofsen, P. A., Encyclopedia of Plant Anatomy, Part 4, The Plant Cell Wall, 3, (Zimmerman, W., and P. G. Ozenda, Eds., Gebrüber Borntraeger, Berlin, Nikolassee, Germany, 335 pp., 1959).

Roland, J.-C., C. A. Lembi and D. J. Morré. 1972. Phosphotungstic acid- chromic acid as a selective electron-dense stain for plasma membrane of plant cells. Stain Technol. 47: 195–200.

Roth, J. 1997. Topology of glycosylation in the Golgi apparatus. In: The Golgi Apparatus. E. G. Berger and J. Roth (eds.). Birkhäuser, Basel. pp. 131–161.

Roth, J. and E. G. Berger. 1982. Immunocytochemical localization of galactosyltransferase in HeLa cells: codistribution with thiamine pyrophosphatase in trans Golgi cisternae. J. Cell Biol. 92: 223–229.

Rothman, J. E. 1981. The Golgi apparatus: Two organelles in tandem. Science 213: 1212–1219.

Rothman, J. E. 1987. Protein sorting by selective retention in the endoplasmic reticulum and Golgi stack. Cell 50: 521–522.

Rothman, J. E. 1994. Mechanisms of intracellular protein transport. Nature 372: 55–63.

Rothman, J. E. and G. Warren. 1994. Implications of the SNARE hypothesis for intracellular membrane topology and dynamics. Curr. Biol. 4: 220–223.

Rothman, J. E. and F. T. Wieland. 1996. Protein sorting by transport vesicles. Science 272: 227–234.

Rothman, J. E., R. L. Miller and L. J. Urbani. 1984. Intercompartmental transport in the Golgi is a dissociative process: facile transfer of membrane protein between two Golgi populations. J. Cell Biol. 99: 260–271.

Rowe, T., M. Aridor, J. M. McCaffrey, H. Plutner, C. Nuoffer and W. E. Balch. 1996. COPII vesicles derived from mammalian endoplasmic reticulum microsomes recruit COPI. J. Cell Biol. 135: 895–911.

Roy, L., J. J. M. Bergeron, C. Lavoie, R. Hendriks, J. Gushue, A. Fazel, A. Pelletier, D. J. Morré, V. N. Subramaniam, W. Hong and J. Paiement. 2000. Role of p97 and syntaxin 5 in the assembly of transitional endoplasmic reticulum. Mol. Biol. Cell 11: 2529–2542.

Ruohola, H., A. K. Kabcenell and S. Ferro-Novick. 1988. Reconstitution of protein transport from the endoplasmic reticulum to the Golgi complex in yeast: the acceptor Golgi compartment is defective in the sec 23 mutant. J. Cell Biol. 107: 1465–1475.

Sager, R. and G. E. Palade. 1957. Structure and development of the chloroplast in Chlamydomonas. J. Biophys. Biochem. Cytol. 3: 463–488.

Sakai, A. and M. Shigenaka. 1967. Behavior of cytoplasmic membranous structures in the spermatogenesis of the grasshopper Atractomorpha bedeli. Bolivar. Cytologia (Tokyo) 32: 72–86.

Salama, N. R., T. Yeung and R. Schekman. 1993. The Sec 13p complex and reconstitution of vesicle budding from the ER with purified cytosolic proteins. EMBO J. 12: 4073–4082.

Salamero, J., E. S. Sztul and K. E. Howell. 1990. Exocytic transport vesicles generated in vitro from the trans-Golgi network carry secretory and plasma membrane proteins. Proc. Natl. Acad. Sci. U.S.A. 87: 7717–7721.

Sanders, E. J. and P. K. Singal. 1975. Furrow formation in Xenopus embryos. Involvement of the Golgi body as revealed by ultrastructural localization of thiamine pyrophosphatase activity. Exp. Cell Res. 93: 219–224.

Saraste, J. and E. Kuismanen. 1984. Pre- and post-Golgi vacuoles operate in the transport of Semliki Forest virus membrane glycoproteins to the cell surface. Cell 38: 535–549.

Saraste, J. and K. Svensson. 1991. Distribution of the intermediate elements operating in ER to Golgi transport. J. Cell Sci. 100: 415–430.

Saraste, J., G. E. Palade and M. G. Farquhar. 1986. Temperature-senstivite steps in the transport of secretory proteins through the Golgi complex in exocrine pancreatic cells. Proc. Natl. Acad. Sci. 83: 6425–6429.

Schachter, H. 1974. The subcellular sites of glycosylation. Biochem. Soc. Symp. 40, 57–71.

Schachter, H. and S. Roseman. 1980. Mammalian glycosgltransferases. Their role in synthesis and function of complex carbohydrates and glycolipids. In: The Biochemistry of Glycoproteins and Proteoglycans. W. J. Lennarz (ed.). Plenum, New York. pp. 85–160.

Schachter, H., I. Jabbal, R. L. Hudgin, L. Pinteric, E. J. McGuire and S. Roseman. 1970. Intracellular localization of liver sugar nucleotide glycoprotein glycosyltransferases in a Golgi-rich fraction. J. Biol. Chem. 245: 1090–1100.

Schatzman, R. C., G. I. Evan, M. L. Privalsky and J. M. Bishop. 1986. Orientation of the v-erb-B gene product in the plasma membrane. Mol. Cell Biol. 6: 1329–1333.

Schekman, R. and L. Orci. 1996. Coat proteins and vesicle budding. Science 271: 1526–1533.

Schilling, E. E., H. Goldenberg, D. J. Morré and F. L. Crane. 1979. Distribution of insulin receptors among mouse liver endomembranes. Biochem. Biophys. Acta 555: 504–511.

Schimmoller, F., B. Singer-Kruger, S. Schroder, U. Kruger, C. Barlowe and H. Riezman. 1995. The absence of Emp24p, a component of ER-derived COPII coated vesicles, causes a defect in transport of selected proteins to the Golgi. EMBO J. 14: 1329–1339.

Schindler, T., R. Bergfeld, M. Hohl and P. Schopfer. 1994. Inhibition of Golgi-apparatus function by brefeldin A in maize coleoptiles and its consequences on auxin-mediated growth, cell-wall extensibility and secretion of cell-wall proteins. Planta 192: 404–413.

Schmitt, H. D., M. Puzicha and D. Gallwitz. 1988. Study of a temperature-sensitive mutant of the ras-related YPT1 gene product in yeast suggests a role in the regulation of intracellular calcium. Cell 53: 635–641.

Schneider, W. C. and E. L. Kuff. 1954. On the isolation and some biochemical properties of the Golgi substances. Am. J. Anat. 94: 209.

Schneider, W. C., A. J. Dalton, E. L. Kuff and M. Felix. 1953. Isolation and biochemical function of the Golgi substance. Nature 172: 161–162.

Schnepf, E. 1961. Quantative Zusammenhange Zwischen der Sekretion des Fangschleimes und den Golgi – strukturen bei Drosophyllum lusitanicum. Naturforsch. 166: 605–610.

Schnepf, E. 1969. Sekretion und exkretion bei pflanzen. Protoplasmatologia, Handbuch der Protoplasmaforschung 8: 1–181.

Schnepf, E. and J. Busch. 1976. Morphology and kinetics of slime secretion in glands of Mimulus tilingii. Z. Pflanzenphysiol. 69: 62–71.

Schnepf, B. and G. Deichgräber. 1986. Inhibition of elongation in Pellia setae by the monovalent ionophore monensin. Bot. Gaz. 147: 252–257.

Schnepf, E., W. Herth and D. J. Morré. 1979. Elongation growth of setae of Pellia (Bryophyta): effects of auxin and inhibitors. Z Pflanzenphysiol 94: 21 1–217.

Schwarz, J. K., J. M. Capasso and C. B. Hirschberg. 1984. Translocation of adenosine 3'-phosphate 5'-phosphosulfate into rat liver Golgi vesicles. J. Biol. Chem. 259: 3554–3559.

Schweizer, A., K. Matter, C. M. Ketcham and H.-P. Hauri. 1991. The isolated ER-Golgi intermediate compartment exhibits properties that are different from ER and cis-Golgi. J. Cell Biol. 113: 45–54.

Seemann, J., M. Pypaert, T. Taguchi, J. Malsam and G. Warren. 2002. Partitioning of the matrix fraction of the Golgi apparatus during mitosis in animal cells. Science 295: 848–851.

Segev, N., J. Mulholland and D. Botstein. 1988. The yeast GTP-binding YPT1 protein and a mammalian counterpart are associated with the secretion machinery. Cell 52: 915–924.

Serafini, T., G. Stenbeck, A. Brecht, F. Lottspiech, L. Orci, J. E. Rothman and F. T. Wieland. 1991. A coat subunit of Golgi-derived non-clathrin-coated vesicles with homology to the clathrin-coated vesicle coat protein beta-adaptin. Nature 349: 214–220.

Sharma, S., C. Birchmeier, J. Nikawa, K. O'Neill, L. Rodgers and M. Wigler. 1989. Characterization of the ros1-gene products expressed in human glioblastoma cell lines. Oncogene Res. 5(2): 91–100.

Shima, D. T., N. Cabrera-Poch, R. Pepperkok and G. Warrant. 1998. An ordered inheritance strategy for the Golgi apparatus: visualization of mitotic disassembly reveals a role for the mitotic spindle. J. Cell Biol. 141: 955–966.

Shorter, J. and G. Warren. 2002. Golgi architecture and inheritance. Annu. Rev. Cell Dev. Biol. 18: 379–420.

Sievers, A. 1967. Elektronenmikroskopische Untersuchungen zur geotropischen Reaction. Protoplasma 64: 225–253.

Silverstein, S. C., R. M. Steinman and Z. A. Cohn. 1977. Endocytosis. Annu. Rev. Biochem. 46: 669–772.

Simons, K. and E. Ikonen. 1997. Functional rafts in cell membranes. Nature 387: 569–572.

Simons, K. and H. Virta. 1987. Perforated MDCK cell support intracellular transport. EMBO J. 16: 2241–2247.

Simons, K. and M. Zerial. 1993. Rab proteins and the road map for intracellular transport. Neuron 11: 789–799.

Singerland, J. and M. Pagano. 2000. Regulation of the cdk inhibitor p^{27} and its deregulation in cancer. J. Cell Physiol. 183: 10–17.

Sitte, P. 1958. Die Ultrastruktur von Wurzelmeristemzellen der Erbse (Pisum sativum). Protoplasma 49: 447–522.

Sjöstrand, F. S. 1956. The ultrastructure of cells as revealed by the electron microscope. Int. Rev. Cytol. 5: 455–533.

Sjöstrand, F. S. 1963. A comparison of plasma membrane, cytomembranes, and mitochondrial membranes with respect to ultrastrucutural features. J. Ultrastruct. Res. 9: 561–580.

Sjöstrand, F. S. 1968. Ultrastructure and function of cellular membranes. In: Ultrastructure in Biological Systems, Vol. 4, The Membranes. A. J. Dalton and F. Haguenau (eds.). Academic Press, New York. pp. 151–210.

Sjöstrand, F. S.and V. Hanzon. 1954. Ultrastructure of Golgi apparatus of exocrine cells of mouse pancreas. Exp. Cell Res. 7: 415–429.

Slingerland, J. and M. Pagano. 2000. Regulation of the cdk inhibitor p27 and its deregulation in cancer. J. Cell Physiol. 183: 10–17.

Smeekens, S. 1994. PC2 and PC3. In: Guidebook to the Secretory Pathway. J. Rothblatt, P. Novick and T. Stevens (eds.). Oxford University Press, London. pp. 197–199.

Smith, C. E. 1980. Ultrastructural localization of nicotinamide adenine dinucleotide phosphatase (NADPase) activity to the intermediate saccules of the Golgi apparatus in rat incisor ameloblasts. J Histochem Cytochem. 28: 16–26.

Söllner, T., S. W. Whiteheart, M. Brunner, H. Erdjument-Bromage, S. Geromanos, P. Tempst and J. E. Rothman. 1993. SNAP receptors implicated in vesicle targeting and fusion. Nature 362: 318–324.

Spiegel, S. and P. H. Fishman. 1987. Gangliosides as bimodal regulators of cell growth. Proc. Natl. Acad. Sci. U.S.A. 84(1): 141–145.

Spiro, R. G. 1994. Golgi endo α-mannosidase. In: Rothblatt, J., P. Novick and T. Stevens (eds) Guidebook to the secretory pathway. Oxford University Press, London, pp. 188–189.

Staehelin, L. A. 1974. Structure and function of intercellular junctions. Int. Rev. Cytol. 39: 191–283.

Staehelin, L. A. and I. Moore. 1995. The plant Golgi apparatus: structure, functional organization and trafficking mechanisms. Ann. Rev. Plant Physiol. 46: 261–288.

Stanley, P. 1994. Genes required for the maturation of N-linked carbohydrates in mammalian cells. In: Guidebook to the Secretory Pathway. J. Rothblatt, P. Novick and T. Stevens (eds.). Oxford University Press, London. pp. 190–194.

Storrie, B., J. White, S. Röttger, E. H. Stelzer, T. Suganuma and T. Nilsson. 1998. Recycling of Golgi-resident glycosyltransferases through the ER reveals a novel pathway and provides an explanation for nocodazole-induced Golgi scattering. J. Cell Biol. 143: 1505–1521.

Strangeways, T. and R. G. Canti. 1927. The living cell in vitro as shown by dark ground illumination and the changes induced in such cells by fixing reagents. Quart. J. Microscop. Sci. 71: 1–14.

Strous, G. J. and H. F. Lodish. 1980. Intracellular transport of secretory and membrane proteins in hepatoma cells infected by vesicular stomatitis virus. Cell 22: 709–717.

Struck, D. K. and W. J. Lennarz. 1980. The function of saccharide-lipids in synthesis of glycoproteins. In: The Biochemistry of Glycoproteins and Proteoglycans. W. J. Lennarz (ed.). Plenum, New York. pp. 35–83.

Sturbois, B., P. Moreau, C. Cassagne and D. J. Morré. 1994. Cell-free transfer of phospholipids between the endoplasmic reticulum and the Golgi apparatus of leek seedlings. Biochim. Biophys. Acta 1189: 31–37.

Subramanian, V. N., F. Peter, R. Philip, S. H. Wong and W. Hong. 1996. GS28, a 38-kilodalton Golgi SNARE that participates in ER Golgi transport. Science 272: 1161–1163.

Sun, I. L., F. L. Crane and D. J. Morré. 1983. Calmodulin-NADH semi-dehydroascorbate oxidoreductases interactions of clathrin-coated vesicles. Biochem. Biophys. Res. Commun. 115: 952–957.

Sun, I., D. J. Morré, F. L. Crane, K. Safranski and E. M. Croze. 1984. Monodehydroascorbate as an electron acceptor for NADH reduction by coated vesicle and Golgi apparatus fractions of rat liver. Biochim. Biophys. Acta 797: 266–275.

Sweiczer, A., B. Novak and J. M. Mitchison. 1996. The size control of fission yeast revisited. J. Cell Sci. 109: 2947–2957.

Swiezewaska, E., G. Dallner, B. Andersson and L. Ernster. 1993. Biosynthesis of ubiquinone and plastoquinone in the endoplasmic reticulum-Golgi membranes of spinach leaves. J. Biol. Chem. 268: 1494–1499.

Szymanski, E. S. and H. M. Farrell. 1982. Isolation and solubilization of casein kinase from Golgi apparatus of bovine mammary gland and phosphorylation of peptides. Biochim. Biophys. Acta 702: 163–172.

Tabas, I. and S. Kornfeld. 1979. Purification and characterization of a rat liver Golgi α-mannosidase capable of processing asparagine-linked oligosaccharides. J. Biol. Chem. 254: 11655–11663.

Taiz, L. 1984. Plant cell expansion: regulation of cell wall mechanical properties. Annu. Rev. Plant Physiol. 35: 585–657.

Takatsuki, A. and G. Tamura. 1985. Protein: intracellular accumulation vesicular stomatitus virus G, Brefeldin A, a specific inhibitor of intracellular translocation of high-mannose type G protein; intracellular accumulation of high-mannose type G protein and inhibition of its cell surface expression. Agric. Biol. Chem. 49: 899–902.

Tandler, B. and D. J. Morré. 1983. The Golgi apparatus of ciliated cells in the cat trachea negatively-stained in situ and in cell fractions. Protoplasma 115: 193–201.

Tartakoff, A. M. 1986. Temperature and energy dependence of secretory protein transport in the exocrine pancreas. EMBO J. 5: 1477–1482.

Tartakoff, A. M. and P. Vassalli. 1977. Plasma cell immunoglobulin secretion: arrest is accompanied by alterations of the Golgi complex. J. Exp. Med. 146: 1332–1345.

Teclebrhan, H., A. Jakobsson-Borin, U. Brunk and G. Dallner. 1995. Relationship between the endoplasmic reticulum-Golgi membrane system and ubiquinone biosynthesis. Biochim. Biophys. Acta 1256: 157–165.

Thomas, G. 1994. Furin. In: Guidebook to the Secretory Pathway. J. Rothblatt, P. Novick and T. Stevens (eds.). Oxford University Press, London. pp. 195–197.

Tokumitsu, S. I. and W. H. Fishman. 1983. Alkaline phosphatase biosynthesis in the endoplasmic reticulum and its transport through the Golgi apparatus to the plasma membrane. J. Histochem. Cytochem. 31: 647–655.

Tooze, J., S. A. Tooze and G. Warren. 1984. Replication of coronavirus MHV-A59 in sac cells: determination of the first site of budding of progeny virions. Eur. J. Cell Biol. 33: 291–293.

Tooze, S. A., J. Tooze and G. Warren. 1988. Site of addition of N-acetylgalactosamine to the E 1 glycoprotein of mouse hepatitis virus-A59. J. Cell Biol. 106: 1475–1487.

Torii, S., K. Murakami and K. Nakayama. 1997. Regulation of association of a 58-kDa peripheral membrane protein (58 K) with the Golgi apparatus: its membrane binding properties related to but distinct from those of COP 1 coat proteins. Biomed. Res. 18: 37–47.

Torii, S., M. Kusakabe, T. Yamamoto, M. Maekawa and E. Nishida. 2004. Sef is a spatial regulator for Ras/MAP kinase signaling. Dev. Cell. 7: 33–44.

Trinchera, M. and R. Ghidoni. 1989. Two glycospingolipid sialyltransferases are localized in different sub-Golgi compartments in rat liver. J. Biol. Chem. 264: 15766–15769.

Tsai, F. M., R. Y. Shyu and S. Y. Jiang. 2006. RIG1 suppresses Ras activation and induces cellular apoptosis at the Golgi apparatus. Cell Signal. 19: 989–999.

Tulsiani, D. R. P., S. C. Hubbard, P. W. Robbins and O. Touster. 1982. α-D-Mannosidases of rat liver Golgi membranes. J. Biol. Chem. 257: 3660–3668.

Turner, F. R. and W. G. Whaley. 1965. Intercisternal elements of the Golgi apparatus. Science 147: 1303–1304.

Tuttle, R. L., N. S. Gill, W. Pugh, J. P. Lee, B. Koeberlein, E. E. Furth, K. S. Polonsky, A. Nan and M. J. Burnbaum. 2001. Regulation of pancreatic β-cell growth and survival by the serine/threonine protein kinase Akt1/PKBα. Nat. Med. 7: 1133–1137.

Twaddle, M. L., R. A. Jersild, Jr., K. N. Dilpeen and D. J. Morré. 1981. Kinetics of appearance of lipoprotein particles in perisinusoidal cisternae of smooth endoplasmic reticulum of isolated rat livers perfused with free fatty acids. Eur. J. Cell Biol. 26: 26–34.

Ueda, K. 1966. Fine structure of Chlorogonium elongatum with special reference to vacuole development. Cytologia (Tokyo) 31: 461–472.

Ulmer, J. B. and G. E. Palade. 1989. Targeting and processing of glycophorins in murine erythroleukemia cells: use of brefeldin A as a perturbant of intracellular traffic. Proc. Natl. Acad. Sci. U.S.A. 86: 6992–6996.

Ungar, D. and F. M. Hughson. 2003. SNARE protein structure and function. Annu. Rev. Cell Dev. Biol. 19: 493–517.

Van Der Woude, W. J., D. J. Morré and C. E. Bracker. 1971. Isolation and characterization of secretory vesicles in germinated pollen in Lilium longiflorum. J. Cell Sci. 8: 331–351.

Van Der Woude, W. J., C. A. Lembi, D. J. Morré, J. A. Kidinger and L. Ordin. 1974. Beta-glucan synthetases of plasma membrane and Golgi apparatus from onion stem. Plant Physiol. 54: 333–340.

Varga, J. M., G. Moellmann, P. Fritsch, E. Godawska and A. B. Lerner. 1976a. Association of cell surface receptors for melanotropin with the Golgi region in mouse melanoma cells. Proc. Natl. Acad. Sci. U.S.A. 73(2): 559–562.

Varga, J. M., M. A. Saper, A. B. Lerner and P. Fritsch. 1976b. Nonrandom distribution of receptors for melanocyte-stimulating hormone on the surface of mouse melanoma cells. J. Supramol. Struct. 4(1): 45–49.

Verbert, A., R. Cacan, P. Debeire and J. Montreuil. 1977. Peculiar behavior of ectosialyltransferase toward exogenous acceptors. FEBS Lett. 74: 234–238.

Verde, C., M. C. Pascale, G. Martire, L. V. Lotti, M. R. Torrisi, A. Helenius and S. Bonatti. 1995. Effect of ATP depletion and DTT on the transport of membrane proteins from the endoplasmic reticulum and the intermediate compartment to the Golgi complex. Eur. J. Cell Biol. 67: 267–274.

Vian, B. 1974. Precisions fournies par le cryodecapage sur la restructuration et l'assimilation au plasmalemme des membranes des derives golgiens. C. R. Acad. Sci. Paris Ser. D. 278: 1483–1486.

Vian, B. and J. C. Roland. 1972. Differenciation des cytomembranes et renouvellement du plasmalemme dans les phenomenes de secretions vegetales. J. Microscop. 13: 119–136.

Vigil, E. L., C. Frantz, H. Goldenberg and F. L. Crane. 1978. Cytochemical demonstration of glutaraldehyde-resistant, NADH-ferricyanide oxido-reductase activities in rat liver plasma membranes and Golgi apparatus. Eur. J. Cell Biol. 18: 213–230.

Virk, S. S., C. J. Kirk and S. B. Shears. 1985. Ca2+ -transport and Ca2+-dependent ATP hydrolysis by Golgi vesicles from lactating rat mammary glands. Biochem. J. 226: 741–748.

Vogt, P. K. 2001. PI 3-kinase, mTOR, protein synthesis and cancer. Trends Mol. Med. 7: 482–484.

Wagner, R. R. and M. A. Cynkin. 1969. Enzymatic transfer of 14C-glucosamine from UDP-N-acetyl-14C-glucosamine to endogenous acceptors in a Golgi apparatus-rich fraction from liver. Biochem. Biophys. Res. Commun. 35: 139–143.

Waits, L., S. Dunkle, F. E. Wilkinson, P. Moreau, K. Safranski, T. Reust, D. M. Morré and D. J. Morré. 1990. Cell-free transfer from dictyosome-like structures to plasma membrane vesicles of guinea pig testes. Protoplasma 154: 8–15.

Walter, P. and V. R. Lingappa. 1986. Mechanism of protein translocation across the endoplasmic reticulum membrane. Ann. Rev. Cell Biol. 2: 499–516.

Walters, J. R. F. and M. M. Weiser. 1984. Characterization of the vitamin D-dependent Ca2+ -binding sites in rat intestinal Golgi-enriched membrane fractions. Biochem J. 218: 347–354.

Ward, R. T. and E. Ward. 1968. The multiplication of Golgi bodies in the oocytes of Rana pipiens. J. Microsc. 7: 1007–1020.

Warnock, D. E., M. S. Lutz, W. A. Blackburn, W. W. Young and J. U. Baenziger. 1994. Transport of newly synthesized glucosylceramide to the plasma membrane by a non-Golgi pathway. Proc. Natl. Acad. Sci. U.S.A. 91: 2708–2712.

Warren, L. and M. C. Glick. 1968. Membranes of animal cells. II. The metabolism and turnover of the surface membrane. J. Cell Biol. 37: 729–746.

Warren, G., P. Woodman, M. Pypaert and E. Smythe. 1988. Cell-free assays and the mechanism of receptor-mediated endocytosis. Trends Biochem. Sci. 13: 462–465.

Watanabe, I. and S. Okada. 1967. Effects of temperature on growth rate of cultured mammalian cells (L5178Y). J. Cell Biol. 32: 309–323.

Waters, M. G., T. Serafini and J. E. Rothman. 1991. 'Coatomer': a cytosolic protein complex containing subunits of non-clathrin-coated Golgi transport vesicles. Nature 349: 248–251.

Wattenberg, B. W. 1990. Glycolipid and glycoprotein transport through the Golgi complex are similar biochemically and kinetically Reconstitution of glycolipid transport in a cell free system. J. Cell Biol. 111: 421–428.

Wattenberg, B. W. 1991. Analysis of protein transport through the Golgi in a reconstituted cell-free system. J. Electron Microsc. Technol. 17: 150–164.

Wattiaux, R., M. Wibo and P. Baudhuin. 1963. Effect of the injection of Triton WR 1339 on the hepatic lysosomes of the rat. In: Ciba Foundation Symposium on Lysosomes. A. V. S. de Reuck and M. P. Cameron (eds.). J. & A. Churchill Ltd. London.

Weibel, E. R. 1969. Stereological principles for morphometry in electron microscopic cytology. Int. Rev. Cytol. 26: 235–302.

Weibel, E. R., W. Staubli, H. Gnagi and F. Hess. 1969. Correlated morphometric and biochemical studies on the liver cell. I. Morphometric model, sterelogic methods, and normal morphometric data for rat liver. J. Cell Biol. 42:68–91.

Weidman, P. J. 1995. Anterograde transport through the Golgi complex: do Golgi tubules hold the key? Trends Biochem. Sci. 5: 302–305.

Weinberg, R. A. 2007. The Biology of Cancer. Garland Scientific, New York. pp. 796.

Weinstein, R. S., F. B. Merk and J. Alroy. 1976. The structure and function of intercellular junctions in cancer. Adv. Cancer Res. 23: 23–79.

Werner, G. 1970. On the development and structure of the neck in urodele sperm. In: Comparative Spermatology. B. Baccetti (ed.). Academic Press, New York. pp. 85–92.

West, D. W. 1981. Energy-dependent calcium sequestration activity in a Golgi apparatus fraction derived from lactating rat mammary glands. Biochim. Biophys. Acta 673: 374–386.

Whaley, W. G. 1975. The Golgi Apparatus. Cell Biology Monographs, Vol. 2. Springer-Verlag, Wien-New York. pp. 190.

Whaley, W. G. and H. H. Mollenhauer. 1963. The Golgi apparatus and cell plate formation – a postulate. J. Cell Biol. 17: 216–221.

Whaley, W. G., M. Dauwalder and J. E. Kephart. 1966. The Golgi apparatus and an early stage of cell plate formation. J. Ultrastruct. Res. 15: 169–180.

Whiteheart, S. W., M. Brunner, D. W. Wilson, M. Wiedmann and J. E. Rothman. 1992. Soluble N-ethylmaleimide-sensitive fusion attachment proteins (SNAPs) bind to a multi-SNAP receptor complex in Golgi membranes. J. Biol. Chem. 267: 12239–12243.

Wilkinson, F. E., D. J. Morré and T. W. Keenan. 1976. Ganglioside biosynthesis. Characterization of uridine diphosphate galactose:GM2 galactosyltransferase in Golgi apparatus from rat liver. J. Lipid. Res. 17: 146–153.

Williams, D. C. 1974. Studies of protistan mineralization. I. Kinetics of coccolith secretion in Hymenomonas carterae. Calcif. Tissue Res. 16: 227–237.

Williamson, F. A. and D. J. Morré. 1976. Distribution of phosphatidylinositol biosynthetic activities among cell fractions from rat liver. Biochem. Biophys. Res. Commun. 68: 1201–1205.

Wilson, E. B. 1925. The Cell in Development and Heredity, 3rd edition. Macmillan. New York.

Wilson, L. A. and D. B. Amos. 1972. Subcellular location of HL-A antigens. Tissue Antigens 2: 105–111.

Wilson, D. W., C. A. Wilcox, G. C. Flynn, E. Chen, W. J. Kuang, W. J. Henzel, M. R. Block, A. Ullrich and J. E. Rothman. 1989. A fusion protein required for vesicle-mediated transport in both mammalian cells and yeast. Nature 339: 355–359.

Wilson, D. W., S. W. Whiteheart, M. Brunner and J. E. Rothman. 1992. A multisubunit particle implicated in membrane fusion. J. Cell Biol. 117: 532–538.

Woodman, P. G. 1997. The roles of NSF, SNAPs and SNAREs during membrane fusion. Biochim. Biophys. Acta 1357: 155–172.

Wu, Y., D. Dowbenko, M. T. Pisabarro, L. Dillard-Telm, H. Koeppen and L. A. Lasky. 2001. PTEN 2, a Golgi-associated testis-specific homologue of the PTEN tumor suppressor lipid phosphatase. J. Biol. Chem. 276: 21745–21753.

Wuestehube, L. J. and R. Schekman. 1992. Reconstitution of transport from the endoplasmic reticulum to the Golgi complex using an ER-enriched membrane fraction from yeast. Methods Enzymol. 219: 124–136.

Xu, R., J. Jin, W. Hu, W. Sun, J. Bielawski, T. Taha and L. M. Obeid. 2006. Golgi alkaline ceramidase regulates cell proliferation and survival by controlling levels of sphingosine and SIP. FASEB J. 20: 1813–1825.

Yamamoto, T. 1963. On the thickness of the unit membrane. J. Cell Biol. 17: 413–422.

Yamamoto, M. 1964. Electron microscopy of fish development. III. Changes in the ultrastructure of the nucleus and cytoplasm of the oocyte during it development in Oryzias latipes. J. Fac. Sci. Tokyo Univ. Sect II. 10: 335–346.

Yeo, K., J. B. Partent, T. K. Yeo and K. Olden. 1985. Variability in transport rates of secretory glycoproteins through the endoplasmic reticulum and Golgi in human hepatomas cells. J. Biol. Chem. 260: 7896–7902.

Younger, J. S., A. W. Scott, J. F. Hollum and W. R. Stinebrng. 1966. Interferon production by inactivated Newcastle disease virus in cell cultures and in mice. J. Bacteriol. 92: 862–868.

Yuan, L., J. G. Barriocanal, J. S. Bonifacino and I. V. Sandoval. 1987. Two integral membrane proteins located in the cis-middle and trans part of the Golgi system acquire sialylated N-linked carbohydrates and display different turnovers and sensitivity to cAMP-dependent phosphorylation. J. Cell Biol. 105: 215–227.

Yunghans, W. N. and D. J. Morré. 1978. Distribution of adenylate cyclase among membrane fractions of rat liver. Cytobiologie 17: 212–231.

Yunghans, W. N., T. W. Keenan and D. J. Morré. 1970. Isolation of a Golgi apparatus-rich fraction from rat liver. III. Lipid and protein composition. Exp. Mol. Path. 12: 36–45.

Yunghans, W. N., J. E. Clark, D. J. Morré and E. D. Clegg. 1978. Nature of the phosphotungstic acid-chromic acid (PACP) stain for plasma membranes of plants and mammalian sperm. Cytobiologie 17: 165–172.

Zambrano, F., S. Fleischer and B. Fleischer. 1975. Lipid composition of the Golgi apparatus of rat kidney and liver in comparison with other subcellular organelles. Biochim. Biophys. Acta 380: 357–369.

Zerial, M., P. Chavrier, L. A. Huber, M. Vingron and K. Simons. 1994. Rab proteins on the exocytic and endocytic pathway. In: Guidebook to the Secretory Pathway. J. Rothblatt, P. Novick and T. Stevens (eds.). Oxford Unviersity Press, London. pp. 202–204.

Zhang, F. and D. L. Schneider. 1983. The bioenergetics of Golgi apparatus function: evidence for an ATP-dependent proton pump. Biochem. Biophys. Res. Commun. 114: 620–625.

Zhang, L. and D. J. Morré. 1993. Isolation and characterization of the principal ATPase of transitional endoplasmic reticulum of rat liver. In: Molecular Mechanisms of Membrane Traffic. D. J. Morré, K. Howell and J. J. M. Bergeron (eds.). Springer, Berlin, Heidelberg, New York, Tokyo (NATO ASI series, series H, Vol. 74), pp. 61–62.

Zhang, L., C. L. Ashendel, G. W. Becker and D. J. Morré. 1994. Isolation and characterization of the principal ATPase associated with transitional endoplasmic reticulum of rat liver. J. Cell Biol. 127: 1871–1883.

Zhao, J., D. J. Morré, M. Paulik, J. Yim and D. M. Morré. 1990. GTP hydrolysis by transitional endoplasmic reticulum from rat liver inhibited by all-trans retinol. Biochim. Biophys. Acta 1055: 230–233.

Zulak, I. M. and T. W. Keenan. 1983. Citrate accumulation by a Golgi apparatus-rich fraction from lactating bovine mammary gland. Int. J. Biochem. 15: 747–750.

Index

A

AAA-ATPase (p97), 216, 234
Acanthamoeba, 133
Acetabularia, 193
N-Acetylglucosamine (galactosyl) transferase
  (*N*-Acetylglucosaminyl transferase), 44,
  91, 142, 183, 245
Acid phosphatase, 43, 101, 139, 181–182
Acinar cells, enzyme and proenzyme secretion
  by, 26, 34, 71, 158–161, 255
Acylation, 177
Adenosine 3′-phosphate 5′-phosphosulfate, 140
Adenosine triphosphatases, 108, 141, 216, 234
ADP-ribosylation factor, 74, 150, 223
Akt, 232–233, 250–251, 273
Albumin
  blocking secretion of, in rat hepatocytes, 222
  formation, Golgi apparatus in, 177–178
Algal cell wall units, formation of, 171–172
Alkaline phosphatase, 106, 109, 139
Apparato reticolare interno, 1
ARF. *See* ADP-ribosylation factor
Arylsulfatases, 106, 139
ATP
  dependent vesicle budding in rat liver,
    215–219
  in ER to Golgi apparatus membrane transfer,
    208–211
  independent vesicle budding, 214
  role in Golgi apparatus, 141–142
ATPases. *See* Adenosine triphosphatases
Autumn Crocus, 226
Avian erythroblastosis virus, 248

B

Baby hamster kidney cell, 222
Bcl-2 proteins, 250
Boulevard périphérique, 19, 31, 69, 255
Brefeldin A (BFA), 72, 74, 194–196, 216, 219,
  222–225, 229, 236, 254–255
  COPI Complex, dissociation, 222
  guanine nucleotide blocking by, 223

  inhibition of transport by, 223, 229
  NADH oxidation, inhibition of, 224–225
  structure, 222–223
Brunner's gland
  Golgi apparatus, 24, 164

C

Cahal, 2
Canals of Holmgren, 4
Cancer and Golgi apparatus, 239–252
  features characteristic of, 245
Cancer phenotype, 239–242
Carbohydrate detection, 99–101, 142
Carbohydrate-processing enzymes, 254
Carbonylcyanide chlorophenyl hydrozone
  (CCCP), 209–210
Caspase, 195, 250–251
Cell-free systems, 197–219
  in ATP-dependent vesicle budding in rat liver,
    215–219
  in cultured cells, 211–213
  development of, 197–198
  Golgi apparatus to plasma membrane transfer,
    211
  NTP dependent ER to Golgi apparatus
    transfer, 208–211
  in plants, 214–215
  in yeast, 213–214
  transfer assay development, 198
  transitional ER to Golgi apparatus transfer,
    198–211
    glycoconjugate processing in, 207–208
    lipid and protein cotransfer, 204–205
    lipid processing in, 205–207
    processing of transferred constituents, 205
    in rat liver and specificity of, 203
    retinol stimulation, 219
    temperature effects in, 203–204
Cell enlargement, 221–237
Cell growth, *See* cell enlargement
  Akt regulation, 232
  coordination, 221

Cell plate formation, 168, 221
Cell proliferation, 221
Cellular proteins, synthesis of, 241
Cell wall matrix polysaccharides, 112, 142
Cell walls and cell walls units secretion, Golgi
    apparatus in, 142, 166–173
Central saccule/saccule, in Golgi apparatus, 12
Ceramide, 81–82, 144–145, 149, 233
Ceramide transfer protein, 82
Cerebrosides, 104
CERT. *See* Ceramide transfer protein
Chinese hamster ovary (CHO) cells, 116, 213
Chromaffin cells of adrenal medulla, 71, 159
Cholesterol, 56, 68, 75, 79–80, 102, 118,
    148–151, 198, 204, 232–233, 245, 251
Chondriome, 4–5
Cis Golgi apparatus network, 37
Chrysophycean algae, 114–115, 126, 134, 171, 181
Cisternae
  appearance of, 189
  changes during replication, 189–191
  defined, 10
  in Golgi apparatus, 10–17
  membranes of, 94
  smooth-surfaced, 193
Cisternal remnants, in Golgi apparatus, 35–36
Cisternal stack. *See* Dictyosome. *See* Stack
    in Golgi apparatus
Clathrin, 12–18, 32–37, 45, 56, 73, 151, 197,
    214, 222, 234, 241, 253
Clathrin coated vesicles, in Golgi apparatus,
    32–37, 234
CMP-NeuNAc transferase, 142, 144
Coated buds, of Golgi apparatus, 71–73
Coatomer proteins. *See* COPs
Colchicine, 32, 113, 178, 180, 222–223, 226
Cognate tethers and scaffold proteins, 255
Colon columnar cells, 165
Condensing vacuoles, in Golgi apparatus,
    33–34, 71, 156
COP. *See* Coatomer proteins (COPs)
β-COP coat protein. *See* COPI
COPI, 72–74, 150, 195, 197, 222–224, 253–256
COPII, 32, 212–214, 216, 253–256, 258
Cytochalasin B, 129, 166, 180, 222, 223
  actin filaments, shortening, 226
  cytoplasmic division inhibition, 225–226
  inhibition, 226
Cytochrome b_5, 105–107, 109, 121–122, 140
Cytochrome P_{450}, 105–107, 121, 122, 140

D
Dictyokinesis, 190
Dictyosome. *See* Stacks
Dictyosome-like structures, 26

Dicumarol, in tubular system breakdown, 73–74
Dipeptidylaminopeptidase (DPP-IV), 208

E
Easter lily, 168, 170
ECTO-NOX protein, 248
EDTA. *See* Ethylenediaminetetraacetic acid
Endomembranes (Endomembrane System)
  biochemical constituents of
    biogenesis, 77–91
    glycolipids, 104
    glycoproteins, 103–104
    lipids, 79–83, 102–103
    oncogene expression, 248
    proteins, 84–88, 103
    sterols, 83
  biogenesis, 77–91, 240
  defined, 29
  differentiation
    biochemical evidence of, 106, 109
    immunological manifestations of, 110–117
  Golgi apparatus, as part of, 29–30
  membrane biogenesis, role in, 240
  signal transduction and oncogene expression
    in cancer, role in, 248–251
Endoplasmic reticulum (ER)
  associations of with Golgi apparatus, 19–21
  cancer phenotype, 239
  cell free analysis of, 198–211
  drug metabolism, 240
  glycoconjugate processing in, 207–208
  lipid and protein cotransfer, 204–205
  lipid processing in, 205–207
  NTP dependent, 208–211
  processing of transferred constituents, 205
  in rat liver and specificity of, 203
  temperature effects in, 203–204
  transfer system, 215–219
  markers, 105
  in mitotic vertebrate cells, 255
  reactions in, 143
  role in membrane biogenesis, 78–79
Endothelial nitric oxidase synthetase (ENOS),
    250
Enzymatic markers, 108–110
Epidermal growth factor (EGF), 248
Essential oils,
  secretion of, 173–175
Ethylenediaminetetraacetic acid, 48
N-Ethylmaleimide
  as inhibitor, 202, 204, 212, 235, 236
  sensitive factor, 50, 183–185, 194–197, 236
  sensitive fusion protein, 150, 183–185
Euglena gracilis, 17, 85
Exo-α–D-mannosidases, 139–140

Exocrine secretion
   defined, 155
   examples. *See* pancreatic and parotid ascinar
      cells
Extrinsic (peripheral) membrane protein, 84,
      151–152

F
FCCP, 227
Flow-differentiation, *See* Membrane
      flow-differentiation
Forkhead (FOXOI), 250
Free coated vesicles, 73
Free-flow electrophoresis
   comparing Golgi apparatus of cancer and
      non cancer cells, 243–247
   in Golgi apparatus subfractionation, 57–61
   theoretical basis for, 60
Fucosyltransferase, 91, 245–246
Furin, 140
Fusiform vesicles, in Golgi apparatus, 35–36

G
Galactosyltransferases, 42–44, 60, 73, 91, 106,
      120, 122, 142, 180, 244–247
Gangliosides, 88, 104, 143–145, 148
GDP-fucose:glycoprotein fucosyltransferase,
      142, 245
GEF. *See* Guanine nucleotide exchange factor
GERL. *See* Golgi apparatus-endoplasmic
      reticulum-lysosome
GERL and lysosome formation, 31, 37, 251,
      261–262
G-glycoprotein processing, 140, 206,
      210–211
Glucose-6-phosphatase, 139
   in rat Golgi apparatus, 43
Glucose-6-phosphate dehydrogenase, 48
Glycerolipids, biosynthetic pathways, 83
Glycerophospholipid biosynthetic enzymes,
      81–82
Glycogen synthase kinase 3β (GSK-3β), 251
Glycolipids, *See* Glycosphingolipids
Glycoproteins
   glycosylation, 176–178
   in Golgi apparatus, 103–104, 143
Glycosphingolipids, 88, 104, 143–144, 148
Glycosylation, of membrane glycoproteins
      and glycolipids, *See* Glycoproteins and
      glycosphingolipids
Glycosyltransferase, 79, 88–91, 120, 142, 144, 180
GM130. *See* Golgi matrix protein
Golgi apparatus
   arrangement of stacks in, 28–29
   associated polyribosomes, 86, 88, 119–120

association with other organelles and cell
      components, 31
association with lysosomes, 31–32, 36
biochemical evidence of membrane
      differentiation, 101–110
biochemistry, 137–153
biosynthetic capabilities of membrane
      biogenesis, 79
characteristics and differences of plant and
      animal, 18, 23–24
coated buds of, 71–73
as control point for tumor cell modification, 240
controversy of, 4–6
definition, 28
dimensional change in heptomas, 241
discovery of, 1–4
distillation tower model, 254
distribution of sugar transferase activities
      among, 246
early model, 66
endomembrane system and functioning of,
      29–31
endoplasmic reticulum-lysosome, 31
and endoplasmic reticulum, association,
      19–21
enzymology of, 138–143
fragmentation and reformation, 194–196
function in secretion, 155–185
free-flow electrophoresis in subfractionation,
      57–61
glycerophospholipids, 80–83
glycosphingolipid synthesis, 143–145
glycosyltransferases distribution in, 90–91
glycosyltransferases of. *See*
      glycosyltransferases
haCER2 in, 233
history of, 1–7, 257
H-Ras and N-Ras localize to, 249
inheritance, 191, 193
inhibitors
   Brefeldin A (BFA), 72, 74, 194–196, 216,
      219, 222–225, 229–236, 254–255
   colchicine, 32, 113, 178, 180, 222–223,
      226–227
   cytochalasin B, 129, 166, 180, 225–226
   FCCP, 227
   in control of secretion, 180
   metabolic, 227
   monensin, 128, 134, 180–181, 222–225,
      231–235, 240
   nocodazole, 195
   in pancreatic ascinar cells, 228
   retinoids, 227
   role in cell enlargement, 228
Intestine absorptive cells, 161

Golgi apparatus (*continued*)
  isolation of
    from mammalian cells, 52–53
    from plant cells, 49–52
    from rodent liver, 40–49
  lipid composition of, 102, 147–148
  maize root, 27–28, 36, 67, 100, 169
  markers, 26, 43, 105–106, 147
  matrix, 11, 24, 188
  matrix protein, 130, 195
  membrane differentiation, mechanisms of,
      118–124
  membrane displacement
    budding of membrane, 229–230
    energy requirement, 233–236
    morphometric quantitation of, 231
    SKERBP and PI3K/Akt pathways, 232
  modern rediscovery of, 6
  morphological basis of membrane
      differentiation
    and cytochemistry, 99–101
    membrane thickness measurements, 94–96
    organization of, 96–98
  nucleotide sugar transporters, 146–147
  phospholipid biosynthesis, 81–83, 148–149
  to plasma membrane transfer, 211–214
  polarity of, 27
  polyribosomes, 85–87, 119
  precisternal stages of, 191–194
  protein composition of, 103, 149–152
  proteins, origins of, 85–88
  purification of, 142
  reassembly, 65, 195
  replication mechanisms of, 187–191
  in secretions
    of cell walls and cell wall units formation,
        166–173
    control of, 179–181
    of enzyme and proenzyme, 158–161
    of hormone, vitamin and essential oils,
        173–175
    large molecules processing by, 175–178
    of lipoprotein particles, 161–162
    of mucin, 162–166
    of protein, sugars, ions and molecules, 175
    role of, 155–156
    saccule. *See* cisternae
    transport vehicles and pathways for,
        156–158
  stacks
    of maize root, 16, 27–28, 67, 169
    from radish root, 12
    of rat epididymis, 10
    from soybean, 14
  sterols, 83

structure of, 9–10, 17–28
subfractionation of, 53–57
tubules
  at Golgi apparatus periphery, 15
  coated buds in, 71–74
  features and functions of, 63–66
  function of, 66–71
  isolation of, 74–76
of tumor cells, 242
vesicles of, 32–37
zone/matrix, 9, 11, 24–25
Golgi, C., 2
Golgi–Cox method, 1
Golgin, 195
gpαF. *See* Glycosylated pro-α-factor
G proteins. *See* Guanine nucleotide binding (G)
    proteins
GRASP65. *See* Golgi reassembly and stack
    protein 65
Growth, 221
GTP, in ER to Golgi apparatus membrane
    transfer, 208–211
GTP-γ-S, 72, 197, 212, 216
Guanine nucleotide binding (G) proteins, 150,
    152, 197–198, 210, 223, 235
Guanine nucleotide exchange factor, 212, 223, 249
Guanosine triphosphatase (GTPase), 124, 195,
    198, 212, 214, 219, 249, 254

H
haCER2, in Golgi apparatus, 233
HeLa cell, haCER2 expression in, 233
Helix pomatia, 23
Helminthosporium dermatioideum, 225
Hemagglutinin antigen (HA), 116
Hepatocyte Golgi apparatus, *See* Rat liver
Herpes virus, 117
Hormones secretion, Golgi apparatus in, 173
Hymenomonas carterae, 134, 172

I
Immunocytochemistry application, 112
Indole-3-acetic acid (IAA), 234
Influenza virus, 116
Integral protein. *See* Intrinsic protein
Intercalary cisternae. *See* Medial cisternae
Intercisternal elements, in Golgi apparatus, 24,
    27–36
Intestinal absorptive cells, 66–70, 161
Intrinsic protein, 84, 87, 151–152

L
Lactate dehydrogenase, 48
Lactosylceramide, 149, 207–208
LDL. *See* Low-density lipoproteins

Lectins, 110–111
Lens culinaris, 110
Lilium longiflorum. See Easter lily
Lipids, in Golgi apparatus, 102–103, 147–149
Lipoprotein
    particles, 19, 22, 31, 45, 55–56, 67–69, 75,
        138, 151–152, 175, 243, 255
    secretion of, 161–162, 265
Liver parenchyma, Golgi apparatus. *See*, Rat liver
Low-density lipoproteins, 67, 69, 161
Low temperature block, 203, 210, 227–228,
    231–232
Lysosomal enzymes
    formation, Golgi apparatus role in, 176
    segregation of, 181–183
Lysosomes and Golgi apparatus, association,
    31–32

M

Maize root, Golgi apparatus of, 27–28, 36,
    67, 108, 169
Mammalian cells, Golgi apparatus isolation
    from, 52–53
Mammary gland, 34, 40, 56, 80–81, 102, 118,
    120–122, 136, 141–142, 174–175, 219
Mannose-6-phosphate receptors, 14
Mannosidases I and II, 139
Medial cisternae, 26, 147
Membrane biogenesis Golgi apparatus in, 77, 91
    biochemical pathways of, 78
    endoplasmic reticulum role in, 78–79
Membrane budding, 229
Membrane differentiation, in Golgi apparatus
    biochemical basis for, 101–102
        constituents concentrated in, 104–110
        survey of, 102–104
    enzyme content and activity of, 107–110, 122
    evidence from cytochemistry, 99
    immunological manifestations of, 110
    in vitro, 123
    mechanisms of
        biosynthetic contributions and, 118–119
        Golgi apparatus polyribosomes, 86, 88,
            119–120
        selective enzyme deletion, 120–121
        selectivity of, 119
    morphological basis for, 94–101
    organization of membrane constituents, 96
Membrane flow hypothesis, 94, 96, 124–125
Membrane flow-differentiation,
        Membrane flow, 63, 67, 74, 93–136,
        156, 196, 223–224, 254
    dynamic aspects, 125
    energetics and regulation, 135
    evidence from induced systems, 134–135

functional significance, 124
kinetics of, 130–133
mechanisms, 122
membrane lipids, bulk flow of, 133–134
morphology of, 126–130
significance, 124
Membrane glycolipids glycosylation, 88,
    143–144, 176–178
Membrane lipids biosynthesis, 80–83
Membrane proteins biosynthesis, 84–85
Membrane sterols biosynthesis, 83–84
Membrane thickness measurements, 94–96
Micrasterias americana, 215
Microfilaments, 32, 35, 222, 236
Microtubules, 32, 37, 122, 226–227
Mimulus tilingii, 166
Mitochondria and Golgi apparatus,
    association, 32
Mitogen-activated protein kinase (MAPK), 249
Mitosis, in Golgi apparatus fragmentation and
    reformation, 194
Monensin
    as cell enlargement inhibitor, 225, 231–235
    evidence in support of flow-differentiation
        hypothesis, 128–129, 134, 240
    function of, 224
    as monovalent polyether antibiotic, 224
    Na^+ ionophore representation by, 224
    as secretion blocker, 180–181, 222–225
Monoamine oxidase, 42–43
Monosialoganglioside G_{MI}, structure of, 145
Morris hepatomas, 239, 241
Mucin secretion, 162–166

N

NADH, Golgi apparatus response to, 211,
    217, 234
NADH dehydrogenase, 139–140
NADH-ferricyanide reductase, 56, 60, 121, 141
NADH oxidase, 219, 224
NAD(P)H oxidoreductases, 105–122
NADPase. *See* NADP phosphatase
NADP phosphatase, 26, 59, 147
Na^+ ionophore representation, Monensin, 224
NEM. *See* N-ethylmaleimide
NEM-sensitive factor, 150, 183–185,
    194–197, 236
NEM sensitive fusion protein, 150, 183–185
Nocodazole, 195
Nonaprenyl-4-hydroxybenzoate transferase, 75
NSF, *See* NEM-sensitive factor
Nucleoside triphosphate, 208, 213, 217
5'-Nucleotidase, 43
Nucleoside diphosphatases, 137, 139
Nucleoside monophosphatases, 139

Nucleoside sugar transporters, in Golgi
    apparatus, 146–147
Nucleoside triphosphatases, 139, 141

O

Ochromonas, 113–115
O-demethylase, in Golgi apparatus, 69–70
Oryzias, 191
Osmium, impregnation of, 27, 59, 96

P

Pancreatic acinar cells, 26, 34, 71, 138,
    158–161, 255
PAPS, *See* adenosine 3′-phosphate 5′-phosphatase
Pancreatic islet cells, 251
Parotid gland ascinar cells, 71, 158–161
Pellia, 231, 234, 235
Peripheral membrane protein. *See* Extrinsic
    membrane protein
Peripheral tubules, in Golgi apparatus, 15–16
    acinar cells and chromaffin cells, 66–71
    coated buds in, 71–73
    functions of, 66–71
Peroxisomes and Golgi apparatus, association
    of, 32
Phenobarbital, 117, 134
Phosphatidylcholine, 79–82, 102, 107, 118,
    148–149, 204–207, 265
Phosphatidylethanolamine, 79–82, 101–102,
    118, 148, 204
Phosphatidylinositol, 79–82, 102, 107, 118, 148
Phosphatidylinositol 4,5-bisphosphate, 250
Phosphatidylinositol 3-kinase (PI3K), 232,
    250–251, 266
Phosphatidylinositol 4-phosphate, 185
Phosphatidylinositol 3-phosphate, 250–251
Phosphatidylserine, 79–82, 118, 148, 204
Phospholipid
    in animal and plant endomembranes, 79–82
    biosynthesis, in Golgi apparatus, 148–149
    in Golgi apparatus, 79–82
    processing of, 205–207
Phosphorylation, 177
Phosphotungstic acid, 15, 50, 65–66, 68,
    72, 99, 100, 123
Physical membrane displacement, 228–229,
    233–235
PIP$_3$, *See* phosphatidylinositol 4,5-bisphosphate
PI4P. *See* Phosphatidylinositol 4-phosphate
Plant cells Golgi apparatus, isolation, 49–52
Plant Golgi apparatus, role of, 173
Plants, cell-free membrane transfer in, 214–215
Plasma membranes
    isolation of, 47–48
    of malignant cells, 245–248
    markers, 106–108
    of tumor cells, 248
Plasma proteins secretion, 222
Plastids and Golgi apparatus, association, 32
Platelet-derived growth factor (PDGF), 248
Pleckstrin homology (PH), 250
Pleurochrysis scherffelii, 114–115, 126, 134
Polyribosomes and Golgi apparatus, 86, 88,
    119–120
Pro-α-factor, 198
Processing, 175–176
α$_1$-Protease blocking, in rat hepatocytes, 222
Protein
    composition, of Golgi apparatus, 131,
        149–152
    in Golgi apparatus, 103
    secretions, in Golgi apparatus, 175
110 kDa Protein dissociation, from membranes,
    222
Proteolytic processing 177–178
PTEN, 250
Pyridine nucleotide reduction, 224
Pythium ultimum, 94, 97

R

Rab GTPases, 150–151, 184–185, 255
Ras, 150, 242, 249–250
Rat hepatocytes. *See* Rat liver
Rat liver
    cell-free systems, 199–211
    cisternae formation and maturation in, 134
    comparing ER, GA and PM of, 106, 108, 109
    cytochrome concentration in, 121
    differentiation of endomembranes in, 96, 121
    endoplasmic reticulum marker for, 105
    enzyme specific activities of endomembrane
        fraction in, 107
    function in lipoprotein secretion, 19, 31,
        66–70, 161–162
    ganglioside biosynthetic pathways of, 145
    Golgi apparatus
        associated polyribosomes, 85–86
        isolated from, 19, 40–44
        morphology, 21, 162
        stacks in, 194
    lipid composition of nuclear envelope and
        endoplasmic reticulum in, 79
    lipid distribution for endomembranes of, 80
    lipid protein ratio, 118
    NADP phosphatase in Golgi apparatus of, 147
    nucleoside diphosphatase, 132–133
    phospholipid composition of animal and plant
        endomembranes comparing, 81
    sugar-nucleotides, distribution of, 89
    terminal enzymes, distribution of, 82

Receptor tyrosine kinases (RTKs), oncogenic activations, 249
Retinol, 148, 173, 217, 219, 227
Retinol stimulation of vesicle budding, in rat liver, 218–219, 227
Retinoids, 227
Rodent liver, Golgi apparatus isolation, 40–49
Rodent liver, reference fractions from, 44–49

S
Saccharomyces, 28
Saccule, in Golgi apparatus, *See* Cisternae
Schizosaccharomyces pombe, 193
Secretagogues, 160
Secretion, control of, 179–181
Secretion granules, in Golgi apparatus, 33–34
Secretory granule. *See* Condensing vacuoles, in Golgi apparatus
Secretory vesicles, in Golgi apparatus, 32–33, 156
  composition of, 56
  isolation of, 54–56
Sendai virus, 116
Semliki Forest Virus, 115
Sialic acid, 57–58, 88–89, 104, 111, 133, 143, 145, 147, 176, 207–208, 244–246
Sialyl transferases, 60, 91, 120, 142, 144, 244–246
Signal hypothesis, 178–179
Slime secretion, plants, 164–166
Smooth endoplasmic reticulum, 241
SNAP receptor proteins, 150, 183–185, 255
SNAPs. *See* Soluble NSF attachment proteins
SNAREs. *See* SNAP receptor proteins
Soluble NSF attachment proteins, 150, 183–185
Sphingomyelin, 80–82, 102–103, 147–149
  synthesis of, 149
SREBP-2. *See* Sterol regulatory element binding protein-2
Sterols, 75, 79, 83, 91, 102, 118–119, 198, 215, 232, 247, 251
Sterol biosynthesis, 83
Sterol regulatory element binding protein-2, 232, 251
Sterology, 42
Succinate dehydrogenase, 42–44
Sucrose gradient procedure, for Golgi apparatus isolation, 46, 50–51, 54
Sugar nucleotides formation, 88–90, 146–147
Sulfation reactions, 90, 176
Sulfotransferases, 79, 177

T
Tetracystis excentrica, 17
Tetrahymena pyriformis, 193
TGN. *See* Trans-Golgi network

Thiamine pyrophosphatase, 101, 105, 108, 138
Tip-growing cells, 168–170, 228
Toxoplasma, 193
Trans-Golgi apparatus reticulum, 241
Trans-Golgi network, 36, 64, 185, 222, 258
Transitional ER to Golgi apparatus, transfer system, 198–202
  glycoconjugate processing, 207–208
  lipid and protein cotransfer in, 204–205
  lipid processing in, 205–207
  processing of transferred constituents, 205
  in rat liver and specificity of, 203
  temperature effects in, 203–204
Trypanosoma, 193
Tubules, in Golgi apparatus. *See* Golgi apparatus tubules
Tumor cell, surface properties, 247
Tyrosylprotein sulfotransferase, 140

U
UDP-galactose, 43, 107, 123, 146, 206, 244, 246
UDP-Gal transferase, 142
UDP-GlcNAc transferas, 142
UDP-*N*-acetylglucosaminyltransferase. *See* *N*-acetylglucosamine transferase
Uric acid oxidase, 42–43
Uridine-5′-diphosphate galactose, 44
Urinary bladder, 35

V
Vacuole, plant lysosome equivalent, 31
Vacuome, 4–5
v-*erb*-B protein, 248
Very low-density lipoproteins, 67, 69, 144, 148, 161
Vesicle transfer complexes, 212
Vesicular stomatitis virus glycoprotein, 116, 135, 198, 205–206, 210–211, 222
Vitamin A and Golgi apparatus. *See* Retinol, 218–219
Vitamins, fat soluble, secretion of, 173–175
VLDL. *See* Very low-density lipoproteins
VTCs. *See* Vesicle transfer complexes

X
Xenopus, 121

Y
Yeast, cell-free membrane transfer, 213–214

Z
Zone of exclusion, 11, 24, 188
  See Golgi apparatus matrix
Zymogen granule, 26, 33, 34, 71, 160–161, 255

Printed in the United States of America